ANALYSIS OF COMPLEX DISEASE ASSOCIATION STUDIES

ANALYSIS OF COMPLEX DISEASE ASSOCIATION STUDIES

A PRACTICAL GUIDE

Edited by

ELEFTHERIA ZEGGINI
Wellcome Trust Sanger Institute, Cambridge, UK

ANDREW MORRIS
Wellcome Trust Centre for Human Genetics, University of Oxford, Oxford, UK

AMSTERDAM • BOSTON • HEIDELBERG • LONDON
NEW YORK • OXFORD • PARIS • SAN DIEGO
SAN FRANCISCO • SINGAPORE • SYDNEY • TOKYO

Academic Press is an imprint of Elsevier

Academic Press is an imprint of Elsevier
32 Jamestown Road, London NW1 7BY, UK
30 Corporate Drive, Suite 400, Burlington, MA 01803, USA
525 B Street, Suite 1800, San Diego, CA 92101-4495, USA

First edition 2011

Notice

No responsibility is assumed by the publisher for any injury and/or damage to
persons or property as a matter of products liability, negligence or otherwise,
or from any use or operation of any methods, products, instructions or ideas
contained in the material herein. Because of rapid advances in the medical
sciences, in particular, independent verification of diagnoses and drug dosages
should be made

British Library Cataloguing-in-Publication Data
A catalogue record for this book is available from the British Library

Library of Congress Cataloging-in-Publication Data
A catalog record for this book is available from the Library of Congress

ISBN: 978-0-12-375142-3

For information on all Academic Press publications
visit our website at www.elsevierdirect.com

Typeset by TNQ Books and Journals
Printed and bound by CPI Group (UK) Ltd, Croydon, CR0 4YY
Transferred to digital print 2013

10 11 12 13   10 9 8 7 6 5 4 3 2 1

Working together to grow
libraries in developing countries

www.elsevier.com | www.bookaid.org | www.sabre.org

ELSEVIER   BOOK AID
International   Sabre Foundation

Table of Contents

List of Contributors

Carl A. Anderson Department of Human Genetics, Wellcome Trust Sanger Institute, Hinxton, Cambridge, UK

Yurii S. Aulchenko Erasmus MC Rotterdam, Rotterdam, The Netherlands

Jennifer H. Barrett Section of Epidemiology and Biostatistics, Cancer Genetics Building, Cancer Research UK Clinical Centre, St James's University Hospital, Leeds, UK

S.S. Cherny Department of Psychiatry and the State Key Laboratory of Brain and Cognitive Sciences LKS Faculty of Medicine, University of Hong Kong, Hong Kong

David Clayton Juvenile Diabetes Research Foundation/Wellcome Trust Diabetes and Inflammation Laboratory Department of Medical Genetics, University of Cambridge, UK

Frank Dudbridge London School of Hygiene and Tropical Medicine, London, UK

Katherine S. Elliott Wellcome Trust Centre for Human Genetics, University of Oxford, UK

David M. Evans MRC Centre for Causal Analyses in Translational Epidemiology, Department of Social Medicine, University of Bristol, UK

Stephen Eyre Epidemiology Unit School of Translational Medicine, University of Manchester, UK

Mark M. Iles Section of Epidemiology and Biostatistics, Cancer Genetics Building, Cancer Research UK Clinical Centre, St James's University Hospital, Leeds, UK

Michael Inouye Immunology Division, The Walter and Eliza Hall Institute of Medical Research, Parkville, Victoria, Australia; Department of Human Genetics, Wellcome Trust Sanger Institute, Hinxton, Cambridge, UK

Christoph Lange Department of Biostatistics, Harvard School of Public Health, Boston, MA, USA; Institute for Genomic Mathematics, University of Bonn, Germany; German Center for Neurodegenerative Diseases (DZNE), BONN, Germany

Jessica Lasky-Su Department of Medicine, Channing Laboratories, Brigham and Women's Hospital and Harvard Medical School, Boston, MA, USA

Jing Li Electrical Engineering and Computer Science Department, Case Western Reserve University, Cleveland, OH, USA

Cecilia Lindgren   Wellcome Trust Centre for Human Genetics, University of Oxford, UK; Oxford Centre for Diabetes, Endocrinology and Metabolism, University of Oxford, UK

Jonathan Marchini   Department of Statistics, University of Oxford, UK

Andrew Morris   Wellcome Trust Centre for Human Genetics, University of Oxford, UK

Amy Murphy   Department of Medicine, Channing Laboratories, Brigham and Women's Hospital and Harvard Medical School, Boston, MA, USA

Vincent Plagnol   Juvenile Diabetes Research Foundation/Wellcome Trust Diabetes and Inflammation Laboratory Department of Medical Genetics, University of Cambridge, UK

P.C. Sham   Department of Psychiatry and the State Key Laboratory of Brain and Cognitive Sciences LKS Faculty of Medicine, University of Hong Kong, Hong Kong

Nicole Soranzo   Wellcome Trust Sanger Institute, Genome Campus, Hinxton, Cambridge, UK and Department of Twin Research & Genetic Epidemiology, King's College London, London, UK

Yik Ying Teo   Departments of Statistics & Applied Probability and Epidemiology & Public Health, National University of Singapore and Genome Institute of Singapore, Singapore

Wendy Thomson   Epidemiology Unit School of Translational Medicine, University of Manchester, UK

Benjamin F. Voight   Medical Population Genetics, The Broad Institute of Harvard and MIT, Cambridge, MA, USA; Center for Human Genetics Research, Massachusetts General Hospital, Cambridge, MA, USA

Wei-Bung Wang   Department of Computer Science and Engineering, University of California-Riverside, Riverside, CA, USA

N. William Rayner   Wellcome Trust Centre for Human Genetics, University of Oxford, Oxford, UK

Sungho Won   Department of Statistics, Chung-Ang University, Seoul, Korea

Krina T. Zondervan   Wellcome Trust Centre for Human Genetics, University of Oxford, Oxford, UK

1

Genetic Architecture of Complex Diseases

P.C. Sham, S.S. Cherny

Department of Psychiatry and the State Key Laboratory of Brain and Cognitive Sciences, LKS Faculty of Medicine, University of Hong Kong, Hong Kong

INTRODUCTION

Most common human diseases, such as coronary heart disease, diabetes, cancers, bipolar affective disorders and schizophrenia, have complex etiology. While they tend to cluster in families, they do not exhibit the characteristic Mendelian segregation ratios of single-gene disorders. These diseases therefore cannot be solely caused by a single genetic mutation, but have a more complex genetic architecture. For any complex disease, many questions about its genetic architecture can be raised. How many genetic variants are involved in individual differences in the propensity to develop the disease (e.g., just a handful, tens, hundreds, or thousands)? Where are these sequence changes located on the 23 chromosomes that constitute the human genome? What is the nature of the sequence changes in these variants (e.g., single base pair changes, copy number changes, etc.)? What are the functional conse-quences of these changes (e.g., change of amino acid sequence and therefore protein structure, or changes in the level or regulation of gene expression)? What are the frequencies and effect sizes of these changes? How important are these changes relative to the environmental variation in explaining individual differences in disease susceptibility? And how do the genetic changes interact with each other and with environmental factors?

This chapter reviews the research approaches that have been used to address some of the above questions and summarizes our current state of knowledge and understanding regarding the genetic architecture of complex diseases that have come out of these studies. The general conclusion is that common diseases are highly heterogeneous, with a small proportion of cases having relatively simple etiology dominated by a single genetic mutation, while the vast majority of cases are caused by the combined effect of multiple genetic and environmental factors each contributing a minor influence.

The genetics approach to the study of complex diseases is comple-mentary to other research paradigms such as the use of cell culture or animal models. The advantages of the genetics approach are that: (1) the finite size and regularity of the genome allows a systematic search for sequence-phenotype relationships, which may unveil novel associations that implicate previously unsuspected biological pathways, and (2) the demonstration of sequence-phenotype relationships offers strong direct evidence for the role of a gene or a pathway in human disease, mini-mizing the need to perform potentially hazardous experiments on humans. On the other hand, the genetics approach is limited in its ability to tease apart detailed molecular mechanisms involved in disease etiology. The genetics approach is therefore a valuable complement, rather than an alternative, to other biological approaches.

GENETIC MODELING: TWIN, ADOPTION AND FAMILY STUDIES

One immediate question regarding the genetic architecture of complex traits is the relative importance of genetic versus environmental factors in explaining the individual differences in disease susceptibility in the population. If genetic factors are relatively unimportant, then further genetic studies may be unwarranted, and research efforts should be directed at environmental factors. On the other hand, if the contribution of genetic factors is substantial, then further genetic studies may help to identify the specific genetic variants involved and elucidate the mechanisms by which these variants influence disease propensity. The proportion of the total variance in disease liability that is explained by genetic (as against environmental) factors is defined as the heritability of the disease. It is important to appreciate that heritability is dependent on the genetic and environmental variations present in a population, so that changes in the variability of genetic or environmental factors can both lead to changes in heritability.

Heritability is typically estimated from twin and adoption studies (see [1] for an overview of methods for estimation of genetic and environmental components of variance). The principle of twin studies is as follows. Identical or monozygotic (MZ) twins share 100% of their genomes, while fraternal or dizygotic (DZ) twins share on average only 50% of their genomes by common descent. On the other hand, for twin pairs who are reared together, their sharing of environmental exposures may be the same regardless of zygosity. Thus, the presence of a greater phenotypic similarity among MZ than DZ twins can be attributed to the greater genetic similarity of MZ than DZ twins. Indeed, if the phenotype is a continuous trait, then the phenotypic similarity within twin pairs can be measured by an intraclass correlation, and the heritability estimated by twice the difference in intraclass correlations in MZ and DZ twins.

The use of twin studies for heritability estimation of disease phenotypes is more complicated because of the dichotomous nature of the phenotype. This is usually done via a liability-threshold model, where the underlying liability is normally distributed in the population, and those individuals with liability above a certain threshold value develop disease. The twin data is then used to estimate the twin correlations for the underlying liability (tetrachoric correlations), and then the heritability can be estimated by twice the difference in these correlations between MZ and DZ twins. As seen in Table 1.1, heritability estimates from twin studies on a number of complex diseases range from 40% to as high as 90%, which are typical of most complex traits.

Heritability can also be estimated by adoption studies, including MZ twins reared apart, whose correlation gives a direct estimate of heritability. In general, the correlations between biological relatives who have been separated by adoption provide estimates of heritability, whereas the correlations between adoptive relatives reared together but are biologically unrelated provide estimates for the influence of the family environment (see Plomin & Loehlin [12] for a discussion of direct estimates of heritability). Arguably the most prominent among such studies,

TABLE 1.1   Heritability of Common Complex Traits and Diseases from Various Twin Studies. Adapted from MacGregor et al. [2]

Trait	Heritability	Reference
Asthma	60	[3]
Blood pressure	40–70	[4]
Bone mineral density	60–80	[5]
Cervical and lumbar disc degeneration	60–80	[6]
Insulin dependent diabetes	70	[7]
Obesity	50–90	[8]
Osteoarthritis	50–70	[9]
Rheumatoid arthritis	60	[10]
Ulcerative colitis	50	[11]

the Minnesota Study of Twins Reared Apart, confirms that practically all complex traits have a substantial genetic component (e.g., [13–15]).

The modeling of twin, adoption and family data can be used to address other important questions about the genetic architecture of complex disorders. For example, two different diseases can be modeled simultaneously, to detect shared genetic influences on the two diseases. In twin studies, shared genetic influences would be indicated by the presence of cross-trait cross-twin correlation (i.e., correlation between disease 1 in twin 1 and disease 2 in twin 2) for both MZ and DZ twins, but which is greater in MZ than in DZ twins. Such studies have indicated substantial genetic sharing for some complex diseases; for example, schizophrenic with manic symptoms [16], and bipolar disorder with unipolar depression [17]. Differences in the genetic influences on disease liability between males and females, for different ages, or under different environmental conditions, can also be modeled in twin data. For example, Kendler et al. [18] found that twin similarity for social phobia was due primarily to genetic influences in males but a result of shared environmental influences in females.

Another type of genetic modeling is aimed not at estimating the relative importance of genetic against environmental factors, but at whether the genetic component is made up of a single genetic factor of major effect (the single major locus [SML] model), or a large number of genetic factors each of small effect (the polygenic model). These are two extreme scenarios, and other possible models include the presence of a major locus on a polygenic background (the mixed model), or a few loci of major effect (the oligogenic model).

One approach to discrimination between different genetic models is to consider the drop-off in recurrence risk of disease as a function of genetic relationship to an affected index case (proband). A polygenic model is predicted to have a steeper drop-off than an SML model; the empirical recurrence risks for schizophrenia appear to be more consistent with a polygenic than an SML model [19]. An alternative, more sophisticated method for discriminating between different genetic models is complex segregation analysis, which uses maximum likelihood on family data to fit model parameters and test different models. Thus, the presence of an SML can be tested by a likelihood ratio test of a mixed model with both SML and polygenic components, against a polygenic model. An alternative test for the presence of an SML considers a generalized transmission model in which genetic transmissions are allowed to deviate from Mendelian proportions, against a model in which genetic transmissions are constrained to Mendelian proportions. Complex segregation analysis has been applied to complex diseases with largely inconclusive results [20, 21]. This is because complex segregation analysis suffers from both low statistical power which throws doubt on negative results, and from numerous possible artifacts which throw doubt on positive results. Thus, a complex segregation analysis showed strong but likely erroneous evidence for an SML effect for medical school attendance. The problem is that the model did not incorporate sibling environment and therefore could account for a higher sibling concordance than parent–offspring concordance for medical school attendance only through a recessive SML [22].

Notable oligogenic models for complex diseases were proposed by Risch [23]. In these models, the effects of the different loci on disease risk can combine in an additive or multiplicative fashion. Risch derived, under each model, how the overall disease prevalence and recurrence risks can be related to the effects of the individual loci. Risch considered that the pattern of recurrence risks in schizophrenia is consistent with an oligogenic model with three to five loci. This would mean that these loci must have quite large effects, providing optimism for studies which aim to identify individual susceptibility loci.

DISEASE GENE MAPPING: LINKAGE STUDIES

Linkage is based on the co-segregation of marker variants and disease in families. In humans, the frequency of crossovers in meiosis is such that each gametic genome has on average only around 35 crossover points. Co-segregation should therefore be detectable for marker loci quite far away from the disease-causing variant. Because linkage operates over long genetic distances, a positional mapping approach based on linkage can cover the entire genome by using a relatively small number of highly polymorphic markers. Standard marker sets for whole-genome linkage scans, based on 200–800 microsatellite polymorphisms, which became available in the 1990s, enabled the successful mapping of hundreds of rare single-gene disorders.

Classical linkage analysis is typically carried out using the lod-score method, which is based on a single major locus parameterized by the disease allele frequency and the penetrances (conditional probability of disease) of the three disease locus genotypes. For Mendelian diseases, the values of these parameters can be easily specified from the results of population prevalence studies and segregation analyses. For complex disease, the SML model is likely to be simplistic, and appropriate values of the model parameters unknown, and indeed may be different for different loci. Nevertheless, classical linkage analysis has been optimistically applied to complex diseases, particularly on pedigrees with an unusually large number of affected individuals. There have been some successes of this approach; for example, the identification of loci responsible for early-onset familial breast cancer [24], maturity-onset diabetes in the young [25] and early-onset Alzheimer's disease [26–28]. However, the patients in these successful linkage studies represent only a very small proportion (<5%) of the overall incidence of each of the complex disease. When successful, the families involved are usually large and have a pattern of inheritance that is very close to autosomal dominant with high penetrance. For collections of smaller families with less clear-cut Mendelian inheritance, the results of classical linkage analysis are much more often unconvincing and difficult to replicate. Examples of non-replicated linkage findings include schizophrenia [29] and bipolar affective disorder [30, 31].

The lack of success in classical linkage analysis for the majority of cases of complex disease suggests that the genetic variants involved in such disorders typically have a small effect on disease risk. Thus multiple genetic variants are often involved in a single family, with the result that the co-segregation between disease and any single variant would be imperfect and different families will show linkage at different loci. It follows that for complex disorders, very large family samples are required to demonstrate conclusive linkage between disease and genetic markers.

A different version of linkage analysis, called non-parametric linkage, is based on excess local allele sharing between the affected relatives, above the level expected for the degree of relationship. For example, sibling pairs are expected to share on average one of the two alleles at any locus, and a locus which shows a significant excess of allele sharing above this level for affected sibling pairs would constitute evidence of linkage with the disease. This method has been considered to be more appropriate for complex diseases as it does not assume an SML. The non-parametric linkage approach, usually based on affected sib-pairs, became popular in the 1990s. The approach was used successfully, for example, on late-onset Alzheimer's disease, to identify a linkage region on chromosome 19 [32], which was subsequently found to contain a major susceptibility variant, the APOE ε4 allele [33]. Other studies using this approach (for example on multiple sclerosis [34, 35], schizophrenia [36–39] and autism [40]) have been less successful. The problem is that, while non-parametric linkage does not require the assumption of a single major locus model, the statistical power to detect linkage is nevertheless highly sensitive to the effect size of the susceptibility variant. For realistic sample sizes, only genetic effects which account for a substantial fraction (e.g., 20%) of the disease heritability are likely to be detected. The lack of success of non-parametric linkage analysis for many complex diseases would exclude the presence of genes of major effect. However, even variants which account for as much as 10% of the disease heritability are likely to go undetected because of inadequate statistical power.

DISEASE GENE MAPPING: ASSOCIATION STUDIES

Association analysis has a potentially far greater power than linkage analysis for detecting variants with modest effect on disease risk, provided that the genetic marker is close enough to exhibit strong linkage disequilibrium (LD) with the functional variant. This fact was recognized by Risch and Merikangas [41], who suggested that the future of complex disease genetics lies with systematic association rather than linkage studies. However, tight LD between polymorphisms requires them to be very close to each other, usually within 50 kb or less. The use of association analysis to identify disease susceptibility variants therefore depends either on prior biological knowledge that points to particular polymorphisms in candidate genes, or requires a very high density of genetic markers in candidate regions suggested by linkage studies or cytogenetic abnormalities, or indeed throughout the entire genome.

Association studies of candidate genes have revealed many susceptibility genes. A comprehensive review of candidate gene association studies, mostly performed when such studies were at their peak of popularity, found that over 600 "significant" associations for complex disease had been reported [42]. However, reviewing the subset of 166 loci which had been tested in three or more studies found that only six were consistently replicated. This suggests that our ability to pick candidate genes may be limited by our current lack of knowledge concerning the biology of complex diseases.

A systematic, genome-wide approach to association studies using dense marker sets is therefore much needed. Single-nucleotide polymorphisms (SNPs) are the most abundant type of sequence variants in the genome, occurring approximately once in every 100 to 300 base-pairs. The systematic cataloging of SNPs began in 1998 when the NIH Human Genome Project set the creation of an SNP map of at least 100,000 markers as one of its objectives. This resulted in the creation of the SNP Consortium, formed in 1999, with the goal of identifying 300,000 SNPs, but ultimately finding 1,400,000 SNPs by the end of 2001 [43]. The International HapMap Project [44], which aimed to create a map of 1 million common SNPs (defined as those where both alleles have frequency at least 5%), with not only their genomic locations but also genotype frequencies and LD relationships among each other, in three populations (Europeans, Africans and East Asians). The project involved the genotyping of enough SNPs to ensure at least one common SNP in every 5 KB bin of the genome, in 270 individuals (90 from each of the three populations). Subsequently, in Phase 2, the HapMap had been extended to include over 3 million SNPs on the same samples [45] and those samples plus additional ones were later genotyped using the latest SNP chip technology (discussed in the following paragraph) from both Affymetrix and Illumina, in Phase 3 of the project. The HapMap provides not only a very high density of common SNPs which can be used as markers in association studies, but also the genotype frequency data to determine the extent to which a set of selected SNPs can serve as surrogates (or tags) for all the other common SNPs in any targeted region of the genome.

The latest technologies for high-throughput SNP genotyping, the Affymetrix SNP Array 6.0 and the Illumina 1M BeadChip, can assay 1 million SNPs in a single reaction. These commercial genotyping products have been evaluated against HapMap data to provide adequate "coverage" for nearly 90% of all common variants in the genomes of European and East Asian populations [46], meaning that nearly 90% of all common variants are either included as one of the genotyped SNPs, or are in such strong LD with one or more of the genotyped SNPs that its genotype can be predicted (and therefore imputed) from those of the genotyped SNPs. Current genotyping technology therefore enables the systematic examination of nearly all common variants in the genome by association analysis, an approach called the genome-wide association study (GWAS). The first successful GWAS, published in 2005, detected a common risk allele for age-related macular degeneration in the gene coding for Complement Factor H [47]. Since then, the GWAS approach has identified multiple risk variants for many complex diseases. For example, over 10 risk variants for breast cancer have been identified, and over 20 for prostate cancer [48]. A consistent finding from all GWAS conducted to date is that nearly all the detected variants have very modest effects on risk of disease (typically with odds ratios, OR, in the 1.2–1.5 range) and explain a very small proportion of the population variance in liability to disease (typically 0.2–0.5%). Even in aggregate, all the risk SNPs identified to date for any complex disease explain a very modest proportion of variance in liability, typically 5–10%. Since the overall heritability of these complex diseases are typically in the range of 40–80%, the GWAS approach has only just begun to characterize the genetic components of these disorders.

These emergent findings from GWAS have important implications for the genetic architecture of complex diseases. First, the number of sequence variants that influence disease risk must be very large, at least in the hundreds if not thousands. Second, since statistical power for detecting an associated variant is determined by both effect size and allele frequency, many risk variants may have escaped detection because of either a very small effect size (OR <1.2) or low allele frequency (<5%). Third, the genome-wide coverage of 90% calculated for current genotyping products applies only to common SNP variants, and it is known that the coverage is much lower for rare SNPs. Lastly, there may be other types of variants for which coverage is lower (e.g., copy number variants, CNV).

There are compelling theoretical reasons why rare variants may be important for complex diseases. Compared to common functional variants, which are likely to have arisen from ancient mutations and have adapted to the rest of the genome (those failing to adapt would have been eliminated by negative selection), rare functional variants have a more recent origin, are more numerous, and are less well-adapted to the rest of the genome. For these reasons, rare functional variants may tend to have a larger effect on disease risk, and collectively explain a substantial proportion of population variance in disease liability [49]. Much evidence has been mounting for the role of rare variants in complex disease. In 2009, a paper in *Science* [50] reported four rare (<3%) variants in the IFIH1 gene which decrease risk for developing type 1 diabetes (T1D). This gene plays a role in the immune response to enterovirus infection and the finding of protective rare variants suggests that T1D may be a result of an overaggressive response to these particular foreign invaders. All four variants discovered are predicted to result in severe functional disruption of the gene. One introduces a premature stop signal, two are found in conserved RNA splicing sites, and the fourth alters an evolutionarily conserved site. One of these variants has a much larger effect than those typically found from GWAS, halving the risk of developing T1D.

For some complex diseases, the combined sample size from all the GWAS is now in the thousands, and yet the associations detected still explain only a small proportion (typically 20% or less) of the total heritability [51]. A number of possible explanations have been proposed for this problem of "missing heritability." First, it has been suggested that heritability estimates for complex diseases may be inflated due to methodological problems. Another possible explanation is that the SNP sets used in current GWAS offer poor tagging, especially for rare variants and structural variations. This would both reduce the number of associations detected, and underestimate the true effect sizes of the detected loci. It is also possible that many susceptibility loci simply have very small effect sizes, so that many have not been detected due to the inadequate statistical power of current studies. Finally, it has been suggested that gene—gene and gene—environment interactions account for a substantial portion of the heritability estimates, but these interactions have been largely neglected in GWAS to date.

CONCLUSION

The rapid development of molecular genetic technologies has allowed highly detailed examination of genome sequence variation, and led to rapid progress in our understanding of the genetic architecture of complex diseases. The field is still rapidly moving, with increasingly higher-powered GWAS being conducted to detect loci with diminishing allele frequency and effect sizes. At the same time, next-generation whole-genome sequencing is becoming less expensive, and it will soon become feasible to examine both known and (previously) unknown sequence variations for association with disease liability. Our picture of genetic architecture of complex diseases will therefore likely change quite rapidly and become much more detailed in the next few years.

Nevertheless, it is useful to consider our current picture of the genetic architecture of complex diseases, taking into account the most recent findings from large-scale association studies. It now appears that most complex diseases are under the influence of a very large number, probably hundreds, of sequence variants. Both common and rare variants are involved, with a wide range of effect sizes. High-penetrance mutations are responsible for some particularly familial and early-onset forms of complex diseases, but these usually account for a very small proportion of cases. Examples include BRCA mutations for early-onset breast cancer, and APP for early-onset familial Alzheimer's disease. A few common genetic polymorphisms have been identified which have moderately strong effects (allelic odds ratio >2); for example, APOE ε4 for late-onset Alzheimer's disease, CFH variant in age-related macular degeneration. It may be that the late age-of-onset of these disorders have reduced selective pressure against these deleterious diseases, and allowed high-risk variants to become quite common in populations. The majority of the genetic factors for complex diseases appear to have very small effect size, with an allelic odds ratio of less than 1.5. At present, the importance of gene–gene and gene–environment interactions is unclear. Although interactions are likely to be widespread, it appears from family data that they probably account for only a modest proportion of the variance in disease liability. However, this remains to be confirmed by studies with larger sample sizes and more comprehensive coverage of sequence variants in the genome. As future studies identify more and more sequence variants that account for individual differences in disease liability, the genes and pathways that determine the development of complex diseases will become clearer. This will enable further studies to focus on the function of these genes and pathways, to elucidate the mechanisms that lead to disease.

References

[1] M.C. Neale, L.R. Cardon, North Atlantic Treaty Organization. Scientific Affairs Division, Methodology for genetic studies of twins and families, Kluwer Academic Publishers, Dordrecht, Boston, 1992.

[2] A.J. MacGregor, H. Snieder, N.J. Schork, T.D. Spector, Twins. Novel uses to study complex traits and genetic diseases, Trends Genet. 16 (3) (2000) 131–134.

[3] D.L. Duffy, N.G. Martin, D. Battistutta, J.L. Hopper, J.D. Mathews, Genetics of asthma and hay fever in Australian twins, Am. Rev. Respir. Dis. 142 (6 Pt 1) (1990) 1351–1358.

[4] H. Snieder, L.J.Pv. Doornen, D.I. Boomsma, Developmental genetic trends in blood pressure levels and blood pressure reactivity to stress, in: J.R. Turner, L.R. Cardon, J.K. Hewitt (Eds.), Behavior Genetic Approaches in Behavioral Medicine, Plenum Press, New York, London, 1995, pp. 105–130.

[5] N.K. Arden, J. Baker, C. Hogg, K. Baan, T.D. Spector, The heritability of bone mineral density, ultrasound of the calcaneus and hip axis length: a study of postmenopausal twins, J. Bone Miner. Res. 11 (4) (1996) 530–534.

[6] P.N. Sambrook, A.J. MacGregor, T.D. Spector, Genetic influences on cervical and lumbar disc degeneration: a magnetic resonance imaging study in twins, Arthritis Rheum. 42 (2) (1999) 366–372.

[7] K.O. Kyvik, A. Green, H. Beck-Nielsen, Concordance rates of insulin dependent diabetes mellitus: a population based study of young Danish twins, BMJ 311 (7010) (1995) 913–917.

[8] H.H. Maes, M.C. Neale, L.J. Eaves, Genetic and environmental factors in relative body weight and human adiposity, Behav. Genet. 27 (4) (1997) 325–351.

[9] T.D. Spector, F. Cicuttini, J. Baker, J. Loughlin, D. Hart, Genetic influences on osteoarthritis in women: a twin study, BMJ 312 (7036) (1996) 940–943.

[10] A.J. MacGregor, H. Snieder, A.S. Rigby, M. Koskenvuo, J. Kaprio, K. Aho, et al., Characterizing the quantitative genetic contribution to rheumatoid arthritis using data from twins, Arthritis Rheum. 43 (1) (2000) 30–37.

[11] C. Tysk, E. Lindberg, G. Jarnerot, B. Floderus-Myrhed, Ulcerative colitis and Crohn's disease in an unselected population of monozygotic and dizygotic twins. A study of heritability and the influence of smoking, Gut 29 (7) (1988) 990–996.

[12] R. Plomin, J.C. Loehlin, Direct and indirect IQ heritability estimates: a puzzle, Behav. Genet. 19 (3) (1989) 331–342.

[13] T.J. Bouchard Jr., D.T. Lykken, M. McGue, N.L. Segal, A. Tellegen, Sources of human psychological differences: the Minnesota Study of Twins Reared Apart, Science 250 (4978) (1990) 223–228.

[14] D.T. Lykken, T.J. Bouchard Jr., M. McGue, A. Tellegen, Heritability of interests: a twin study, J. Appl. Psychol. 78 (4) (1993) 649–661.

[15] K.E. Markon, R.F. Krueger, T.J. Bouchard Jr., Gottesman II, Normal and abnormal personality traits: evidence for genetic and environmental relationships in the Minnesota Study of Twins Reared Apart, J. Pers. 70 (5) (2002) 661–693.

[16] A.G. Cardno, F.V. Rijsdijk, P.C. Sham, R.M. Murray, P. McGuffin, A twin study of genetic relationships between psychotic symptoms, Am. J. Psychiatry 159 (4) (2002) 539–545.

[17] P. McGuffin, F. Rijsdijk, M. Andrew, P. Sham, R. Katz, A. Cardno, The heritability of bipolar affective disorder and the genetic relationship to unipolar depression, Arch. Gen. Psychiatry 60 (5) (2003) 497–502.

[18] K.S. Kendler, K.C. Jacobson, J. Myers, C.A. Prescott, Sex differences in genetic and environmental risk factors for irrational fears and phobias, Psychol. Med. 32 (2) (2002) 209–217.

[19] M. McGue, I.I. Gottesman, D.C. Rao, Resolving genetic models for the transmission of schizophrenia, Genet. Epidemiol. 2 (1) (1985) 99–110.

[20] P.C. Sham, N.E. Morton, W.J. Muir, M. Walker, A. Collins, D.C. Shields, et al., Segregation analysis of complex phenotypes: an application to schizophrenia and auditory P300 latency, Psychiatr. Genet. 4 (1) (1994) 29–38.

[21] G.P. Vogler, I.I. Gottesman, M.K. McGue, D.C. Rao, Mixed-model segregation analysis of schizophrenia in the Lindelius Swedish pedigrees, Behav. Genet. 20 (4) (1990) 461–472.

[22] P. McGuffin, P. Huckle, Simulation of Mendelism revisited: the recessive gene for attending medical school, Am. J. Hum. Genet. 46 (5) (1990) 994–999.

[23] N. Risch, Linkage strategies for genetically complex traits. I. Multilocus models, Am. J. Hum. Genet. 46 (2) (1990) 222–228.

[24] J.M. Hall, M.K. Lee, B. Newman, J.E. Morrow, L.A. Anderson, B. Huey, et al., Linkage of early-onset familial breast cancer to chromosome 17q21. Science 250 (4988) 1684–1689.

[25] G.I. Bell, K.S. Xiang, M.V. Newman, S.H. Wu, L.G. Wright, S.S. Fajans, et al., Gene for non-insulin-dependent diabetes mellitus (maturity-onset diabetes of the young subtype) is linked to DNA polymorphism on human chromosome 20q. Proc. Natl. Acad. Sci. USA 88 (4) 1484–1488.

[26] R.E. Tanzi, J.F. Gusella, P.C. Watkins, G.A. Bruns, P. St George-Hyslop, M.L. Van Keuren, et al., Amyloid beta protein gene: cDNA, mRNA distribution, and genetic linkage near the Alzheimer locus, Science 235 (4791) (1987) 880–884.

[27] R. Sherrington, E.I. Rogaev, Y. Liang, E.A. Rogaeva, G. Levesque, M. Ikeda, et al., Cloning of a gene bearing missense mutations in early-onset familial Alzheimer's disease, Nature 375 (6534) (1995) 754–760.

[28] R.F. Clark, M. Hutton, M. Fulder, S. Froelich, E. Karran, C. Talbot, et al., The structure of the presenilin 1 (S182) gene and identification of six novel mutations in early onset AD families. Alzheimer's Disease Collaborative Group, Nat. Genet. 11 (2) (1995) 219–222.

[29] R. Sherrington, J. Brynjolfsson, H. Petursson, M. Potter, K. Dudleston, B. Barraclough, et al., Localization of a susceptibility locus for schizophrenia on chromosome 5, Nature 336 (6195) (1988) 164–167.

[30] E.I. Ginns, J. Ott, J.A. Egeland, C.R. Allen, C.S. Fann, D.L. Pauls, et al., A genome-wide search for chromosomal loci linked to bipolar affective disorder in the Old Order Amish. Nat. Genet. 12 (4) 431–435.

[31] J.A. Egeland, D.S. Gerhard, D.L. Pauls, J.N. Sussex, K.K. Kidd, C.R. Allen, et al., Bipolar affective disorders linked to DNA markers on chromosome 11. Nature 325 (6107) 783–787.

[32] M.A. Pericak-Vance, J.L. Bebout, P.C. Gaskell Jr., L.H. Yamaoka, W.Y. Hung, M.J. Alberts, et al., Linkage studies in familial Alzheimer disease: evidence for chromosome 19 linkage, Am. J. Hum. Genet. 48 (6) (1991) 1034–1050.

[33] W.J. Strittmatter, A.M. Saunders, D. Schmechel, M. Pericak-Vance, J. Enghild, G.S. Salvesen, et al., Apolipoprotein E: high-avidity binding to beta-amyloid and increased frequency of type 4 allele in late-onset familial Alzheimer disease, Proc. Natl. Acad. Sci. USA 90 (5) (1993) 1977–1981.

[34] S. Kuokkanen, M. Gschwend, J.D. Rioux, M.J. Daly, J.D. Terwilliger, P.J. Tienari, et al., Genomewide scan of multiple sclerosis in Finnish multiplex families, Am. J. Hum. Genet. 61 (6) (1997) 1379–1387.

[35] The Transatlantic Multiple Sclerosis Genetics Cooperative, A meta-analysis of genomic screens in multiple sclerosis. Mult. Scler. 7 (1) (2001) 3–11.

[36] J.L. Blouin, B.A. Dombroski, S.K. Nath, V.K. Lasseter, P.S. Wolyniec, G. Nestadt, et al., Schizophrenia susceptibility loci on chromosomes 13q32 and 8p21, Nat. Genet. 20 (1) (1998) 70–73.

[37] H. Coon, S. Jensen, J. Holik, M. Hoff, M. Myles-Worsley, F. Reimherr, et al., Genomic scan for genes predisposing to schizophrenia, Am. J. Med. Genet. 54 (1) (1994) 59–71.

[38] H.W. Moises, L. Yang, H. Kristbjarnarson, C. Wiese, W. Byerley, F. Macciardi, et al., An international two-stage genome-wide search for schizophrenia susceptibility genes, Nat. Genet. 11 (3) (1995) 321–324.

[39] S. Wang, C.E. Sun, C.A. Walczak, J.S. Ziegle, B.R. Kipps, L.R. Goldin, et al., Evidence for a susceptibility locus for schizophrenia on chromosome 6pter-p22, Nat. Genet. 10 (1) (1995) 41–46.

[40] A. Philippe, M. Martinez, M. Guilloud-Bataille, C. Gillberg, M. Rastam, E. Sponheim, et al., Genome-wide scan for autism susceptibility genes. Paris Autism Research International Sibpair Study, Hum. Mol. Genet. 8 (5) (1999) 805–812.

[41] N. Risch, K. Merikangas, The future of genetic studies of complex human diseases, Science 273 (5281) (1996) 1516–1517.

[42] J.N. Hirschhorn, K. Lohmueller, E. Byrne, K. Hirschhorn, A comprehensive review of genetic association studies, Genet. Med. 4 (2) (2002) 45–61.

[43] R. Sachidanandam, D. Weissman, S.C. Schmidt, J.M. Kakol, L.D. Stein, G. Marth, et al., A map of human genome sequence variation containing 1.42 million single nucleotide polymorphisms, Nature 409 (6822) (2001) 928–933.

[44] D. Altsuler, L.D. Brooks, A. Chakravarti, et al., A haplotype map of the human genome, Nature 437 (7063) (2005) 1299–1320.
The International HapMap Project, Nature 426 (6968) (2003) 789–796.

[45] K.A. Frazer, D.G. Ballinger, D.R. Cox, D.A. Hinds, L.L. Stuve, R.A. Gibbs, et al., A second generation human haplotype map of over 3.1 million SNPs, Nature 449 (7164) (2007) 851–861.

[46] C. Li, M. Li, J.R. Long, Q. Cai, W. Zheng, Evaluating cost efficiency of SNP chips in genome-wide association studies, Genet. Epidemiol. 32 (5) (2008) 387–395.

[47] R.J. Klein, C. Zeiss, E.Y. Chew, J.Y. Tsai, R.S. Sackler, C. Haynes, et al., Complement factor H polymorphism in age-related macular degeneration, Science 308 (5720) (2005) 385–389.

[48] T.A. Manolio, L.D. Brooks, F.S. Collins, A HapMap harvest of insights into the genetics of common disease, J. Clin. Invest. 118 (5) (2008) 1590–1605.

[49] W. Bodmer, C. Bonilla, Common and rare variants in multifactorial susceptibility to common diseases, Nat. Genet. 40 (6) (2008) 695–701.

[50] S. Nejentsev, N. Walker, D. Riches, M. Egholm, J.A. Todd, Rare variants of IFIH1, a gene implicated in antiviral responses, protect against type 1 diabetes, Science 324 (5925) (2009) 387–389.

[51] T.A. Manolio, F.S. Collins, N.J. Cox, D.B. Goldstein, L.A. Hindorff, D.J. Hunter, et al., Finding the missing heritability of complex diseases, Nature 461 (7265) (2009) 747–753.

[52] D. Aetsuler, L.D. Brook, A. Chakravarthi, et al., A haplotype map of the human genome, Nature 437 (2005) 1299–1320.

Population Genetics and Linkage Disequilibrium

Jeffrey C. Barrett

Wellcome Trust Sanger Institute, Genome Campus,
Hinxton, Cambridge, UK

OUTLINE

THE ORIGIN AND STRUCTURE OF VARIATION IN THE HUMAN GENOME

All genetic studies are aimed at understanding the connection between observed differences in *phenotype* and underlying variation in *genotype*. While there are many ways of approaching this problem, all are influenced by the patterns of variation in the human genome, and an understanding of how they have arisen. Two forces — mutation and recombination — are continually increasing the variation among genome sequences within a population, and this variation is acted upon by drift, population demography and selection to influence phenotypic traits in that population.

While mutation and recombination both contribute to the overall variability within a population of genomes, they do so in different ways. Novel mutations create entirely new sequences, and arise in each generation via a number of processes, ranging from cosmic radiation to errors generated during DNA replication or meiosis. Some of these mutations rise to appreciable frequency over time, either stochastically (i.e., genetic drift) or because selection favors the new allele over the old one. In the absence of recombination these mutations could be visualized as a tree (Fig. 2.1) with the root representing the ancestral sequence and each individual in a population at the tip of a branch containing all the mutations that had accumulated through the generations of its lineage. Ancient mutations (shared across many of the extant individuals) would lie near the root and more recent mutations would be close to the tips of the branches. Recombination, in contrast, does not introduce any new sequence-level variation to the population but creates mixtures from the existing pool. Individuals no longer have a single lineage in the tree, but instead "jump" from one branch to another at the point of crossing over in every ancestral generation (Fig. 2.1).

In order to understand the effects of these processes at the population level (i.e., all mutation and recombination in the historical generations which led to extant individuals) it is helpful to consider a single generation. During meiosis, pairs of homologous chromosomes cross over to generate a new "recombinant" chromosome consisting of the sequence of one parental chromosome up to the point of crossover, where the sequence transitions to the other parental chromosome. Thus, variants on either side of the crossover point will no longer be inherited together, but

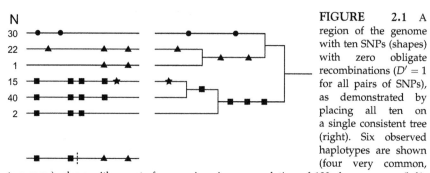

FIGURE 2.1 A region of the genome with ten SNPs (shapes) with zero obligate recombinations ($D' = 1$ for all pairs of SNPs), as demonstrated by placing all ten on a single consistent tree (right). Six observed haplotypes are shown (four very common, two rarer), along with counts from an imaginary population of 100 chromosomes (left). SNPs on the same branch have $r^2 = 1$, and SNPs denoted by the same shape (circle, triangle, square, star) have $r^2 > 0.8$, and would thus be captured by a single tag. The haplotype shown at bottom would represent an obligate recombination between branches containing squares and triangles, located approximately at the dashed line. If such a haplotype had been observed in the population, $D' < 1$.

mixed across the daughter chromosomes. There is therefore a relationship between the physical proximity of two variant positions on a chromosome and how likely they are to be inherited together (i.e., that the meiotic recombination did not occur between them). This concept, called linkage, can be quantified by the *recombination fraction*, which is simply the fraction of meioses resulting in a recombinant set of alleles at the two variants, ranging from 0 ("linked") for extremely proximal loci which are nearly always inherited together to 0.5 ("unlinked") loci at opposite ends of one chromosome, or on separate chromosomes.

Whereas linkage refers to the observation that two loci are inherited together (rather than being separated by recombination) in a single generation, the somewhat confusingly named *linkage disequilibrium* refers to the pattern of correlation between loci at a population level. An imaginary population with no new mutations and an infinite number of generations would have accumulated many historical recombinations at every position in the genome, resulting in linkage equilibriums, and a complete breakdown of correlation between mutations. By contrast, mutation and recombination in real human populations have occurred a finite number of times in particular locations in the genome, and the age, size and demography of these populations have left nearby loci correlated. This residual correlation, not yet ablated by historical recombination, is linkage disequilibrium (LD).

The following sections present a discussion of pair-wise measures of LD, their application to complex disease association studies, and developments of large-scale collaborative projects which have generated genome-wide catalogues of LD and enabled genome-wide association studies.

PAIR-WISE MEASUREMENT OF LD

The most straightforward way to summarize LD in a particular region is via simple statistics calculated for each pair of single-nucleotide polymorphisms (SNPs). Such an approach will generate a matrix of values that are both individually informative and from which broader patterns can be deduced. Consider a pair of SNPs, where the first has alleles A and a, and the second has alleles B and b. The allele frequencies of these alleles are f_A, $f_a = 1 - f_A$, f_B and $f_b = 1 - f_B$, respectively. We will define a *haplotype* as a particular series of alleles on one chromosome at consecutive variant sites. So if an individual had genotype AA at the first SNP and Bb at the second, the two haplotypes (corresponding to two chromosomes) would be AB and Ab. If f_{AB} represents the frequency of the AB haplotype, equation (1) shows the standard coefficient of LD between the two loci. Because the expected haplotype frequency in the absence of LD is the product of the marginal allele frequencies, D_{AB} represents the departure from the uncorrelated state.

$$D_{AB} = f_{AB} - f_A f_B \tag{1}$$

There are a number of problems with this statistic, however. Equation (1) focuses on D_{AB}, but it can be computed for any combination of alleles, and an algebraic rearrangement shows that $D_{AB} = -D_{aB} = -D_{Ab} = D_{ab}$. The sign of D thus depends on an arbitrary coding of the alleles. Furthermore, the range of D is highly dependent on the specific values of the allele frequencies, which makes it difficult to compare LD among many pairs of markers with diverse frequencies. For this reason, two derived LD statistics which both are frequency normalized and always positive are more commonly used [1].

The first, D' [2], is defined in equation (2) as the absolute value of the ratio of the observed value of D to the most extreme possible value given the allele frequencies. The second, r^2, is the squared Pearson correlation coefficient, given in equation (3). Each of these two statistics has particular properties and situations in which they are useful (see next section) but it is worthwhile to consider how they relate to the two generative forces of variation: recombination and mutation.

$$|D'| = \begin{cases} \dfrac{-D_{AB}}{\min(f_A f_B, f_a f_b)} & D_{AB} < 0 \\[2ex] \dfrac{D_{AB}}{\min(f_A f_b, f_a f_B)} & D_{AB} > 0 \end{cases} \tag{2}$$

$$r^2 = \frac{D_{AB}^2}{f_A f_a f_B f_b} \tag{3}$$

D' is sensitive only to recombination, and its principal usage is in characterizing historical patterns of recombination. The upper part of Fig. 2.1 shows a region of the genome with a number of SNPs that can all be placed on a single, consistent tree. As noted above, the existence of a consistent tree implies the absence of historical recombination. In such circumstances $D' = 1$ between all pairs of SNPs, providing a useful way to translate from pair-wise statistics into a better understanding of continuous LD (as reflected by the large dark swathes in the bottom part of Fig. 2.2). If one were to observe a haplotype similar to that in the bottom of Figure 2.1, however, it would be impossible to recreate a consistent tree to explain all the haplotypes, instead requiring an obligate recombination between two branches (squares and triangles, in this case) and $D' < 1$.

Whereas many historical mutations in a recombination-free region have $D' = 1$, both mutation history and recombination drive $r^2 = 1$ (those

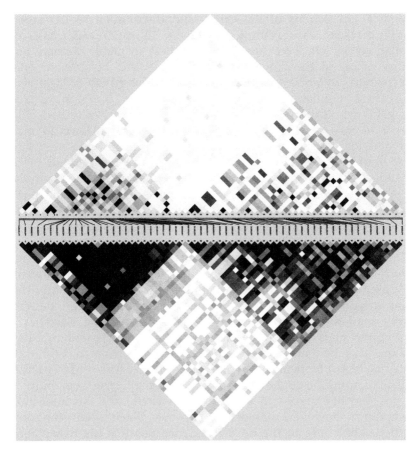

FIGURE 2.2   LD plots from Haploview [16] from HapMap data in a 100 kb region on chromosome 7. The middle bar shows the location of all genotyped SNPs in the region, the upper triangle shows pair-wise r^2 values, the lower triangle shows pair-wise D' (white = 0, black = 1). While the D' is high in two large, mostly contiguous blocks, r^2 is close to 1 only in sparse fragments within those blocks.

with the same shape in Fig. 2.1 have $r^2 > 0.8$). Recombination will clearly reduce r'; it is mutations on a historical branch, rather than a physical segment, which dictate whether r^2 are always within a region of high D', but can otherwise be non-contiguous (Fig. 2.2).

PREDICTED AND OBSERVED PATTERNS
OF RECOMBINATION

Given the importance of LD patterns in designing association studies (especially genome-wide), there was great interest in evaluating theoretical predictions of the extent and distribution of LD among nearby markers. For example, Kruglyak [3] presented coalescent simulations which were based on the best available human population model at the time (an ancient population of constant effect size followed by an exponential increase approximately 100,000 years ago). These simulations predicted that even for common variants, strong LD would only persist at an extremely local scale (dwindling to zero in less than 100 kb). Such extremely local LD would make genome-wide association studies extremely difficult, as a very large number of SNPs would be required to capture most or all of the variation; Lander and Schork commented that "Association studies are not well suited to whole-genome searches in large, mixed populations" [4]. For this reason, many believed that focused gene-based studies would be feasible far earlier than unbiased, truly genome-wide scans.

These predictions were contradicted, however, by an intriguing pattern discovered by Daly *et al.* [5] in a region of chromosome 5q31, as part of a study of inflammatory bowel disease. Rather than seeing the predicted smooth, rapid decline from perfect LD between nearby markers to zero after a few tens of kb, long stretches of consecutive markers were highly correlated, punctuated by sharp breakpoints where even adjacent SNPs were uncorrelated. Within the blocks of high LD, nearly all common variation was represented by only a handful of common haplotypes (despite the possibility of 2N possible haplotypes in a set of N markers). At the inter-block breakpoints, however, haplotypes in one block were randomly shuffled with those in the next block. Because this discovery could have significant implications for association study design, it was followed up by several groups [6, 7] who consistently observed the same pattern of blocks of high LD with punctate breakpoints (Fig. 2.2).

An explanation for these observations was posited as a result of unrelated work which examined individual recombination events observed in sperm meioses [8]. Nearly all detected recombinations were confined to narrow hotspots rather than being homogeneously distributed across the sequence examined. If generalized to all recombinations that have occurred historically in a population across the whole genome, this would explain the discrepancy from theoretical predictions, which assumed a constant rate of local recombination. Long stretches of the genome are characterized by low levels of historical recombination broken by intense, narrow recombination hotspots. This hotspot view did not, however, win immediate widespread acceptance, nor was the block view of LD considered to fully capture the subtlety of the underlying pattern of variation.

THE INTERNATIONAL HAPMAP PROJECT

It became clear that a full resolution of these issues would require a large-scale project to investigate the nature of LD across the entire human genome. The International HapMap Project was formally initiated in October 2002 and included researchers from 20 groups in six countries with the aim of characterizing millions of DNA sequence variants, their frequencies and the correlations between them in samples from populations with ancestry from Africa, Asia and Europe [9].

Anonymized samples were selected from four populations: 90 Yoruba (30 parent-parent-offspring trios) from Ibadan, Nigeria (abbreviation: YRI); 90 individuals (30 trios) of European ancestry from Utah, collected in 1980 by the Centre d'Etude du Polymorphisme Humain (CEU); 45 unrelated Han Chinese from Beijing (CHB); and 45 unrelated Japanese from Tokyo (JPT). Pilot studies and simulations indicated that complete ascertainment in 45 unrelated individuals would represent 99% of variation with MAF >5% in the populations from which the samples were drawn.

When confronted with the choice of SNPs to genotype, the first challenge for the project was the need for a much larger set of variants than existed at the time, with good coverage in as large a fraction of the genome as possible. Building upon the previous work of the SNP Consortium and the International Human Genome Sequencing Consortium [10], large-scale shotgun sequencing was undertaken to increase the catalog of observed SNPs. Because the project was focused on common variation, "double-hit" SNPs (i.e., those where the non-reference allele had been seen at least twice) were preferentially selected with the aim of generating a map with a common (MAF >5%) SNP in each population every five kilobases.

The resulting first generation HapMap [11] yielded insight into a wide variety of areas of human genetics research. Two of the more prosaic, but nonetheless important, advances were the validation of a large number of common SNPs (about 1.2 million unique SNPs passed quality control measures) and the opportunity to develop medium- and high-throughput genotyping technologies which enabled the subsequent wave of genome-wide association studies (see chapters on SNP selection chapter 4, study design chapter 3 and case studies chapters 18 and 19). The project also verified that the previously observed block-like pattern of LD generalized to the entire genome, and was not an artifact of small sample size or lower marker density. Furthermore, population genetics analysis [12] demonstrated that recombination hotspots are ubiquitous in the genome and are the driving force behind the observed LD patterns (with blocks of high LD corresponding to inter-hotspot intervals and sharp breakdowns of LD mapping to hotspots). Recombination rate maps resulting from these models have largely supplanted the earlier block-like models, and provide a useful continuous variable summary of LD. Finally, the genome-wide LD map made it straightforward to select a subset of common markers that captures all (or nearly all) the information contained in the full set. This process of thinning out markers based on r^2, or "tagging" (see Chapter 4 on SNP selection), made association studies more comprehensive and efficient.

A second generation map was published in 2007 [13] with over two million additional SNPs which provided better tag SNP selection, more precise estimates of local recombination rates and new insights into natural selection. The open availability of all project data (Fig. 2.2) has been a key to the design of complex disease association studies. The project is now being extended (HapMap3) to include samples from a number of additional populations with the hope of providing a comprehensive variation and LD resource to researchers in as many parts of the world as possible. Finally, many of the same partners are involved in the 1000 Genomes project, which pushes the HapMap paradigm into the analysis of rarer variation. Deep resequencing (including in the same HapMap samples) aims to provide a similar catalogue of all variants with MAF $> 1\%$, including copy number variation and short insertion/deletion polymorphisms in addition to SNPs. These international efforts, beginning with the sequencing of the human genome, extending into the study of variation via the HapMap, and culminating in the ongoing generation of thousands of complete genomes, have transformed human genetics and demonstrated the importance of large-scale collaboration.

CONCLUSION

The study of linkage disequilibrium is not only important in the design of efficient, powerful association studies, but has also yielded new insights into the history of human populations [14] and the biological mechanisms relating to recombination [15]. This chapter aims to provide an introduction to the important concepts related to LD, so that they can serve as a foundation for exploring human disease via genetic association studies. The fundamental goal of deciphering the biological underpinnings of human disease (via association studies or any other experimental design) can only be achieved by first understanding the forces which create and shape variation in human populations, and the patterns which result from those forces.

References

[1] P.W. Hedrick, Gametic disequilibrium measures: proceed with caution, Genetics 117 (1987) 331–341.

[2] R.C. Lewontin, The interaction of selection and linkage. I. General considerations; heterotic models, Genetics 49 (1964) 49–67.

[3] L. Kruglyak, Prospects for whole-genome linkage disequilibrium mapping of common disease genes, Nat. Genet. 22 (2) (1999) 139–144.

[4] E.S. Lander, N.J. Schork, Genetic dissection of complex traits, Science 265 (1994) 2037–2048.

[5] M.J. Daly, J.D. Rioux, S.F. Schaffner, T.J. Hudson, E.S. Lander, High-resolution haplotype structure in the human genome, Nat. Genet. 29 (2) (2001) 229–232.

[6] S.B. Gabriel, S.F. Schaffner, H. Nguyen, J.M. Moore, J. Roy, B. Blumenstiel, et al., The structure of haplotype blocks in the human genome, Science 296 (5576) (2002) 2225–2229.

[7] D.E. Reich, S.F. Schaffner, M.J. Daly, G. McVean, J.C. Mullikin, J.M. Higgins, et al., Human genome sequence variation and the influence of gene history, mutation and recombination, Nat. Genet. 32 (1) (2002) 135–142.

[8] A.J. Jeffreys, L. Kauppi, R. Neumann, Intensely punctate meiotic recombination in the class II region of the major histocompatibility complex, Nat. Genet. 29 (2) (2001) 217–222.

[9] International HapMap Consortium, The International HapMap Project, Nature 426 (2003) 789–796.

[10] R. Sachidanandam, D. Weissman, S.C. Schmidt, J.M. Kakol, L.D. Stein, G. Marth, et al., A map of human genome sequence variation containing 1.42 million single nucleotide polymorphisms, Nature 409 (6822) (2001) 928–933.

[11] International HapMap Consortium, A haplotype map of the human genome, Nature 437 (7063) (2005) 1299–1320.

[12] G.A. McVean, S.R. Myers, S. Hunt, P. Deloukas, D.R. Bentley, P. Donnelly, The fine-scale structure of recombination rate variation in the human genome, Science 304 (2004) 581–584.

[13] International HapMap Consortium, A second generation human haplotype map of over 3.1 million SNPs, Nature 449 (2007) 851–861.

[14] S.R. Grossman, I. Shylakhter, E.K. Karlsson, E.H. Byrne, S. Morales, G. Frieden, et al., A composite of multiple signals distinguishes causal variants in regions of positive selection, Science 327 (2010) 883–886.

[15] S. Myers, R. Bowden, A. Tumian, R.E. Bontrop, C. Freeman, T.S. MacFie, et al., Drive against hotspot motifs in primates implicates the PRDM9 gene in meiotic recombination, Science 327 (2010) 876–879.

[16] J.C. Barrett, B. Fry, J. Maller, M.J. Daly, Haploview: analysis and visualization of LD and haplotype maps, Bioinformatics 21 (2) (2005) 263–265.

3

Genetic Association Study Design

Krina T. Zondervan

Wellcome Trust Centre for Human Genetics, University of Oxford, UK

The era of genetic association studies started in the late 1990s, after linkage studies in families — which had been very successful in mapping genetically homogeneous rare Mendelian disorders — proved to be unsuccessful for mapping heterogeneous complex traits [1]. Population-based (i.e., non-family-based) genetic association studies in particular were heralded as the answer to gene mapping in complex traits [2]. The basis for this optimism was the belief that the genetic origin of complex traits would include common variants with modest effects on risk, also termed the "common disease–common variant" hypothesis [3, 4]. Each of these common variants would increase susceptibility to disease, rather than be a direct cause (the situation for rare variants underlying Mendelian disorders). This hypothesis was soon met with considerable skepticism when findings from most population-based association studies focusing on biologically plausible candidate genes could not be replicated [5]. Some scientists interpreted this as evidence against the basic premise of the study design, and argued that many rare, rather than few common, variants of small effect underlie complex traits [6, 7]. Population-based genetic association studies from even several thousand individuals had little power to detect such effects [8, 9]. Others argued that many of these candidate gene studies had a very low biological *a priori* probability of success, and that this was compounded by a low statistical probability of finding true effects due to limited adherence to basic study design principles, such as use of adequate sample sizes [8, 10, 11]. Only when the first genome-wide association (GWA) studies — not based on any biological hypotheses and with thousands of individuals sufficiently powered to detect common variants of modest effect — produced the first replicated successes, did confidence in the population-based association study return [12—17].

Ten years on from the advent of association studies, and the field has taken a middle ground, generally accepting that most complex diseases are likely caused by varying spectras of both common and rare genetic variations, while recognizing that identifying all variations contributing to the heritability of a trait remains a major challenge [9, 18, 19]. The past has learnt that a good understanding of study design principles optimizes an investigator's chances of finding genetic variation underlying complex disease. Focusing on population-based designs, this chapter discusses these principles. In Chapter 14, family-based association designs are discussed.

CONCEPTS AND SCOPE OF ASSOCIATION STUDIES

Association studies can be performed using any type of genetic polymorphism, but the one most commonly used is the single nucleotide polymorphism (SNP). This is because of its abundance across the genome, comprising $\sim 90\%$ of all human variation [20]. Unless targeting a specific, known polymorphism, all genetic association studies utilize one important population-genomic feature in their design: linkage disequilibrium (LD). LD is the allelic association between polymorphisms located near each other on the genome in a population, a feature shaped by genomic ancestral history influenced by founder origin, population bottlenecks, recombination, and genetic drift (see Chapter 2). LD is an extremely useful feature, as it means investigators do not need to genotype all polymorphisms in a region of interest. Instead, they can select a subset of SNPs that are proxies for the majority of all common genetic variation nearby (so-called "tagSNPs"). After a first version of the Human Genome sequence was completed [21], the International SNP Consortium published information on the location of millions of SNPs for potential use in association studies [22]. The International HapMap Project [23] subsequently set out to provide a map of LD between these SNPs across the genome of four populations of different ethnic origin (Caucasians of Northern and Western European origin, Japanese from Tokyo, Han Chinese from Beijing and Yoruba from Nigeria). Phase II of this map provided information on ~3.2 million common SNPs [24]; Phase III extends this effort to include seven further populations but for a more limited number of SNPs (http://hapmap.ncbi.nlm.nih.gov/). This wealth of information provided the backbone of all modern genetic association studies, allowing investigators to select tagSNP sets that provide optimal coverage of all common variations in a region of interest. Together with the rapid development of high-throughput genotyping technologies, available at ever-decreasing costs, it allowed investigators for the first time to expand their investigations from small-scale candidate gene studies in a few hundred individuals, to large-scale studies across the genome in thousands of individuals.

Candidate Gene Studies

The first type of genetic association study to be conducted, before the availability of dense SNP LD maps, was the candidate gene study. Numerous candidate gene studies were (and continue to be) published, but the results of very few could be replicated, suggesting that most reported findings were false positives. In 2002, a review of 603 candidate gene case-control studies from 1986 to 2000 was published, which found that results of only six were independently replicated [5]. A crude tally of original studies in humans published up to 2009 with "candidate gene" in the title, abstract or keywords shows that — despite lack of success — their popularity has certainly not diminished (Fig. 3.1). Candidate gene studies to date have mainly followed biological hypotheses, usually generated because of experimental evidence that particular biological pathways are likely to be involved in disease onset. Such *de novo* candidate gene studies have an intrinsically very low chance of success, given the many genes that are involved in often highly complex and interacting biological pathways. Moreover, our knowledge of pathways and their gene membership is currently still limited although this situation will no doubt change in the future (see, for instance, the *Kyoto Encyclopaedia of Genes and Genomes* — KEGG [25]). Candidate gene studies that have the highest

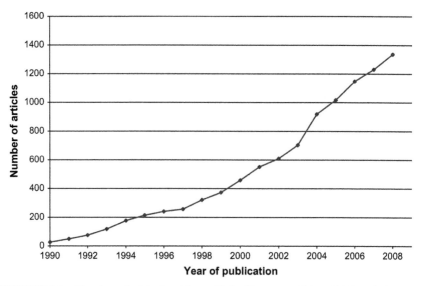

FIGURE 3.1   Number of original articles published between 1990 and 2009 with the term "candidate gene" in title, abstract, or keywords, retrieved using PubMed.

chance of success, therefore, are those that aim to replicate initial findings of hypothesis-free association studies that implicate the gene or region surrounding it.

A second reason for the failure of many candidate gene studies to date is that many have focused only on a small selection of SNPs in coding or regulatory sequences, rather than conducting an exhaustive search using tagSNPs to cover the entire gene (where necessary supplemented by less well-tagged coding SNPs) [26]. However, a recent analysis of all GWA studies published to date showed that 88% of signals were either inter-genic (43%) or intronic (45%) [27]. This clearly indicates that limiting marker selection to SNPs with a functional implication is unlikely to provide the best approach, at least for the identification of common variants. A more detailed overview of SNP selection aspects is provided in Chapter 4 "Tag SNP Selection".

Lastly, most candidate gene studies have suffered from limitations in study design, including inappropriate case definition and control selection and − most importantly − vastly inadequate sample sizes, aspects that will be discussed later in this chapter.

Candidate Region Studies

Candidate region studies distinguish themselves from candidate gene studies by focusing on a larger genomic region containing several, to even 100s, of genes. They tend to be based on a statistical rather than a biological hypothesis, as the region is selected on the basis of *a priori* statistical evidence, usually significant linkage. Often, the region is first characterized through bioinformatic approaches, and candidate genes of particular interest may be highlighted. However, because of the above-described issues with defining candidate genes of interest, an association study across the entire region is usually conducted. As such, candidate region studies can be seen as a hybrid between candidate gene and GWA studies, and are seemingly a potentially attractive strategy. Their main drawback, however, is the basis for selection of the region, which is most often the observation of significant linkage in the area. Given that the variants underlying linkage signals are most likely to be rare, and population-based association studies are only able to detect the effects of common variants, it is questionable whether candidate region studies are able to identify the variants responsible for the linkage peak [28]. Often the argument behind the approach is that the same genes harboring rare variants impacting on the complex trait may also contain common variation that is of influence. To date, few association studies following linkage evidence have uncovered such evidence of common variation [29, 30], with some notable examples [31, 32]. As resequencing strategies are becoming more cost-effective for targeted regions, new methodological approaches developed to investigate cumulative evidence of rare variants [19, 33] are likely to increase the popularity of the candidate region study.

GWA Studies

Despite early concerns about multiple testing issues and power, GWA studies have brought the gene mapping successes for complex traits, which candidate gene studies were largely unable to deliver. The historical first successful GWA was conducted in age-related macular degeneration [12], a study notable for its extremely small sample size (<100 cases) and the use of only 100,000 SNPs randomly distributed across the genome, and successful because of a fortuitous selection of SNPs and the uncharacteristically large effect size of the variant in question. One of the first large-scale GWA studies was conducted by the Wellcome Trust Case Control Consortium — WTCCC [14]. With a sample size of 2,000 cases per disease and 3,000 common controls of similar ethnic origin, the WTCCC identified a total of 23 significantly associated common variants for six out of seven selected common diseases; 22 of these were replicated in independent studies. Other GWA studies showed similar successes [16]. To date, hundreds of GWA studies have been conducted in over 80 diseases and traits; published studies are summarized in a continuously updated online catalog: http://www.genome.gov/gwastudies/ [27].

GWA studies were made possible by the publication of the International HapMap [23] providing SNP LD information across the human genome, and the simultaneous development of genotyping chip arrays based on this information, allowing the investigation of hundreds of thousands of SNPs in thousands of individuals. Rapid reductions in high-throughput genotyping costs subsequently made GWA studies financially feasible. The latest genotyping arrays provide up to 1 million common SNPs — either selected entirely on the basis of LD coverage or based on a mixture of random and LD-based selection — supplemented by selected sets to examine copy number variants [26]. At the time of writing, average cost per array currently approximates $400, a figure that will no doubt continue to decrease. It is likely, however, that in future GWA studies based on genotyping arrays will be superseded by those based on whole-genome resequencing, with the "$1,000 dollar genome" eagerly anticipated ([34]; http://www.nature.com/ng/qoty/index.html). In the meantime, current ongoing efforts such as the 1000 Genomes project (http://www.1000genomes.org/page.php), which aims to resequence the genomes of 1,000 individuals around the world, will help investigators understand to what extent GWA genotyping arrays cover all genetic variation across the genome and what is being missed [19, 35].

POPULATION-BASED STUDY DESIGNS

The most popular population-based study design for binary traits (whether focusing on candidate genes, regions, or the entire genome) is the case–control study. In a case–control study, a set of cases is identified and their genotyping information compared to a set of suitable controls in order to find genetic variants associated with the trait under study. Cases and controls are usually sampled independently from the same ethnic source population. An alternative approach involves sampling from a cohort of individuals that is being followed up prospectively (nested case–control or case–cohort study designs), such as provided by large-scale biobanking initiatives [36].

For quantitative traits, individuals are either sampled from the two tails of the distribution and their genotyping information is compared — similar to a case–control study — or they are sampled randomly across the spectrum of trait values in a population; association is investigated by analyzing mean trait values between genotypes categories [37]. Although the remainder of this chapter will focus on case–control studies — as the most popular design currently being adopted — implications for the design of quantitative trait association studies will be briefly discussed towards the end.

The population-based genetic case—control study has been criticized for two main reasons: (1) the fact that it cannot generally detect rare genetic variation underlying complex disease — unless their effects are very large [6, 38]; and (2) its susceptibility to spurious findings (false positives) through confounding by population origin [39, 40]. The first criticism is undeniably true; with sample sizes in excess of 1,000 individuals, genetic case—control studies in unrelated individuals are designed to detect relatively common variants of modest effect. They are generally unable to detect variants with allele frequencies less than ~1—5% in the general population [9]. When employing LD mapping, the lack of power is further compounded under a scenario of allelic heterogeneity where multiple rare variants arising on different ancestral backgrounds effectively cancel out each other's signals [8, 38]. Provided that there are common variants that underlie complex traits — and the results of recent GWA studies indicate that this is so — this is no argument against case—control studies *per se*. Rather, it shows that population-based case—control studies will only be able to detect a proportion of all genetic variation underlying complex traits.

Confounding by population origin — or population stratification — is an important and often quoted potential drawback of population-based genetic case—control studies. Population stratification arises when (different proportions of) cases and controls are sampled from genetically different underlying populations, thus causing any associations found to be due to sampling differences rather than the disease of interest. Concerns for confounding in case—control studies are not new. In epidemiology, it has long been recognized that case—control studies can be particularly prone to bias and confounding, and thus producing spurious results, when little or no attention is based on study design [41, 42]. However, the design of currently published genetic case—control studies can sometimes be a far cry from the original epidemiological principles intended to minimize problems of bias and confounding. To what extent does this matter? Some of these principles were designed to prevent information biases related to changeable non-genetic risk factors under study, and not necessarily to non-changeable genetic factors. Furthermore, investigators have argued that as long as populations are reasonably matched on origin [43, 44], residual population stratification in GWA data can be adjusted for in downstream analysis using various methods of genomic control (see Chapter 9) [45—49] — although it remains unclear to what extent fine-scale substructure can be adjusted for in ever-increasing sample sizes utilized in GWAs [50]. It is important, however, to realize the epidemiological origins of the case—control design before considering its implications in a genetic research setting. Moreover, as soon as genetic association studies are extended to include environmental factors and their interaction, an appreciation of classic epidemiological study design becomes crucially important.

Case–control Study Design – Theoretical Basis

The case–control study was first adopted in 1926 [51], with methodology fully employed in the early 1950s in a seminal paper on the association between smoking and lung cancer [52]. Its design provided a cost-effective alternative to the prospective cohort study, allowing the investigation of many different risk factors in relation to a single, specifically defined, disease. The prospective cohort study is the gold standard in epidemiological study design, and aims to identify unbiased risk estimations of disease for exposures (which can be environmental or genetic) in populations at risk. In a cohort study, a representative subsample of a population free of the disease of interest, but exposed and unexposed to the exposure(s) of interest, is followed up to determine who develops disease. By definition, cases – recruited following a precise clinical definition – are incident (rather than prevalent) and absolute risk of disease is expressed in terms of incidence rates among exposed and unexposed [53]:

$$I_{exposed} = \frac{D_{exposed}}{T_{exposed}} \quad \text{and} \quad I_{unexposed} = \frac{D_{unexposed}}{T_{unexposed}} \tag{1}$$

where I = incidence rate, D = the number of individuals developing disease and T = the amount of person-time spent by individuals in the exposed or unexposed group.

Relative risk of disease is then calculated as the incidence rate ratio among exposed over unexposed:

$$RR = \frac{I_{exposed}}{I_{unexposed}} = \frac{D_{exposed} \cdot T_{unexposed}}{D_{unexposed} \cdot T_{exposed}} \tag{2}$$

In its originally intended form, the case–control study aims to derive a disease risk estimate for an exposure that is as close as possible to the estimate that would have been derived if a prospective cohort study had been performed. This means that a case–control study should include the same individuals as cases that would have been sampled in a prospective cohort study. In this situation, the number of cases among unexposed can be directly approximated by the number of exposed among cases in the case–control study:

$$D_{exposed} \approx E_{cases} \quad \text{and} \quad D_{unexposed} \approx U_{cases} \tag{3}$$

where E = exposed and U = unexposed individuals.

The ratio of person times spent in the unexposed and exposed groups in a cohort study can be approximated in a case–control study by the ratio of unexposed and exposed individuals among controls, provided that these are sampled from the same population as cases, and sampling is independent of exposure status:

$$\frac{T_{exposed}}{T_{unexposed}} = \frac{E_{controls}}{U_{controls}} \tag{4}$$

Combining Equations (2), (3) and (4) gives the equation for the risk estimate obtained in case–control studies, the odds ratio (OR):

$$OR = \frac{E_{cases} \cdot U_{controls}}{E_{controls} \cdot U_{cases}} \tag{5}$$

So, the two key principles that allow the OR to provide a good, unbiased approximation of the RR are: (1) sampling of incident cases; and (2) consecutive sampling of controls from the same at-risk population from which cases were sampled. This sampling scheme has an additional important advantage for epidemiological studies aimed at investigating "environmental" (i.e., non-genetic) risk factors (e.g., through question-naires or interviews completed by study participants). The recruitment of incident (rather than prevalent) cases and controls simultaneously minimizes differential information biases between the two groups, which could result in spurious associations. For instance, it is likely that prevalent cases may have changed lifestyle habits in response to the disease, and efforts by investigators to address this by assessing risk-factor exposure prior to disease onset may be influenced by recall bias.

Sampling of controls from the same source population from which cases arose is a key principle in case–control design. Depending on how cases were sampled, this source population may not be easy to define. For instance, cases can be sampled from a population-based registry identifying all cases in a certain region or they may have been sampled from a range of different clinics. It is important that controls should have had the same opportunity as cases to develop the disease of interest, and — had they done so — the same opportunity to be diagnosed with the disease and to be recruited into the study. This ensures that their exposure profile does not differ systematically from cases simply because of sampling differences. As an example, consider an epidemiological study of the effect of alcohol use on pre-eclampsia — a complex and dangerous condition in pregnant women involving hypertension and proteinurea [54]. Controls selected needed the same opportunity as cases to have developed the condition — i.e., been pregnant and recruited from the same area/clinic as cases. This is because alcohol use is likely to differ markedly with pregnancy and gestational status as well as with, for example, geographical catchment area (and a host of socio-economic indicators), and any design not allowing for this would cause spurious associations between alcohol use and pre-eclampsia. In practice, studies of non-genetic risk factors can employ a range of control selection schemes — depending on the disease of interest and the case sampling scheme used — varying from community-based controls (random selection; "friend" controls) to clinic-based controls diagnosed with other conditions [42]. When the case source population is difficult to define, multiple control groups recruited from different settings are sometimes used.

One approach to improve the similarity between cases and controls for potential confounding factors (and thereby increase the probability that they were sampled from the same theoretical source population) is matching. Matching refers to a selection of controls (that are as similar as possible) to cases with respect to the distribution of one or more potential confounding variables that are difficult to measure or for which the distribution is likely to be extremely dissimilar between cases and controls (thus inhibiting the potential for statistical adjustment at the analysis stage) [55]. In the example of the study relating alcohol use to pre-eclampsia, additional potential confounding factors, such as age and gestational stage, could be taken account of through matching of controls to cases at the time of recruitment. Matching is not without disadvantages, however, as matching variables need to be used as stratification variables in the analysis, and can no longer be investigated as an exposure of interest. Furthermore, matching on variables that are unlikely to be confounders ("overmatching") will reduce power and efficiency [55, 56]. In the pre-eclampsia study, it might be a better option to allow for age in the analysis phase, as long as its distribution among cases and controls is not very dissimilar.

Genetic vs. Environmental Case—control Study Design

How do principles of epidemiological case—control design, originally intended to reduce bias in environmental case—control studies, impact on studies which aim to investigate the influence of genetic variants alone? The key difference between genetic and environmental association studies is that in genetic studies no allowance has to be made for biases inherent to the changeable nature of exposure information — our DNA remains the same over our lifetime. Because of the non-changeable nature of genetic variants in individuals, most of the biases described for classical environmental case—control studies will not be applicable to genetic association studies. One important bias remains, however, which is confounding due to sampling from different source populations when these populations have different genetic profiles — population stratification. But there are other, more subtle, differences between the two fields that have influenced the manner in which genetic association study design has evolved.

First, genetic association studies — most of which are designed to uncover *de novo* associations — are mainly focused on hypothesis testing rather than additional effect size estimation. The reason for this is that due to the vast numbers of statistical tests being conducted in a single study, the first aim is to find a true association. This means that sample size is one of the most important factors in study design, and both incident and prevalent cases are included in recruitment (as there is generally no concern for the OR to be a close estimate of the relative risk in the underlying population, and inclusion of both will not result in spurious results due to information biases such as differential recall). Also, study designs can include cases with a family history of the disease of interest, or multiple cases from the same family (allowing for the genetic correlation between them), in an effort to enrich cases for shared genetic variants, and thus artificially increase the effect size in the study sample [8, 57]. Only after an association is found, and replicated in independent studies, is there an interest in investigating to what extent the polymorphism in question accounts for the total phenotypic variance observed in the population — requiring more accurate effect size estimation.

A second subtle difference is that genetic association studies of complex disease aim to find common variants with small effects, with ORs in the 1.1–1.5 range, requiring very large sample sizes. This is true by definition, since a common variant with a much larger relative risk would result in a large attributable risk for that variant with respect to the disease, i.e., the variant would explain a very large proportion of the causality of the disease, making it resemble a Mendelian rather than a complex condition. Although it is recognized that large sample sizes are also required to investigate non-genetic factors underlying complex traits, relative risks of 1.5 and over are not uncommon. So, whereas sample sizes of a few hundred cases and controls could potentially still provide information on environmental exposures with large effects, they will certainly not be sufficient to uncover the vast majority of genetic associations.

The next section discusses the practical design implications for genetic association studies in more detail.

Designing a Genetic Case–Control Study

There are several steps in the design of a population-based genetic case–control study — be it a candidate gene, candidate region, or genome-wide association study. A paper outlining these, and providing a detailed protocol with practical recommendations, was published in 2007 [58], and

BOX 3.1

CHECKLIST OF DESIGN CONSIDERATIONS FOR POPULATION-BASED GENETIC CASE–CONTROL STUDIES

1. *Define disease/trait of interest accurately and specifically.* Non-specific definitions will increase heterogeneity of underlying causal factors and decrease power of the study.
2. *Consider evidence of the heritability of the disease/trait.* Low heritability values mean a larger number of individuals will need to be recruited to have sufficient power of detection. However, high heritability values do *not* guarantee that genetic variants will be easier to detect. The (unknown) allele spectrum of the trait will be crucial.
3. *Is a population-based design appropriate?* Consider the feasibility of recruiting thousands of cases required to detect common variants of modest effect. Does the trait show Mendelian-type inheritance patterns? Is there *a priori* evidence that the variants to be detected are likely to be rare?
4. *Select controls derived from the same ethnic (sub)population as cases.* This requires an understanding of how cases were sampled. Are there any other covariates that may give rise to population stratification? Even when matching controls to cases on ethnicity, analytical methods will likely need to be adopted in GWA studies to adjust for more subtle population substructure.
5. *Conduct power calculations to determine minimum required sample size.* At least 1000 cases and 1000 controls will be needed for candidate-gene studies involving a small number of polymorphisms. In determining appropriate sample size, parameters such as case definition and prevalence, the extent of genomic region to be investigated, investigation of direct vs. indirect (LD-based) association, and the inclusion of phenotypic measurement error, need to be considered.
6. *Is it a de novo or a replication study?* Note that effect sizes in previous (small) studies tend to be biased upwards and need to be reduced for sample size calculations in replication studies. Replication studies need to be conducted in the same ethnic (sub)population as the original study to provide evidence of replication.

others have provided similar guidelines [59]. The key points are summarized below and in Box 3.1. Noteworthy here is also the recent development of guidelines for the reporting of genetic association studies, which aims to enhance transparency of study reporting (design, conduct, and analysis): the STrengthening the REporting of Genetic Association studies (STREGA) initiative [60].

The first step in the design of any case—control study is to *define the disease or phenotype of interest* as accurately and specifically as possible. This is important because non-specific disease definitions will increase both genetic and environmental causal heterogeneity. This will reduce the power of detection of a true effect, and the ability to replicate findings in other studies. Often a compromise will have to be made between using a phenotype definition that is as specific as possible — and of optimal biological relevance — and one that is deemed to be clinically relevant. Moreover, such definitions are subject to change over time, as knowledge of the phenotype in question increases. Multi-center recruitment may result in case definitions being relaxed in an effort to acquire sufficiently large sample sizes and increase power. In reality, this may result in a smaller increase in power than anticipated, and could even result in a reduction in power. In practice, the aim should be to define cases using a definition that minimizes likely causal heterogeneity based on all existing clinical and biological evidence. In genetic association studies, both incident and prevalent cases can be included, as their focus tends to be on hypothesis testing rather than effect estimation. For those complex traits that are self-limiting (non-chronic), this can have potential consequences for interpretation of the results. In theory, results of such studies including incident cases would measure genetic susceptibility to disease onset, whereas those including prevalent cases could measure susceptibility to both disease onset and prolonged maintenance.

Although seemingly obvious, the next step is to consider to what extent the disease or phenotype of interest — under the decided-upon definition — is *heritable*. Evidence from family-based (particularly twin) studies may indicate a relatively low heritability (e.g., 10—20%), which means that large sample sizes are likely to be required to detect any common variants contributing to the genetic variance. It should be noted, however, that a high heritability does not necessarily mean genetic variants will be more easily detectable, as most of this genetic variation could be due to rare variants of small effect. Poignant examples of this are hypertension and psychiatric conditions such as bipolar disorder, for which heritabilities in excess of 80% have been reported [61]. Although both suffer from complications surrounding heterogeneity of case definition that could have negatively impacted on power, it was still surprising — given the high heritability observed — that underlying variants for hypertension/blood pressure were not identified until recently through a very large meta-analysis of studies involving >70,000 individuals [62], when the first two genes underlying bipolar disorder were revealed after combining evidence from three separate GWA studies

[63]. Indeed, a recent GWA analysis of schizophrenia involving ~3,300 cases and 3,500 controls, which did not find any genome-wide significant signals, provided evidence of a large polygenic component involving thousands of common alleles of very small effect, and showed that this component also contributed to the risk of bipolar disorder [64]. As such, the allelic frequency spectrum of a trait override's the importance of heritability values, but the problem is that the former is most often unknown until evidence from different GWA studies accumulates.

Next, it is important to weigh up whether a *population-based* case—control study is the *appropriate* design for the research question. To attain sufficient levels of power to detect common variants with odds ratios in the 1.3–1.5 range, thousands of cases need to be recruited. Is this feasible for the complex disease under study? One way of potentially decreasing the number of cases required is through a selection of cases with a family history of the condition (or even by selecting multiple cases from families while adjusting for familial correlation in analysis) — also termed "enrichment sampling" [57]. The hypothesis behind this approach is that familial cases may be more similar in genetic etiology, effectively increasing the effect size of the underlying genetic variants to be detected. Reducing the number of cases to be recruited on this basis, however, may be very risky, as familial aggregation could also be due to shared environmental factors, or — in the case of candidate gene/region studies — to genetic variants not under study. Moreover, if the case definition involves a relatively rare subphenotype that shows clear familial aggregation in families (and such multiple families can be recruited and genotyped), a family-based approach is likely to be more suitable than a population-based study.

Once a case definition and appropriateness of population-based study design is decided upon, the next step is to consider *control selection*. For reasons previously outlined, the general rule for controls is that they should be selected from the same population of individuals who are at risk of becoming — and being selected as — cases according to the case definition and recruitment strategies for the study. Applying this guideline minimizes the chance of false positive results due to various biases and confounding data. In genetic association studies many of the information biases that are problematic in environmental risk-factor studies do not apply, because of the non-changeable nature of the genetic exposure. The most important type of bias is confounding by population origin, or "population stratification," a situation in which cases and controls differ in terms of allele frequency of genetic variants not because of the disease under study, but because of having been sampled from populations with different ancestral histories. Population stratification is minimized when controls are matched to cases on ethnicity, or when cases and controls are restricted to a particular ethnic group. For candidate gene/region studies, the design stage is often the only opportunity available to minimize confounding by population origin. In GWA studies, further detailed investigation of genome-wide genotypes during the analysis stage, combined with external genotyping information from defined populations of different ethnic origin, can help in identifying and correcting for more subtle population substructure (see Chapter 9) [45, 46, 49]. It should be noted here that such correction for population substructure may become more difficult as sample sizes increase (and the influence of fine-scale substructure becomes more difficult to adjust for [50]), and that these approaches adjust for structure involving common variations. In the analysis of rarer variants, matching on common variant structure may not be sufficient and chromosomal region-specific matching may need to be considered [19]. However, as genomic control approaches are dependent on the availability of GWA information from subpopulations potentially impacting on the results in question, it remains important to avoid any potential stratification problems in control selection at the design stage.

Whether or not further matching on other covariates, such as sex, is necessary and reduces population stratification is a question for debate. It is likely to depend on the disease in question, how controls were selected, and on the extent in which phenotypic and covariate data are available for these. In situations where there are very pronounced gender differences in disease prevalence (e.g., gender-specific or autoimmune conditions), and very large studies are conducted, differences in gender distribution between cases and controls could potentially result in confounding results by genomic associations from other gender-related traits. Here, matching on sex could be considered, though a more efficient approach would likely involve simple adjustment for sex in the analysis stage. When a covariate has a marked influence on disease risk but is unlikely to be a confounder, such as age, some degree of matching may still improve power of the study by ensuring that controls had a similar opportunity to be diagnosed with disease as cases had. This only applies, however, where disease information is available for controls and they have been screened to be unaffected. For example, selecting controls that are considerably younger than cases for an age-dependent disease may reduce power because young individuals exposed to risk genotypes who may develop disease in future would be included as "unaffected" among controls. It is important to note though that "overmatching" on unnecessary variables carries a penalty in terms of power, as matching variables need to be adjusted for in the analysis stage [55].

To what extent controls should be "healthy" or include individuals with a mixture of other conditions is an interesting topic, and one that may cause less concern for genetic association studies than it does for studies focusing on environmental risk factors. Nevertheless, selective inclusion of other diseases among controls has the potential to increase the false positive rate, particularly for GWA studies of ever-increasing size. In addition, pleiotropic effects (influence of the same genetic variant on different traits) could potentially reduce effect size and increase false negative rate. If case datasets of other, seemingly unrelated, conditions are included as controls, extreme care has to be employed in the analysis stage to investigate to what extent association signals are driven by particular incorporated datasets [65]. An increasingly common practice for GWA studies is to use already genotyped "common" controls, such as those provided by the Wellcome Trust Case Control Consortium [14]. However, phenotypic and demographic information on these individuals is often very limited, which minimizes opportunities of excluding individuals with certain conditions to increase power or adjusting for covariates in analyses. Large-scale cohort biobanking studies, rich in such phenotypic information, should in future be able to provide such well-characterized control sets [36].

Adequate sample size is one of the most important considerations in analysis of genetic association studies [66]. If cases are sampled from the general population (i.e., without employing enrichment strategies), then the ORs to be detected for common variants with allele frequencies >0.2 in a case–control study are likely to be in the region of 1.1–1.5. For variants that are of lower frequency, though not rare (i.e., with allele frequencies between 0.05 and 0.2), ORs of up to ~3.0 are possible. As a bare minimum, sample sizes of at least 1,000 cases and 1,000 controls are required to detect ORs of ~1.5 with at least 80% power when testing a limited number of polymorphisms, but the required size will depend on whether: (1) the analysis will include case subgroups; (2) the analysis focuses on candidate genes with a limited number of independent tests or GWAs with many thousands such tests; (3) the study is designed to replicate a previously found association for a polymorphism with a known allele frequency (direct association), or whether it aims to find a novel polymorphism of unknown frequency and effect size through LD mapping (indirect association). Figure 3.2a and b show the number of cases and controls required for a candidate gene and GWA study of a disease with 5% prevalence in the population, to detect an SNP with allelic odds ratios varying from 1.1 to 1.7 with 80% power. A multiplicative risk model (where $GRR_{AA} = (GRR_{Aa})^2$) is assumed. For the candidate gene study, a scenario of 20 tagSNPs is assumed, resulting in an SNP-based significance threshold of $\alpha = 0.0025$ after Bonferroni correction of the target overall type I error of 0.05. For the GWA study, an SNP-based significance threshold of $\alpha = 5.0 \times 10^{-7}$ is assumed, as adopted previously and estimated to correspond to a genome-wide type I error of 0.05 [14]. Often recruitment and genotyping of cases is more expensive than that of controls (particularly if the latter are publicly available) and power can be increased by increasing the case:control ratio up to ~3 to 4 controls per case [41]. Figure 3.2a and b clearly show this power increase, particularly for SNPs with small effect (OR = 1.1–1.3) or with lower allele frequency in the population (<0.2). These simple calculations, however, do not take account of phenotypic assessment error often present in practical study settings. A recent paper describes the development of a web-based simulation program to calculate realistic sample sizes for case–control studies under various scenarios of measurement error: ESPRESSO (www.p3gobservatory.org/powercalculator.htm) [66].

If genome-wide genotyping of thousands of cases and controls is prohibitively expensive, a two-stage design can be more economical [67]. In such a design all SNPs are tested for association in a subset of cases

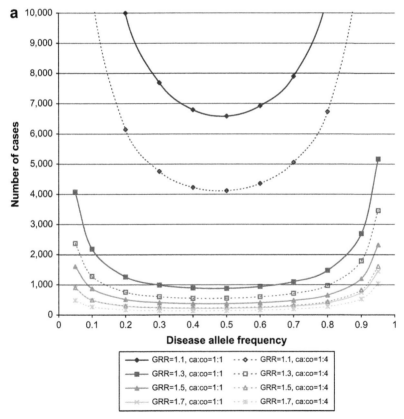

FIGURE 3.2 Number of cases required to detect a variant conferring genotype relative risks between 1.1 and 1.7 (under a multiplicative model) to a disease with a population prevalence of 5%, with 80% power, for (a) a candidate gene study involving 20 independent tagSNPs (SNP-based type I error = 0.0025); (b) a GWA study (SNP-based type I error = 5.0×10^{-7}. Adapted from Zondervan & Cardon [58]. Please refer to color plate section.

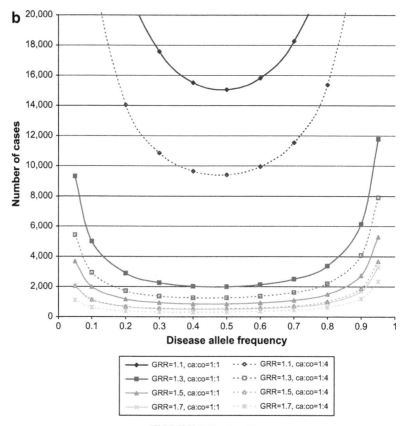

FIGURE 3.2—*Cont'd.*

and controls, and those exceeding a predetermined liberal significance threshold are genotyped in the remainder of the study sample; subsequent analysis is carried out for the two stages combined to maintain sufficient power [68]. The optimal two-stage design depends on the genotyping costs in stages 1 and 2, but — importantly — also on the underlying (unknown) disease model. With genotyping costs continuing to decrease, one-stage designs are now generally preferred over two-stage designs.

A final consideration in genetic case–control design is whether a *de novo* or a replication study is involved. If a replication study is being designed, it is important to note that the effect size found in the original study is likely to be biased upward, the so-called "Winner's curse" [69, 70]. Sample size calculations for a replication study should therefore assume a smaller effect size than originally reported. Another important point is that a true replication study will involve analysis of the same polymorphism with the same direction of effect in the same ethnic population measured on the same phenotype as the original study. As causal pathways and the relative contribution of a polymorphism may differ between populations, failure to observe a significant association in a different ethnic population does not constitute a failure to replicate, but only provides information on the lack of a similarly sized effect.

Quantitative Trait Study Design

Studies investigating genetic association for quantitative, rather than binary, generally adopt two strategies [37]. Either, they use phenotypic values from all individuals randomly sampled from a particular population (clinical or otherwise); or they sample individuals from the two extreme tails of the distribution, in an effort to increase power of the study while reducing genotyping cost [71]. The latter approach has been a natural follow-on from linkage studies, which had shown that including extreme phenotypic individuals could increase power of linkage detection [72]. Also, being able to use individuals already ascertained in this manner for (family-based) association studies appeared attractive. However, this type of sampling tended to enrich the detection of rare, deleterious poly-morphisms underlying extreme trait values originating from Mendelian-type transmission — exactly the scenario linkage studies had the most power to detect. In a population-based association setting, aiming to uncover a common variation, sampling from the extreme tails of a quan-titative trait distribution is unlikely therefore to be a useful strategy [19].

Many of the concerns that relate to appropriate case–control study design are also useful considerations in the design of quantitative trait genetic association studies. Careful consideration of a specific phenotypic definition, to optimize chances of finding genes underlying the trait, remains important, as is its application to a genetically homogeneous population to avoid population stratification issues. Moreover, recent GWA studies have reported that extremely large sample sizes exceeding tens of thousands individuals are often needed to detect common varia-tion underlying normally distributed quantitative traits [62, 73–76]. Even with these extremely large sample sizes, the polymorphisms found tend to explain only small proportions (typically <10%) of phenotypic variance [19, 35].

CONCLUSIONS

The aim of the present chapter was to review various aspects of population-based study design, aimed at uncovering common genetic variations underlying complex traits. With the advent of large-scale GWA studies in recent years, many successes have been — and continue to be — made in establishing common variants underlying many different complex conditions for the first time. With these successes have also come some sobering thoughts. First, there is the realization that many of these variants lie in what are currently considered as gene deserts, and their biological function is not obvious. Second, these variants tend to explain only very small proportions of overall heritability, leaving many wondering where the missing heritability can be found [18, 19]. Undoubtedly, rare variants will play a role in many conditions, but unfortunately our current arsenal of (population-based) study design tools — as reviewed here — is not well equipped to uncover these. Another question that remains to be fully explored is to what extent are other types of genetic variation (e.g., structural variation such as copy number variants [77]) important.

In order to understand the influence of implicated genetic variants, both biologically and from a clinical/public health point of view, future studies will need to incorporate genetic/molecular as well as environmental exposures in assessing disease risk. Incorporation of environmental exposure assessment in case–control studies will require a more careful consideration of epidemiological study design than adopted in many genetic association studies to date. Large-scale biobanking initiatives involving cohorts for which detailed biological as well as epidemiologically sound disease and exposure information is collected, and in which nested case–control studies can be conducted, will help towards these efforts — at least for conditions for which they are able to accrue sufficient numbers of cases. For conditions that are less common, or not generally prioritized for studying these settings, dedicated well-designed case–control studies collecting comprehensive biological and environmental data are likely to remain important in the years to come.

References

[1] J. Altmuller, L.J. Palmer, G. Fischer, H. Scherb, M. Wjst, Genome-wide scans of complex human diseases: true linkage is hard to find, Am. J. Hum. Genet. 69 (2001) 936–950.

[2] N. Risch, K. Merikangas, The future of genetic studies of complex human diseases, Science 273 (1996) 1516–1517.

[3] D.E. Reich, E.S. Lander, On the allelic spectrum of human disease, Trends Genet 17 (2001) 502–510.

[4] F.S. Collins, M.S. Guyer, A. Charkravarti, Variations on a theme: cataloging human DNA sequence variation, Science 278 (1997) 1580–1581.

[5] J.N. Hirschhorn, K. Lohmueller, E. Byrne, K. Hirschhorn, A comprehensive review of genetic association studies, Genet. Med. 4 (2002) 45–61.

[6] K.M. Weiss, J.D. Terwilliger, How many diseases does it take to map a gene with SNPs? Nat. Genet. 26 (2000) 151–157.

[7] A.F. Wright, N.D. Hastie, Complex genetic diseases: controversy over the Croesus code, Genome Biol. 2 (2001) COMMENT2007.1-2007.8.

[8] K.T. Zondervan, L.R. Cardon, The complex interplay among factors that influence allelic association, Nat. Rev. Genet. 5 (2004) 89–100.

[9] M.I. McCarthy, G.R. Abecasis, L.R. Cardon, D.B. Goldstein, J. Little, J.P. Ioannidis, et al., Genome-wide association studies for complex traits: consensus, uncertainty and challenges, Nat. Rev. Genet. 9 (2008) 356–369.

[10] N.J. Risch, Searching for genetic determinants in the new millennium, Nature 405 (2000) 847–856.

[11] L.R. Cardon, J.I. Bell, Association study designs for complex diseases, Nat. Rev. Genet. 2 (2001) 91–99.

[12] R.J. Klein, C. Zeiss, E.Y. Chew, J.Y. Tsai, R.S. Sackler, C. Haynes, et al., Complement factor H polymorphism in age-related macular degeneration, Science 308 (2005) 385–389.

[13] R. Sladek, G. Rocheleau, J. Rung, C. Dina, L. Shen, D. Serre, et al., A genome-wide association study identifies novel risk loci for type 2 diabetes, Nature 445 (2007) 881–885.

[14] The Wellcome Trust Case Control Consortium, Genome-wide association study of 14,000 cases of seven common diseases and 3,000 shared controls, Nature 447 (2007) 661–678.

[15] E. Zeggini, M.N. Weedon, C.M. Lindgren, T.M. Frayling, K.S. Elliott, H. Lango, et al., Replication of genome-wide association signals in UK samples reveals risk loci for type 2 diabetes, Science 316 (2007) 1336–1341.

[16] S.F. Kingsmore, I.E. Lindquist, J. Mudge, D.D. Gessler, W.D. Beavis, Genome-wide association studies: progress and potential for drug discovery and development, Nat. Rev. Drug Discov. 7 (2008) 221–230.

[17] K.C. Seng, C.K. Seng, The success of the genome-wide association approach: a brief story of a long struggle, Eur. J. Hum. Genet. 16 (2008) 554–564.

[18] B. Maher, Personal genomes: the case of the missing heritability, Nature 456 (2008) 18–21.

[19] T.A. Manolio, F.S. Collins, N.J. Cox, D.B. Goldstein, L.A. Hindorff, D.J. Hunter, et al., Finding the missing heritability of complex diseases, Nature 461 (2009) 747–753.

[20] L. Kruglyak, D.A. Nickerson, Variation is the spice of life, Nat. Genet. 27 (2001) 234–236.

[21] E.S. Lander, L.M. Linton, B. Birren, C. Nusbaum, M.C. Zody, J. Baldwin, et al., Initial sequencing and analysis of the human genome, Nature 409 (2001) 860–921.

[22] R. Sachidanandam, D. Weissman, S.C. Schmidt, J.M. Kakol, L.D. Stein, G. Marth, et al., A map of human genome sequence variation containing 1.42 million single nucleotide polymorphisms, Nature 409 (2001) 928–933.

[23] The International HapMap Consortium, A haplotype map of the human genome, Nature 437 (2005) 1299–1320.

[24] K.A. Frazer, D.G. Ballinger, D.R. Cox, D.A. Hinds, L.L. Stuve, R.A. Gibbs, et al., A second generation human haplotype map of over 3.1 million SNPs, Nature 449 (2007) 851–861.

[25] M. Kanehisa, S. Goto, KEGG: Kyoto Encyclopedia of Genes and Genomes, Nucleic Acids Res. 28 (2000) 27–30.

[26] F.H. Pettersson, C.A. Anderson, G.M. Clarke, J.C. Barrett, L.R. Cardon, A.P. Morris, et al., Marker selection for genetic case-control association studies, Nat. Protoc. 4 (2009) 743–752.

[27] L.A. Hindorff, P. Sethupathy, H.A. Junkins, E.M. Ramos, J.P. Mehta, F.S. Collins, et al., Potential etiologic and functional implications of genome-wide association loci for human diseases and traits, Proc. Natl. Acad. Sci. USA 106 (2009) 9362–9367.

[28] J.K. Pritchard, N.J. Cox, The allelic architecture of human disease genes: common disease-common variant...or not? Hum. Mol. Genet. 11 (2002) 2417–2423.

[29] I. Prokopenko, E. Zeggini, R.L. Hanson, B.D. Mitchell, N.W. Rayner, P. Akan, et al., Linkage disequilibrium mapping of the replicated type 2 diabetes linkage signal on chromosome 1q, Diabetes 58 (2009) 1704–1709.

[30] G.B. Ehret, A.A. O'Connor, A. Weder, R.S. Cooper, A. Chakravarti, Follow-up of a major linkage peak on chromosome 1 reveals suggestive QTLs associated with essential hypertension: GenNet study, Eur. J. Hum. Genet. 17 (2009) 1650–1657.

[31] J.P. Hugot, M. Chamaillard, H. Zouali, S. Lesage, J.P. Cezard, J. Belaiche, et al., Association of NOD2 leucine-rich repeat variants with susceptibility to Crohn's disease, Nature 411 (2001) 599–603.

[32] J. VanderMolen, L.M. Frisse, S.M. Fullerton, Y. Qian, L. Bosque-Plata, R.R. Hudson, et al., Population genetics of CAPN10 and GPR35: implications for the evolution of type 2 diabetes variants, Am. J. Hum. Genet. 76 (2005) 548–560.

[33] A.P. Morris, E. Zeggini, An evaluation of statistical approaches to rare variant analysis in genetic association studies, Genet. Epidemiol. 34 (2009) 188–193.

[34] E.R. Mardis, Anticipating the 1,000 dollar genome, Genome Biol. 7 (2006) 112.

[35] D.B. Goldstein, Common genetic variation and human traits, N. Engl. J. Med. 360 (2009) 1696–1698.

[36] B.M. Knoppers, I. Fortier, D. Legault, P. Burton, The Public Population Project in Genomics (P3G): a proof of concept? Eur. J. Hum. Genet. 16 (2008) 664–665.

[37] J. Barrett, Association studies, in: N. Camp, A.J. Fox (Eds.), Methods in Molecular Biology. Quantitative Trait Loci – Methods and Protocols, Humana Press, New Jersey, 2002, pp. 3–12.

[38] J.K. Pritchard, Are rare variants responsible for susceptibility to complex diseases? Am. J. Hum. Genet. 69 (2001) 124–137.

[39] L.R. Cardon, L.J. Palmer, Population stratification and spurious allelic association, Lancet 361 (2003) 598–604.

[40] P. Gorroochurn, S.E. Hodge, G. Heiman, D.A. Greenberg, Effect of population stratification on case–control association studies. II. False-positive rates and their limiting behavior as number of subpopulations increases, Hum. Hered. 58 (2004) 40–48.

[41] J.J. Schlesselman, Case-control Studies: Design, Conduct and Analysis, First ed., Oxford University Press, Oxford, 1982.

[42] K.J. Rothman, S. Greenland, Case-control studies, in: K.J. Rothman, S. Greenland (Eds.), Modern Epidemiology, Lippincott-Raven, Philadelphia, 1998, pp. 93–114.

[43] S. Wacholder, N. Rothman, N. Caporaso, Counterpoint: bias from population stratification is not a major threat to the validity of conclusions from epidemiological studies of common polymorphisms and cancer, Cancer Epidemiol, Biomarkers Prev. 11 (2002) 513–520.

[44] K. Yu, Z. Wang, Q. Li, S. Wacholder, D.J. Hunter, R.N. Hoover, et al., Population substructure and control selection in genome-wide association studies, PLoS One 3 (2008) e2551.

[45] B. Devlin, K. Roeder, Genomic control for association studies, Biometrics 55 (1999) 997–1004.

[46] K.G. Ardlie, K.L. Lunetta, M. Seielstad, Testing for population subdivision and association in four case-control studies, Am. J. Hum. Genet. 71 (2002) 304–311.

[47] N.A. Rosenberg, J.K. Pritchard, J.L. Weber, H.M. Cann, K.K. Kidd, L.A. Zhivotovsky, et al., Genetic structure of human populations, Science 298 (2002) 2381–2385.

[48] B. Devlin, S.A. Bacanu, K. Roeder, Genomic control to the extreme, Nat. Genet. 36 (2004) 1129–1130.

[49] A.L. Price, N.J. Patterson, R.M. Plenge, M.E. Weinblatt, N.A. Shadick, D. Reich, Principal components analysis corrects for stratification in genome-wide association studies, Nat. Genet. 38 (2006) 904–909.

[50] J. Marchini, L.R. Cardon, M.S. Phillips, P. Donnelly, The effects of human population structure on large genetic association studies, Nat. Genet. 36 (2004) 512−517.

[51] N. Paneth, E. Susser, M. Susser, Origins and early development of the case−control study: Part 1, Early evolution, Soz. Praventivmed. 47 (2002) 282−288.

[52] R. Doll, A.B. Hill, A study of the aetiology of carcinoma of the lung, Br. Med. J. 2 (1952) 1271−1286.

[53] K.J. Rothman, S. Greenland, Measures of disease frequency, in: K.J. Rothman, S.G. Greenland (Eds.), Modern Epidemiology, Lippincott-Raven, Philadelphia, 1998, pp. 29−46.

[54] B. Sibai, G. Dekker, M. Kuperminc, Pre-eclampsia, Lancet 365 (2005) 785−799.

[55] K.J. Rothman, S. Greenland, Matching, in: K.J. Rothman, S. Greenland (Eds.), Modern Epidemiology, Lippincott-Raven, Philadelphia, 1998, pp. 147−162.

[56] B.C. Choi, G.R. Howe, Methodological issues in case-control studies: II. Test statistics as measures of efficiency, Int. J. Epidemiol. 13 (1984) 229−234.

[57] A.C. Antoniou, D.F. Easton, Polygenic inheritance of breast cancer: implications for design of association studies, Genet. Epidemiol. 25 (2003) 190−202.

[58] K.T. Zondervan, L.R. Cardon, Designing candidate gene and genome-wide case−control association studies, Nat. Protoc. 2 (2007) 2492−2501.

[59] C.I. Amos, Successful design and conduct of genome-wide association studies. *Hum. Mol. Genet*, 16 Spec No 2 (2007) R220−R225.

[60] J. Little, J.P. Higgins, J.P. Ioannidis, D. Moher, F. Gagnon, E. von Elm, et al., STrengthening the REporting of Genetic Association Studies (STREGA) − an extension of the STROBE statement, Genet. Epidemiol. 33 (2009) 581−598.

[61] M.C. O'Donovan, N.J. Craddock, M.J. Owen, Genetics of psychosis; insights from views across the genome, Hum. Genet. 126 (2009) 3−12.

[62] C. Newton-Cheh, T. Johnson, V. Gateva, M.D. Tobin, M. Bochud, L. Coin, et al., Genome-wide association study identifies eight loci associated with blood pressure, Nat. Genet. 41 (2009) 666−676.

[63] M.A. Ferreira, M.C. O'Donovan, Y.A. Meng, I.R. Jones, D.M. Ruderfer, L. Jones, et al., Collaborative genome-wide association analysis supports a role for ANK3 and CACNA1C in bipolar disorder, Nat. Genet. 40 (2008) 1056−1058.

[64] S.M. Purcell, N.R. Wray, J.L. Stone, P.M. Visscher, M.C. O'Donovan, P.F. Sullivan, et al., Common polygenic variation contributes to risk of schizophrenia and bipolar disorder, Nature 460 (2009) 748−752.

[65] J.J. Zhuang, K. Zondervan, F. Nyberg, C. Harbron, A. Jawaid, L.R. Cardon, et al., Optimising the power of genome-wide association studies by using publicly available reference samples to expand the control group, Gen. Epidem. 34 (2010) 319−326.

[66] P.R. Burton, A.L. Hansell, I. Fortier, T.A. Manolio, M.J. Khoury, J. Little, et al., Size matters: just how big is BIG?: quantifying realistic sample size requirements for human genome epidemiology, Int. J. Epidemiol. 38 (2009) 263−273.

[67] D. Thomas, R. Xie, M. Gebregziabher, Two-stage sampling designs for gene association studies, Genet. Epidemiol. 27 (2004) 401−414.

[68] A.D. Skol, L.J. Scott, G.R. Abecasis, M. Boehnke, Joint analysis is more efficient than replication-based analysis for two-stage genome-wide association studies, Nat. Genet. 38 (2006) 209−213.

[69] K.E. Lohmueller, C.L. Pearce, M. Pike, E.S. Lander, J.N. Hirschhorn, Meta-analysis of genetic association studies supports a contribution of common variants to susceptibility to common disease, Nat. Genet. 33 (2003) 177−182.

[70] C. Garner, Upward bias in odds ratio estimates from genome-wide association studies, Genet. Epidemiol. 31 (2007) 288−295.

[71] N.J. Schork, S.K. Nath, D. Fallin, A. Chakravarti, Linkage disequilibrium analysis of biallelic DNA markers, human quantitative trait loci, and threshold-defined case and control subjects, Am. J. Hum. Genet. 67 (2000) 1208–1218.

[72] G.R. Abecasis, W.O. Cookson, L.R. Cardon, The power to detect linkage disequilibrium with quantitative traits in selected samples, Am. J. Hum. Genet. 68 (2001) 1463–1474.

[73] M.N. Weedon, H. Lango, C.M. Lindgren, C. Wallace, D.M. Evans, M. Mangino, et al., Genome-wide association analysis identifies 20 loci that influence adult height, Nat. Genet. 40 (2008) 575–583.

[74] C.M. Lindgren, I.M. Heid, J.C. Randall, C. Lamina, V. Steinthorsdottir, L. Qi, et al., Genome-wide association scan meta-analysis identifies three loci influencing adiposity and fat distribution, PLoS Genet 5 (2009) e1000508.

[75] J.R. Perry, L. Stolk, N. Franceschini, K.L. Lunetta, G. Zhai, P.F. McArdle, et al., Meta-analysis of genome-wide association data identifies two loci influencing age at menarche, Nat. Genet. 41 (2009) 648–650.

[76] Y.S. Cho, M.J. Go, Y.J. Kim, J.Y. Heo, J.H. Oh, H.J. Ban, et al., A large-scale genome-wide association study of Asian populations uncovers genetic factors influencing eight quantitative traits, Nat. Genet. 41 (2009) 527–534.

[77] J.M. Kidd, G.M. Cooper, W.F. Donahue, H.S. Hayden, N. Sampas, T. Graves, et al., Mapping and sequencing of structural variation from eight human genomes, Nature 453 (2008) 56–64.

Tag SNP Selection

Jing Li[1], Wei-Bung Wang[2]

[1] Electrical Engineering and Computer Science Department,
Case Western Reserve University, Cleveland, OH, USA
[2] Department of Computer Science and Engineering,
University of California-Riverside, Riverside, CA, USA

INTRODUCTION

Deciphering the genetic and molecular basis underlying phenotypic variation is critical in the understanding of biological systems and their evolution. More importantly, it has many practical applications in disease diagnosis, drug target identification, molecular and personalized medicine. Marker-based gene mapping approaches including linkage and association studies have existed for many years. Using these methods, researchers have successfully identified many genes responsible for rare monogenic diseases. With advances in new technology, the number of markers in genome-wide analysis has dramatically increased from dozens (e.g., microsatellite) to millions (e.g., single nucleotide polymorphisms − SNPs, the NCBI dbSNP database has ~7 million validated SNPs). However, given the huge number of SNPs, it is too costly and inefficient to type all of them for genome-wide association studies (GWAS). Instead, a few panels with reasonable coverage will greatly facilitate the automatic manufacture of marker sets and high-throughput genotyping of large samples. Therefore, one important question is how to select subsets of markers to faithfully represent all known SNPs as much as possible. In addition to cost and efficiency reasons, a set of reduced but informative SNPs will also help the downstream computational analysis. By removing redundant SNPs, statistical power of association analyses can be improved since the number of tests, which is usually proportional to the total number of SNPs, will be greatly reduced.

There are many considerations in selecting an "ideal" set of SNPs in genome-wide association studies. Based on the "common disease− common variant" hypothesis, GWAS has mainly targeted at identifying common variants that are associated with an increased risk of diseases through an unbiased essay across the genome. Therefore the overall genomic coverage of a potential SNP set, in comparison with a large reference set of common SNPs, has been commonly used in evaluating different sets of SNPs and different selection approaches. In most cases, common SNPs are defined based on the frequency estimation from HapMap samples, which consist of individuals from China (CHB), Japan (JPT), Nigeria (YRI) and US residents with ancestries from Northern and Western Europe (CEU) [1, 2]. Experimental results from the HapMap project [1, 2] have shown that, although a great majority of SNPs are shared by different populations, there are significant differences in terms of allele frequency distributions and local linkage disequilibrium structures in different populations. An SNP selection procedure also needs to consider coverage for different populations. In addition to frequency, another important consideration is the existing knowledge of gene structures and other functional annotations about the human genome. SNPs that are found in these important regions or SNPs that have known functional roles (e.g., non-synonymous SNPs) should be included. At the same time, to make the set of SNPs available either as a predefined or a customized chip, there are other practical limitations such as array efficiency, call rate and call accuracy.

In this chapter, we primarily focus on existing computational approaches for the selection of a "smallest" set of tagSNPs with least information loss. A SNP in a reference panel is covered by the set of tagSNPs if it is either included in the tagSNP set or it can be approximated by SNP(s) in the set. In most cases, proximity is measured by linkage disequilibrium (LD). Selections according to other criteria (such as inclusion of functional SNPs, or consideration of array efficiency) may not involve much computation or may not have many alternative solutions. Therefore they are beyond the scope of this chapter.

Generally speaking, one can tag a SNP based either on pair-wise correlations or on correlations of multiple SNPs (i.e., haplotypes). We denote such approaches as haplotype-independent and haplotype-based approaches, respectively. This distinction is partly due to historical reasons because some of the earliest work started with known haplotypes obtained through experimental approaches. One should note that some of those haplotype-based approaches were later extended to directly handle genotype data. On the other hand, haplotype-independent approaches, such as methods which totally rely on pair-wise LD, also need haplotype frequency information of two loci, because the basic concept of pair-wise LD is defined as the deviation of the haplotype frequency and the product of allele frequencies of the two loci. Methods of both categories will be discussed. This chapter is organized as follows. In the first section, we introduce commonly used pair-wise LD measures, followed by discussions of haplotype-based and haplotype-independent approaches. Selection approaches that target multiple populations will then be briefly introduced. We conclude this section by briefly discussing a closely related topic, i.e., SNP selections in multi-stage designs. In the following section, a list of commonly used computational tools to select tagSNPs and their internet links will be provided with brief descriptions. We finish the chapter by summarizing the coverage of several popular existing commercial genotyping platforms.

APPROACHES

Linkage disequilibrium, which refers to the non-random association among alleles within short distances, holds the key to association mapping as well as tagSNP selection. Given our intensive use of LD, we first give mathematical definitions of a few commonly used pair-wise LD measures and briefly introduce algorithms to estimate them. Consider two bi-allelic SNPs A and B, with possible alleles A/a and B/b, respectively. The uppercase letters represent the major alleles and the lowercase letters represent the minor alleles. Let h be an allele or a haplotype with SNPs A and B, and p_h be the probability that h is observed in the population. LD can be measured by D:

$$D = p_{AB} - p_A \, p_B$$

Because D depends on allele frequencies, a more commonly used measure is the normalized D':

$$D' = \begin{array}{ll} D/D_{\max}, & D \geq 0 \\ D/D_{\min}, & D < 0 \end{array}$$

where $D_{\max} = \min\{p_A \, p_b, \ p_a \, p_B\}$ and $D_{\min} = \max\{-p_A \, p_B, -p_a \, p_b\}$. $D_{\max}$ and $D_{\min}$ are theoretical maximum and minimum value of D, therefore D' always ranges in $[0,1]$. The higher the value is, the more correlation there is between the two SNPs. When $D' = 1$, the two alleles are in *complete LD* (i.e., the strongest possible correlation for the given allele frequencies). It also implies that there has been no historical recombination between these two loci under the assumption of no recurrent mutations.

LD can also be measured by the correlation coefficient, which is denoted by r^2:

$$r^2 = D^2/(p_A p_a p_B p_b)$$

The value of r^2 always ranges in $[0,1]$. When $r^2 = 1$, the two SNPs are "perfectly" correlated to each other. For two perfectly correlated SNPs, the minor allele frequencies of two SNPs must be the same. SNPs with $r^2 = 1$ can predict each other perfectly when one of them is unknown. Because r^2 is directly related to the statistical power of association mapping, the tagSNPs selected using r^2 can be effective in disease association mapping studies.

Measures of linkage disequilibrium D, D' and r^2 all require haplotype frequencies across the two loci. If haplotype information is not available, two-locus haplotype frequencies can still be estimated from genotype data assuming the population is randomly mating [3]. Briefly, let f_h be the frequency of h in a population, and $\hat{f}_h$ be the maximum likelihood estimate of f_h. Let n_g be the number of times that genotype g is observed in the sample, and N be the number of individuals in the sample. For SNPs A and B, the maximum likelihood estimate of haplotype frequency f_{AB} satisfies the following equation under the assumption of the Hardy-Weinberg equilibrium:

$$\hat{f}_{AB} = \frac{1}{2N}\left(2n_{AABB} + n_{AABb} + n_{AaBB} + n_{AaBb}\frac{\hat{f}_{AB}\hat{f}_{ab}}{\hat{f}_{AB}f_{ab} + \hat{f}_{Ab} + \hat{f}_{aB}}\right)$$

Similar equations can be set up for haplotypes Ab, aB and ab. These equations can be solved by the standard expectation maximization (EM) algorithm.

Studies have shown that LD patterns vary greatly for different genomic regions and different populations. Furthermore, experimental results published in 2002 [4] have shown that the human genome can be partitioned into segments of haplotype blocks, within which there is little evidence for historical recombination, separated by recombination hotspots. For each block, only a few common haplotypes have been observed from the HapMap samples. Therefore, to capture all common haplotypes in a population, only a small fraction of SNPs are needed within each haplotype block. Given haplotype blocks, the problem of selecting a subset of SNPs to distinguish all common haplotypes within each block is known as the *minimum test set* problem in computer science. It is unlikely to find efficient (i.e., polynomial) algorithms for the problem in general because it is known that the problem is NP-complete [5]. In practice, because both the number of common haplotypes and the number of common SNPs within a block are expected to be small, in many cases it is still feasible to find exact solutions or close to optimal solutions when blocks and haplotypes are given. However, for the tag SNP selection problem in general, neither haplotypes nor blocks are readily available in most cases. Existing high-throughput genotyping technologies can only output genotypes, not haplotypes. Experimental methods such as sperm typing to determine haplotypes are costly. In terms of haplotype block structures, population genetics tells us that many evolutionary factors can contribute to their formation. There is no way one can directly observe the exact boundaries of haplotype blocks.

In what follows, we will first introduce selection approaches assuming that haplotypes are given. In reality, haplotypes can be obtained through experimental approaches. In addition, for tightly linked markers, family-based haplotype inference approaches can usually return reliable haplotype results with great efficiency [6]. Haplotypes and their frequencies can also be estimated based on population genetics models using unrelated individuals, which have been intensively investigated (e.g., Stevens & Smith [7]). Many earlier tagSNP selection approaches have been developed based on the assumption that individual haplotypes are given. These approaches are denoted as haplotype-based approaches. It should be noted that some of these approaches were extended later to directly handle genotype data.

Haplotype-based Approaches

To evaluate haplotype structure, diversity and to select tagSNPs, Patil et al. [8] relied on experimentally determined haplotypes and considered the problem of tagSNP selection and haplotype block partition simultaneously. They proposed a computational framework with the aim to minimize the number of tag SNPs required to distinguish all common haplotypes within each block. The authors also proposed a greedy algorithm to solve the problem. Subsequently, such a framework has been adopted by other researchers. For example, Zhang et al. [9] proposed a dynamic programming algorithm to minimize the number of tagSNPs required to capture common haplotypes within each block. For different block partitions with the same number of tagSNPs, the algorithm picks the partition with the smallest number of blocks. Briefly, assume that there are n consecutive SNPs and K haplotypes. For a block that consists of the set of consecutive SNPs from i to j $(i \le j)$, let $f(i, i + 1, ..., j)$ denote the smallest number of tagSNPs required to distinguish common haplotypes within the block. Define S_j to be the optimal number of tagSNPs required for SNPs from 1 to j for all possible partitions. S_j can be calculated recursively:

$$S_j = \min\{S_{i-1} + f(i, i + 1, ..., j)\}(1 \le i \le j), S_0 = 0$$

When multiple block partitions with the same minimum number of tagSNPs exist, the partitions with the minimum number of blocks can also be calculated using a dynamic programming algorithm. Though the problem of finding the smallest number of tagSNPs to distinguish all common haplotypes (i.e., to calculate $f(i, i + 1, ..., j)$) is theoretically hard, optimal solutions can usually be obtained through an exhaustive search given the small size of the problem in practice. Zhang et al. [9] have shown that their algorithm identified smaller numbers of tagSNPs as well as smaller numbers of blocks than those returned by the greedy algorithm in Patil et al. [8].

To directly apply these haplotype-based approaches to genotype data, one has to infer haplotypes beforehand. For SNPs with high densities, haplotype assignments of each individual can be reliably obtained when family genotype data are available [10]. Many computational approaches have been developed to infer haplotypes for such data [6]. For unrelated individuals, haplotypes of short regions can also be inferred computationally, mainly based on the assumption of limited haplotype diversity. However, the obvious two-step approach (i.e., infer haplotypes first and then partition them into blocks and select tag SNPs within each block) is not optimal in the sense that long-range haplotypes are hard to infer computationally, especially when they cross block boundaries. Zhang et al. [11] further extended their haplotype-based dynamic programming algorithm to deal with genotype data by incorporating an expectation-maximization (EM)-based haplotype inference algorithm into their block partition framework. The algorithm is essentially heuristic. It examines the SNPs sequentially and infers haplotype frequencies along the way. As long as the resulting haplotypes still support the current segment to be within one block, it will try to add one new SNP to it. The current block ends when adding a new

SNP will result in too many haplotype diversities. The algorithm then selects the minimum number of tagSNPs for the current block and starts a new block.

Different to Patil et al. [8] and Zhang et al. [9], in which blocks were used as a means to select the minimum number of tagSNPs, many other researchers define block boundaries naturally using LD patterns and/or the concept of recombination hotspots. For example, in Daly et al. [10], haplotypes were inferred first based on parents' genotype data and then a hidden Markov model (HMM) was used to estimate the historical recombination frequency between each pair of adjacent markers. Blocks were then defined based on the estimated historical recombination frequency using a threshold approach. In addition to pair-wise LD or recombination fraction measures of adjacent markers, one can also define a haplotype block as a contiguous set of markers in which the average LD is greater than some predetermined threshold [12]. TagSNPs can then be selected block by block. Alternatively, the four-gamete test has also been used to define haplotype blocks [13]. The rationale of the four-gamete test lies in the fact that under the assumption of no recurrent or backward mutations, and no genotyping or haplotype inference errors, one can only observe three different haplotypes between two loci. There must be at least one historical recombination event when observing all four gametes. Blocks can then be defined as a set of consecutive markers with no recombination.

None of the block definitions is perfect. Though the concept of haplotype block is reasonably robust, there is substantial variability in inferred block boundaries using different definitions, which motivates the tagSNP selection method without using the block concept. For example, Bafna et al. [14] defined the informativeness of an SNP s with respect to an SNP t based on the probability that two haplotypes are not the same at SNP s given they are not the same at SNP t. They extended the definition of informativeness to a set of SNPs and their goal was to find a smallest subset with the largest informativeness. The key to avoid the use of blocks here is to use only SNPs that are in close proximity to tag unseen SNPs. However, this formulation cannot totally solve the problem because one has somehow to choose the size of a neighborhood.

In cases when a block partition is not necessary or naturally defined (e.g., candidate gene studies), there are many ways to select tagSNPs for each of such regions. For example, Sebastiani et al. [15] proposed an algorithm to remove those SNPs that can be represented by Boolean functions of tagSNPs. To quantify haplotype diversities, researchers [16] have used the concept of entropy. Their goal was to select the minimal number of SNPs so that the entropy based on haplotype frequencies is the same or the change is minimal. Lin and Altman [17] formulated the tagSNP selection problem as the dimension-reduction problem and employed a well-known statistical approach called principal component analysis (PCA), which is a powerful tool to extract features (i.e., variables) for multivariate analysis. Briefly, given a set of variables, PCA finds a set of ordered orthogonal variables, which are linear combinations of the original variables, with decreasing empirical variance. Linear algebra shows that the kth new variable is precisely the kth eigenvector of the input matrix. For the SNP selection problem, the PCA-based approach [17] assumed that each SNP is a variable, and selected the most significant eigenSNPs, which are linear combinations of the original SNP that retain the maximum variability. The eigenSNPs will then be mapped to the original SNPs according to their correlation to the eigenSNPs.

Haplotype-independent Approaches

The haplotype-independent approaches do not require multi-maker haplotypes as inputs. Instead, they usually consider pair-wise correlations among SNPs across the entire genome. Most approaches use LD to evaluate the quality of tagSNP sets [18–23]. But there are several alternatives, including entropy [24, 25] and prediction accuracy [26]. We primarily focus on LD measures and briefly discuss other criteria.

For many approaches using pair-wise LD measures (such as D, D' or r^2) in selecting tagSNPs, there is usually a predetermined threshold. Two or more SNPs are considered, being correlated only if the measure of LD between these SNPs exceeds the threshold. One typical selection criterion is to choose a minimum set of SNPs to be tagSNPs such that each SNP is either a tagSNP or is correlated with a tagSNP. Some other variations include: selecting a fixed number of tagSNPs to cover as many SNPs as possible, or selecting a minimum set of tagSNPs to cover a certain fraction of SNPs. A SNP that is correlated to a (subset of) tagSNP is *covered* by the (subset of) tagSNP. If a non-tagSNP is not correlated to any tagSNPs, then it is *uncovered*. When the selection criteria of tagSNPs is to cover all SNPs and the measurement of LD is r^2 of pair-wise SNPs, it is equivalent to the dominating set problem in computer science, which is NP-complete [5]. Briefly, the dominating set problem on a graph $G = (V,E)$ is to find a subset V' of V such that every vertex not in V' is joined to at least one vertex in V' by some edge(s). For a tag SNP selection problem using pair-wise r^2, one can construct a graph $G = (V,E)$ with each vertex v_i representing an SNP s_i. Two vertices v_i and v_j are connected if and only if two corresponding SNPs s_i and s_j are correlated. The tagSNP set that covers all SNPs is therefore the dominating set of the graph G. Using this framework, Carlson et al. [18] have developed a greedy algorithm to select the maximally informative tagSNPs out of a set of common SNPs. The selection criterion is that all common SNPs are either directly assayed or correlated to a tagSNP. The algorithm first evaluates r^2 values between all pairs of

SNPs. In each round of selection, the SNP which is correlated with the maximum number of uncovered SNPs is picked as a tagSNP. The algorithm stops when all non-tagSNPs are covered.

In addition to pair-wise LD, researchers consider LD among a small group of SNPs that are in close proximity. The idea originated from the concept of *haplotype blocks* [4]. Strong LD between two distant SNPs is often considered an artifact because of insufficient sample sizes. Although true strong LD may occur between two significantly distant SNPs, in practice, it is commonly ignored when the distance between the two SNPs is greater than a threshold. Qin et al. [22] have proposed a comprehensive search algorithm called FESTA, which capitalized the block structure of strong LD. FESTA first partitions SNPs into groups, called *precincts*, such that there is only low LD between SNPs in different groups. Tag SNP selection can then be performed independently in each precinct. FESTA evaluates the cost of an exhaustive search for the minimum tagSNP selection. If the cost is below a threshold, an exhaustive search is then performed. Otherwise FESTA performs a search in a smaller space, then applies a greedy algorithm to select additional tagSNPs. Therefore, FESTA can be viewed as a mixture of an exhaustive search and a heuristic algorithm. Researchers [20, 23] have also proposed multi-marker tagSNP selection approaches based on multi-marker LD measures. They consider a two- or three-locus haplotype as a "compound" allele of an artificial "compound" SNP. The r^2 value between an SNP and a compound SNP is evaluated. It has been shown that multi-marker LD usually enhances the prediction power of tagSNPs and reduces the number of tagSNPs significantly.

In addition to traditional LD measures, researchers use the concept of relative entropy to evaluate the quality of tagSNP sets. The calculation of entropy and relative entropy requires haplotype frequencies. When such information is not available, approximations have to be sought. For a locus X with k alleles with frequencies p_i ($i = 1, ..., k$), the entropy $H(X)$ of X is defined as:

$$H(X) = -\sum_{i=1}^{k} p_i \log_2 p_i$$

The joint entropy of two loci X and Y is defined as the entropy of corresponding two-locus haplotype frequencies. The conditional entropy $HXY = XXY - H(Y)$ is the entropy of X when Y is given. Therefore, the information of Y about X can be represented as the deviation of the two entropies of X before and after Y is given, which is defined as $IXY = HX + HY - H(XY) = IYX$. Given the purpose of disease gene mapping, the ideal objective of SNP selection is to find a set of loci $X_1, X_2, ..., X_k$ that maximizes $IZX_1, X_2, ..., X_k$, where Z is the "unknown" disease locus. Since Z is unknown, the practical objective function is defined either within the tagSNP set [16] or the whole SNP sets [24, 25].

One can also measure the quality of tagSNPs by evaluating their ability in predicting other SNPs. For example, Halperin et al. [26] have proposed a framework to minimize the overall prediction error using genotypes of tagSNPs to predict genotypes of other SNPs. The prediction error is evaluated by:

$$\eta = \sum_{j=1}^{m} \Pr[f_j(z_T(g)) \neq g(j)]$$

where $g(j)$ is the jth genotype in a sample, $z_T(\cdot)$ eliminates non-tagSNPs, and $f(\cdot)$ is the prediction function using only tagSNPs. The authors proposed two algorithms to tackle the problem, one was based on random sampling and the other was an exact algorithm [26].

There are some other measures of tagSNPs. Stram et al. [27] proposed a formal measure, squared correlation R_h^2, between true and predicted haplotype frequencies from given genotype data. The squared correlation is to characterize the uncertainty in the prediction of haplotypes from genotype data and use it for tagSNP selection. Weale et al. [28] formulated haplotype R_2 as the coefficient of determination. In statistics, the coefficient of determination provides a measure of how well future outcomes are likely to be predicted.

SNP Selections for Multiple Populations

Most approaches discussed so far consider the tagSNP selection problem within one population. Although results from HapMap have shown that the majority of SNPs are shared by different populations, their frequencies and haplotype structures can be quite different. Therefore, intuitively, a set of tagSNPs extracted from one population may not be a good representative of another population. However, it is not practical to obtain reference panels for all populations. For an association study in a population without a reference panel itself, reference panels from other populations must be used. And the primary existing reference panel is the HapMap samples. There has been some research on how tagSNPs in one population may be transferred to different populations based on HapMap samples [19, 29, 30]. They generally concluded that HapMap samples can be used to select tags for genome-wide association studies in different populations around the world.

A few computational approaches have been proposed to address tagSNP selection problems from panels consisting of multiple populations. Howie et al. [31] have developed an algorithm called MultiPop-TagSelect to select tagSNPs from multiple populations. The algorithm works by binning tagSNPs selected within each population. However, the quality of common tagSNPs relies on the quality of bins of each population. Liu et al. [21] have worked on a formulation called the minimum common tagSNP selection (MCTS) problem for universal tagSNPs of multiple populations. The MCTS problem is defined as follows. Let $r_i^2(s_1,s_2)$ be the correlated coefficient r^2 between s_1 and s_2 in population i. Given a set of SNPs V and a collection of populations, each has its LD pattern $E_i = \{(s_1,s_2)|r_i^2(s_1,s_2) \geq r_0\}$ where r_0 is the pre-determined threshold, find the smallest subset of SNPs to be tagSNPs, so that for each population i, for each non-tag SNP s_1, there is a tag SNP s_2 such that $s_1,s2 \in E_i$. Not surprisingly, the MCTS problem has been shown to be NP-hard. Liu et al. [21] have developed two algorithms to solve the MCTS problem. Both algorithms first calculate LD separately for each population. Heuristics are then used on each connected component of the union graph from all populations. The first algorithm applies a simple greedy idea to the problem after reduction. The second algorithm transforms the problem to an integral linear programming instance and solves it by Lagrangian relaxation techniques, which outperforms the greedy algorithm and MultiPop-TagSelect algorithm. An advantage of the algorithm is that it deals with multiple populations directly, and its result does not rely on quality of single population tagSNP selections.

SNP Selections in Multi-stage Designs

One related but different problem is to select SNPs in a multi-stage design. Studies have shown that a two-stage or multi-stage design might be able to achieve a similar power but with much reduced genotyping costs, compared to a single-stage design [32–34]. The purpose of SNP selection in a multi-stage design is to select a much smaller but promising subset of SNPs from all available SNPs for further tests in later stages. The set of SNPs selected not only need to be representative, but also need to be informative in terms of revealing genotype-phenotype associations. The simplest method for SNP selection in a multi-stage design is to utilize a simple statistic to test the association of each SNP with the disease/phenotype, and adopt a liberal significance level per test. All SNPs that pass the test will advance to the next stage. An issue with this approach is that when a SNP shows moderate association with the disease at stage 1 and is selected for stage 2, it is very likely that other SNPs in high correlations with this SNP will also be selected for stage 2. Furthermore, such a strategy relies only on marginal effects and causal SNPs with haplotype or epistatic effects but small marginal effects may be filtered out. Li [34] suggested a procedure consisting of three phases to select a subset of highly discriminative SNPs. In the first phase, all the SNPs are ranked based on their associations with the disease as usual.

Correlations among SNPs will then be explored using a clustering algorithm in Phase 2. In the third phase, potential haplotype effects and/or gene—gene interactions will be considered using an information theory-based approach. A SNP will be genotyped and tested in stage 2 only if it passes all three phases. Experimental results show that the approach can effectively eliminate irrelevant SNPs and always has a higher power than the original two-stage method.

TOOLS

There are some tagSNP selection tools available online. We introduce some popular ones (Table 4.1) that provide various features and options for users.

TABLE 4.1   Popular Tag SNP Selection Tools and their Characteristics

	Operating system	**Linux (x86), Macintosh OS X**
FESTA	Features	Optional threshold bounding execution time Select included/excluded SNPs in advance Optional criteria Double coverage option Optional MAF
	Comments	No GUI
	Reference	Qin et al. [22]
	Availability	http://www.sph.umich.edu/csg/qin/FESTA/
	Operating system	**Web server (Tagger), Windows**
Haploview (Tagger)	Features	LD and haplotype block analysis Optional block definition LD plot and haplotypes display Permutation testing for association significance Multi-marker LD
	Comments	Tagger is the core of Haploview GUI
	Reference	Barrett et al. [35], de Bakker et al. [36]
	Availability	http://genecruiser.broadinstitute.org/mpg/ haploview (Haploview) http://genecruiser.broadinstitute.org/mpg/ tagger/ (Tagger)

(Continued)

TABLE 4.1 Popular Tag SNP Selection Tools and their Characteristics—cont'd

	Operating system	**Windows XP/2000 or NT**
	Features	Customized adjacent SNP masking Various visualization of SNPs Assess statistical power Select haplotype tagging SNPs in real time
SNPbrowser	Comments	Registration required GUI
	Reference	De La Vega et al. [37]
	Availability	http://www.allsnps.com/snpbrowser
	Operating system	**Unix, Linux, Windows 95, NT, 2000, XP**
	Features	Optional evaluation Various input data types, including genotype data
HapBlock	Comments	No GUI
	Reference	Zhang et al. [9]
	Availability	http://www.cmb.usc.edu/msms/HapBlock/

FESTA

FESTA stands for Fragmented Exhaustive Search for TAgSNPs. It is the implementation of the comprehensive search algorithm in [22]. FESTA was developed by the Center for Statistical Genetics at the Department of Biostatistics of the University of Michigan. The measurement is based on r^2. As discussed earlier, the key idea of the algorithm in FESTA is the mixture of a comprehensive search algorithm and a greedy algorithm. Users may change the search depth depending on the power of their machines. The selection criterion is to cover all SNPs. When there are alternative solutions of the same size, four different additional criteria can be entertained:

1. Maximize average r^2 value between tagSNPs and non-tagSNPs.
2. Maximize lowest r^2 between tagSNPs and non-tagSNPs.
3. Minimize average r^2 value among tagSNPs.
4. Maximize average r^2 value among tagSNPs.

Criteria 1 and 2 try to identify tagSNP sets that have strongest connections with non-tagSNPs. Criterion 3 is to find a tagSNP set whose members are as independent as possible which minimizes overlap between tagSNPs. Criteria 4 may increase redundancy and robustness to genotype failure.

HapBlock

HapBlock is a set of dynamic programming algorithms for haplotype block partitioning and tagSNP selection. HapBlock was developed by Zhang et al. [9] in the Molecular and Computational Biology Program of the Department of Biological Sciences at the University of Southern California. HapBlock provides various definitions of block and selection criteria for users to choose. Three definitions of haplotype block have been included:

1. Coverage of common haplotypes. At least a certain percentage of observed haplotypes must be common.
2. LD-based blocks. In each block, a certain proportion of SNPs must be in strong LD.
3. No historical recombination. A set of consecutive SNPs without historical recombination events.

Five different criteria for tagSNP selection are implemented:

1. Common haplotypes. The minimum tagSNP set that can uniquely distinguish a percentage of all haplotypes.
2. Haplotype diversity. The minimum tagSNP set that can account for a percentage of overall haplotype diversity.
3. Haplotype entropy. The minimum tagSNP set that can account for a percentage of overall haplotype entropy.
4. Haplotype determination coefficient R_h^2 [27]. The minimum tagSNP set such that the overall prediction strength exceeds a predetermined threshold.
5. LD measure r^2. The minimum tagSNP set such that the overall prediction power exceeds a predetermined threshold.

Haploview

Haploview is a visualization tool for the analysis of LD patterns [35]. A graphical genome browser allows researchers to navigate to a particular region of the genome. Haploview was developed in and is maintained by Mark Daly's lab at the Broad Institute. The algorithm of tag SNP selection is "best N method" in de Bakker et al. [36]. It allows two measurements of LD: D' and r^2. The algorithm Tagger also allows tagging with 2-marker and 3-marker haplotypes. Four haplotype block definitions have been included:

1. Confidence intervals. This definition is given in Gabriel et al. [4]. A block is created if the one-sided upper 95% bound on D' is >0.98 and the lower bound is above 0.7.
2. Four-gamete rule. A set of consecutive SNPs without historical recombination events. Recombination events are tested by the four-gamete rule.

3. Solid spine of LD. The first and last markers in a block are in strong LD with all intermediate markers while the intermediate markers are not necessarily in LD with each other.
4. Custom. Users may click and draw along the marker number row to define haplotype blocks.

SNPbrowser

SNPbrowser was developed by Applied Biosystems [37] and can provide query, visualization of SNPs, gene annotations, power, haplotype blocks and LD map coordinates. TagSNP selection can be based on genotype correlations, pairwise r^2, or haplotype R^2 [28]. The tagSNP selection algorithm is a dynamic programming algorithm proposed by Halldórsson et al. [38]. Users may change the threshold of LD and minor allele frequencies. SNPbrowser will select and visualize tagSNPs accordingly in real time.

GENOTYPING PLATFORMS

After intensive investigation of the tagSNP selection problem for GWAS, researchers eventually have to carry out their studies using a handful of commercial SNP chips, many of which have incorporated the tagSNP concepts. Therefore, it is important to know differences in coverage for different chips, which may result in power variation and potential bias in association analysis [39–41]. The most commonly used criterion for SNP chip evaluation is the global coverage, defined as the fraction of common SNPs that are tagged by the SNPs on the chip [39–41]. Other important considerations include coverage for functional regions (such as gene regions) and coverage for different populations.

There are two major vendors on the market that provide commercial SNP chips, i.e., Illumina and Affymetrix. Popular SNP chips from Illumina include Human1M, HumanHap550 and HumanHap300 (Table 4.2). In addition, Illumina also provides a specially designed chip Human-Hap650Y, which intentionally adds more SNPs to tag common SNPs in African populations. The Illumina SNP chips include LD-based tagSNPs derived from over 2 million common SNPs (minor allele frequency greater than 0.05) in the HapMap data. Affymetrix pipelines include SNP Array 6.0, SNP Array 5.0 and Gene Chip Human Mapping 100K. Prior to Array 6.0, Affymetrix SNP chips include randomly selected SNPs based on their physical positions and call rates, without considering LD patterns [39]. The additional SNPs in SNP Array 6.0 are mostly tagSNPs. In addition to these SNP chips that provide genome-wide coverage, both companies provide SNP chips that target at

TABLE 4.2 Number of SNPs and their Genomic Coverage of Common SNPs at $r_2 = 0.8$ for Different Populations from HapMap Samples. Results were Mainly Obtained from Li et al. [41] (Entries Labeled with [1] are from Barrett and Cardon [39])

SNP chips	# of SNPs	CEU	CHB + JPT	YRI
HumanHap300	317,511	75%[1], 77%	63%[1], 66%	28%[1], 29%
HumanHap550	555,352	87%	83%	50%
HumanHap650Y	660,917	87%	84%	60%
Human1M	1,072,820	93%	92%	68%
Gene Chip 100K	116,204	31%[1]	31%[1]	15%[1]
SNP Array 5.0	500,568	64%	66%	41%
SNP Array 6.0	934,968	83%	84%	62%

functional SNPs, especially non-synonymous SNPs (nsSNPs) in protein coding regions. For example, the MegAllele system marketed by Affymetrix consists of 12,000 nsSNPs. Illumina's Human-1 BeadChip broadly targets at gene regions.

A few studies [39–41] have explicitly evaluated the coverage of different SNP chips over genomic regions and gene regions, as well as coverage over different populations. Generally speaking, for products from the same company, chips with large numbers of SNPs certainly provide better coverage because they usually contain all SNPs from chips with smaller numbers as subsets. But chips with similar number of SNPs from different companies may be quite different. More specifically, SNP sets on Illumina chips (Human1M, HumanHap550, HumanHap300) were selected using LD-based tagSNP selection approaches to maximize the coverage based on HapMap samples. However, Affymetrix basically used randomly selected sets of SNPs in their pipelines, with the only exception that their latest product SNP Array 6.0 has also incorporated tagSNPs on top of the randomly selected SNPs from SNP Array 5.0. Therefore, in terms of global coverage, chips with smaller numbers of SNPs from Illumina may have better coverage than chips with larger numbers of SNPs from Affymetrix for European populations. For example, the genomic coverage of HumanHap300 for common SNPs at $r^2 > 0.8$ in Phase II HapMap Caucasian samples is 75%, compared to 65% of Affymetrix SNP Array 5.0, although SNP Array 5.0 has significantly more SNPs than HumanHap300. Both companies have improved their coverage for their later products (e.g., HumanHap500: 87%, Human 1M: 93%, Array 6.0: 83%). On the other hand, one should also notice that the global coverage is usually different for different populations. For example, for African populations, the coverage of Affymetrix SNP Array

5.0 is 41% while HumanHap300 is only about 28%. More detailed results about genomic and population coverage are summarized in Table 4.2.

In terms of local coverage, defined as the percentage of tagged common SNPs in HapMap samples at $r^2 > 0.8$ for a fixed window size of 1 Mbps, Li et al. [41] concluded that Human1M has the best coverage among all the chips for all four populations. Like the result for global coverage, the local coverage of the HumanHap550 is almost always better than SNP Array 6.0 for the CEU samples, despite the fact that the latter chip has about 380,000 more SNPs. Similar results were observed when comparing HumanHap300 to SNP Array 5.0. For the YRI sample, the coverage of HumanHap650Y has significantly improved compared to Human-Hap550. One should also notice that local coverage of all chips varies greatly and the variation patterns show high correlations across chips, which probably reflects gaps in the reference sequence, low call rates in some genomic locations, and/or low LD in some regions. Li et al. [41] also calculated the gene coverage, defined as the percentage of tagged common SNPs in the HapMap samples at $r^2 > 0.8$ for each gene. A reasonable fraction of genes (89.1%) in the CEU samples have coverage >80% for Human1M. The fractions of genes covered by other panels range from 81.4% (Human-Hap650Y) to 34.2% (SNP Array 5.0). A slightly smaller number of genes have been covered for the CHB + JPT samples, but it drops substantially for the YRI samples. There are also a significant portion of genes (9.5−51.8% for different populations) that have coverage <80% for all chips. Some genes, including those that have been associated with pathways or diseases in previous studies may have very low coverage. These results can guide investigators when they interpret their GWAS results and when they design follow-up studies.

Acknowledgments

JL's research is supported by National Institutes of Health/National Library of Medicine grant LM008991, and National Center for Research Resources grant RR03655.

References

[1] T.I.H. Consortium, The International HapMap Project, Nature 426 (6968) (2003) 789−796.

[2] K.A. Frazer, D.G. Ballinger, D.R. Cox, et al., A second generation human haplotype map of over 3.1 million SNPs, Nature 449 (7164) (2007) 851−861.

[3] W.G. Hill, Estimation of linkage disequilibrium in randomly mating populations, Heredity 33 (2) (1974) 229−239.

[4] S.B. Gabriel, S.F. Schaffner, H. Nguyen, et al., The structure of haplotype blocks in the human genome, Science 296 (5576) (2002) 2225−2229.

[5] M.R. Garey, D.S. Johnson, Computers and Intractability: A Guide to the Theory of NP-Completeness, W.H. Freeman, San Francisco, 1979. x, 338 pp.

[6] J. Li, T. Jiang, A survey on haplotyping algorithms for tightly linked markers, J. Bioinform. Comput. Biol. 6 (1) (2008) 241–259.

[7] M. Stephens, N.J. Smith, P. Donnelly, A new statistical method for haplotype reconstruction from population data, Am. J. Hum. Genet. 68 (4) (2001) 978–989.

[8] N. Patil, A.J. Berno, D.A. Hinds, et al., Blocks of limited haplotype diversity revealed by high-resolution scanning of human chromosome 21, Science 294 (5547) (2001) 1719–1723.

[9] K. Zhang, M. Deng, T. Chen, et al., A dynamic programming algorithm for haplotype block partitioning, PNAS 99 (11) (2002) 7335–7339.

[10] M.J. Daly, J.D. Rioux, S.F. Schaffner, et al., High-resolution haplotype structure in the human genome, Nat. Genet. 29 (2) (2001) 229–232.

[11] K. Zhang, Z. Qin, J. Liu, et al., Haplotype block partitioning and tag SNP selection using genotype data and their applications to association studies, Genome Res. 14 (5) (2004) 908–916.

[12] D.E. Reich, M. Cargill, S. Bolk, et al., Linkage disequilibrium in the human genome, Nature 411 (6834) (2001) 199–204.

[13] N. Wang, M.J. Akey, K. Zhang, et al., Distribution of recombination crossovers and the origin of haplotype blocks: the interplay of population history, recombination, and mutation, Am. J. Hum. Genet. 71 (5) (2002) 1227–1234.

[14] V. Bafna, B.J. Halldórsson, R. Schwartz, et al., Haplotypes and informative SNP selection algorithms: don't block out information, in: Proceedings of the Seventh Annual International Conference on Computational Molecular Biology, ACM Press, Berlin, Germany, 2003.

[15] P. Sebastiani, R. Lazarus, S.T. Weiss, et al., Minimal haplotype tagging, Proc. Natl. Acad. Sci. USA 100 (17) (2003) 9900–9905.

[16] J. Hampe, S. Schreiber, M. Krawczak, Entropy-based SNP selection for genetic association studies, Hum. Genet. 114 (1) (2003) 36–43.

[17] Z. Lin, R.B. Altman, Finding haplotype tagging SNPs by use of principal components analysis, Am. J. Hum. Genet. 75 (5) (2004) 850–861.

[18] C.S. Carlson, M.A. Eberle, M.J. Reider, et al., Selecting a maximally informative set of single-nucleotide polymorphisms for association analyses using linkage disequilibrium, Am. J. Hum. Genet. 74 (1) (2004) 106–120.

[19] P.I. de Bakker, N.P. Burt, R.R. Graham, et al., Transferability of tag SNPs in genetic association studies in multiple populations, Nat. Genet. 38 (11) (2006) 1298–1303.

[20] W.B. Wang, T. Jiang, A new model of multi-marker correlation for genome-wide tag SNP selection, Genome Inform. 21 (2008) 27–41.

[21] L. Liu, Y. Wu, S. Lonardi, et al., Efficient algorithms for genome-wide tagSNP selection across populations via the linkage disequilibrium criterion, Comput. Syst. Bioinformatics Conf. 6 (2007) 67–78.

[22] Z.S. Qin, S. Gopalakrishnan, G.R. Abecasis, An efficient comprehensive search algorithm for tagSNP selection using linkage disequilibrium criteria, Bioinformatics 22 (2) (2006) 220–225.

[23] K. Hao, Genome-wide selection of tag SNPs using multiple-marker correlation, Bioinformatics 23 (23) (2007) 3178–3184.

[24] Z. Liu, S. Lin, Multilocus LD measure and tagging SNP selection with generalized mutual information, Genet. Epidemiol. 29 (4) (2005) 353–364.

[25] Z. Liu, S. Lin, M. Tan, Genome-wide tagging SNPs with entropy-based Monte Carlo method, J. Comput. Biol. 13 (9) (2006) 1606–1614.

[26] E. Halperin, G. Kimmel, R. Shamir, Tag SNP selection in genotype data for maximizing SNP prediction accuracy, Bioinformatics 21 (Suppl. 1) (2005) i195–i203.

[27] D.O. Stram, C.A. Haiman, J.N. Hirschhorn, et al., Choosing haplotype-tagging SNPS based on unphased genotype data using a preliminary sample of unrelated

subjects with an example from the Multiethnic Cohort Study, Hum. Hered. 55 (1) (2003) 27–36.

[28] M.E. Weale, C. Depondt, S.J. Macdonald, et al., Selection and evaluation of tagging SNPs in the neuronal-sodium-channel gene SCN1A: implications for linkage-disequilibrium gene mapping, Am. J. Hum. Genet. 73 (3) (2003) 551–565.

[29] R. Magi, L. Kaplinski, M. Remm, The whole genome tagSNP selection and transferability among HapMap populations, Pac. Symp. Biocomput. (2006) 535–543.

[30] A.C. Need, D.B. Goldstein, Genome-wide tagging for everyone, Nat. Genet. 38 (11) (2006) 1227–1228.

[31] B.N. Howie, C.S. Carlson, M.J. Rieder, D.A. Nickerson, Efficient selection of tagging single-nucleotide polymorphisms in multiple populations, Hum. Genet. 120 (1) (2006) 58–68.

[32] J.N. Hirschhorn, M.J. Daly, Genome-wide association studies for common diseases and complex traits, Nat. Rev. Genet. 6 (2) (2005) 95–108.

[33] A.D. Skol, L.J. Scott, G.R. Abecasis, M. Boehnke, Optimal designs for two-stage genome-wide association studies, Genet. Epidemiol. 31 (7) (2007) 776–788.

[34] J. Li, Prioritize and select SNPs for association studies with multi-stage designs, J. Comput. Biol. 15 (3) (2008) 241–257.

[35] J.C. Barrett, B. Fry, J. Maller, M.J. Daly, Haploview: analysis and visualization of LD and haplotype maps, Bioinformatics 21 (2) (2005) 263–265.

[36] P.I. de Bakker, R. Yelensky, I. Pe'er, et al., Efficiency and power in genetic association studies, Nat. Genet. 37 (11) (2005) 1217–1223.

[37] F.M. De La Vega, H.I. Isaac, C.R. Scafe, A tool for selecting SNPs for association studies based on observed linkage disequilibrium patterns, Pac. Symp. Biocomput. (2006) 487–498.

[38] B.V. Halldórsson, V. Bafna, R. Lippert, et al., Optimal haplotype block-free selection of tagging SNPs for genome-wide association studies, Genome Res. 14 (8) (2004) 1633–1640.

[39] J. C. Barrett, L. R. Cardon, Evaluating coverage of genome-wide association studies. Nat. Genet. 38(6) 659-662.

[40] R. Mägi, A. Pfeufer, M. Nelis, et al., Evaluating the performance of commercial whole-genome marker sets for capturing common genetic variation, BMC Genomics 8 (2007) 159.

[41] M. Li, C. Li, W. Guan, Evaluation of coverage variation of SNP chips for genome-wide association studies, Eur. J. Hum. Genet. 16 (5) (2008) 635–643.

Genotype Calling

Michael Inouye[1,2]*, Yik Ying Teo*[3,4]

[1] Immunology Division, The Walter and Eliza Hall Institute of Medical Research, Parkville, Victoria, Australia
[2] Department of Human Genetics, Wellcome Trust Sanger Institute, Hinxton, Cambridge, UK
[3] Departments of Statistics & Applied Probability and Epidemiology & Public Health, National University of Singapore, Singapore
[4] Genome Institute of Singapore, Singapore

BIAS AND ERROR IN GENOTYPE CALLING

The poor cost efficiency and potential sources of bias when manually determining genotypes are obvious for large datasets. Computational algorithms alleviate this burden by being both fast and robust, but even so, any systematic bias in calling can reduce both the power and coverage of a study. This is especially true for case—control studies where informative missingness, differential patterns of missing genotype calls between cases and controls can lead to an artificial inflation in the number of "significant associations" [1—3]. To identify potential informative missingness or other areas of bias, quantile-quantile plots of observed and expected test statistics are typically generated and the degree of inflation determined and the genotype calling algorithm designed by Plagnol and colleagues directly targets this inflation [4].

There are different strategies one can take to reduce bias. As Clayton and colleagues note, when there are cluster shifts between cases and controls (or indeed for any kind of batch processing), calling the cases and controls separately can alleviate test statistic inflation [1]. However, for algorithms which are dependent on pooling data from multiple arrays and do not have strong priors for cluster positions, it is important to call in large batches since low frequency SNPs will have relatively few data points per cluster therefore making it harder to cluster. Additionally, genotype imputation can offer a powerful solution to informative missingness as well as increase the power and coverage of a study [5, 6].

GENOTYPING PLATFORMS

Of the many genotyping platforms currently available, the two that have by far generated the most data for genetic association studies are Affymetrix (Santa Clara, CA, USA) and Illumina (San Diego, CA, USA). Since genotype calling algorithms have been most widely used and have been at their most useful on these platforms, here we focus on Affymetrix and Illumina; however, the principles covered in this chapter can be readily applied to almost any genotyping method.

Microarrays from Affymetrix assess each SNP at the middle position of a 25 bp oligonucleotide probe of predefined sequence. However, there are two major design differences surrounding the Genome-wide Human SNP Array 5.0. Microarrays designed before the 5.0, e.g. the Mapping 500K Array Set, use probes which are either perfectly complementary to each allele of a bi-allelic SNP and its proximal sequence (perfect match, PM) or complementary to the proximal sequence but not to the SNP alleles (mismatch, MM). The probe types are intended to give estimates of foreground signal and background signal (e.g. cross-hybridization), respectively. In addition, for each probe type there are up to six offset positions where the SNP position is shifted toward either the 3' or 5' end of the probe. The Affymetrix 5.0 and 6.0 microarrays reduce this probe hierarchy, keeping only the perfect matches, thus freeing up space on the array to add more SNP and copy-number probes [7]. For all Affymetrix arrays, binding of a probe to a target sequence elicits a fluorescent signal which, in the case of an SNP probe, measures the relative amount of allele A or allele B in a particular sample. Illumina microarrays contain probe-coated beads organized into bead types, each of which assesses a single SNP locus. The beads are randomly allocated to a microarray and each bead type is typically represented at multiple locations on the array. Therefore each SNP is measured with a certain amount of redundancy. On each bead, 50-bp oligonucleotide probes bind to the sequence directly adjacent to an SNP position, after which a single-base extension with a fluorescent ddNTP then leads to a signal for allele A or allele B [8].

NORMALIZATION ALGORITHMS

In studies that involve the processing of many samples, the incorporation of non-biological variation is inevitable. This artifactual variation can come from any number of sources; for example, batch processing, differential reagent use or DNA extraction protocols. Therefore, algorithms that seek to reduce or remove these effects, normalization algorithms, can be immensely important to genotype calling and thus the success of the study. Since many genotyping platforms (Affymetrix and Illumina included) generate a signal measure of allele A and allele B, the following normalization algorithms are highly portable. It should, however, be noted that many normalization algorithms exist and no one algorithm will produce an optimal result for every dataset; we present here only two of the most widely used.

Quantile normalization seeks to minimize probe intensity variation between arrays [9]. The algorithm does this by forcing the distribution of probe intensities for all arrays in a set to be the same:

1. Given n arrays of length p (e.g., p could be the number of SNP probes), form X of dimension $p \times n$ where each array is a column.
2. Sort each column of X to give X_{sort}.
3. Take the means across rows of X_{sort} and assign this mean to each element in the row to get X'_{sort}.
4. Get $X_{normalized}$ by rearranging each column of X'_{sort} to have the same ordering as original X.

The median polish algorithm [10, 11] is intended to protect against failed or outlier probes. Let $y_{k,ij}$ denote the logarithm of the normalized PM intensity for allele k in probe j of array i. For SNP l, a linear model of the form:

$$y^l_{k,ij} = \theta^l_{k,i} + \beta^l_{k,j} + \varepsilon^l_{k,ij}$$

is fitted to the log-normalized intensities, where $\theta^l_{k,i}$ and $\beta^l_{k,i}$ denote the effects for allele k for SNP l from array i and probe j, respectively, and $\varepsilon^l_{k,ij}$ is a normally distributed error centered around zero. As the intensity data can be considered as an $n_l \times N$ matrix, where n_l denotes the number of probes for SNP l, the model parameters $\theta^l_{k,i}$ and $\beta^l_{k,i}$ are estimated by the following iterative procedure:

1. Compute the row median of each row and compute the median of all the row medians. Record this value as the overall effect.
2. Subtract each row median from the corresponding row, and subtract the overall effect from the row medians.
3. Compute the column median of each column and add the median of all column medians to the overall effect.
4. Subtract each column median from the corresponding column, and subtract the median of all column medians from the column medians.
5. Compute the row median of each row and add the median of all column medians to the overall effect.
6. Subtract each row median from the corresponding row, and subtract the median of all row medians from the row medians.
7. Iterate between (3) and (6) until there are no changes with the row and column medians.

The sum of the overall and column effects yields summary estimates of the log-normalized intensities for alleles A and B of SNP l for chip i, written as:

$$\theta^l_i = (\theta^l_{A,i}, \theta^l_{B,i}),$$

and the procedure is applied to all the microarrays. This effectively summarizes the information across the probes to yield an overall signal for each chip which eliminates effects that may be introduced by particular problematic probes.

Generally, after allelic signals have been suitably normalized, the available algorithms for assigning genotypes can be classified into two categories:

1. Assigning genotypes for each sample individually.
2. Assigning genotypes to multiple samples simultaneously by pooling information from these samples.

Assigning the genotypes for each sample independently means that there is the advantage of not having to rely on achieving a target sample size or having to combine samples, which can introduce additional complexities if there are variations in hybridization profiles as a result of assaying the samples on different plates or different batches. However, when the data across multiple samples are suitably standardized, there exist significant advantages of assigning genotypes concurrently to a collection of samples, as this enables sophisticated statistical algorithms to utilize the overall characteristics of the hybridization profiles across multiple samples for accurately calibrating the background hybridization. Regardless of this distinction, an algorithm must be robustly designed in order to handle a wide variety of cluster characteristics (Fig. 5.1).

In the following sections, we provide an overview of the popular genotype calling algorithms used for Affymetrix and Illumina microarrays. We first provide a review of the popular dynamic model mapping (DM) algorithm used for Affymetrix microarrays which assigns genotypes to each sample independently [12]. We then provide an introduction to two clustering-based algorithms that perform concurrent genotype calling of multiple samples, the Bayesian Robust Linear Modeling with Mahalanobis distance (BRLMM) algorithm [13] for Affymetrix microarrays and the ILLUMINUS algorithm [14] for Illumina microarrays. Finally, we discuss other popular genotype calling algorithms.

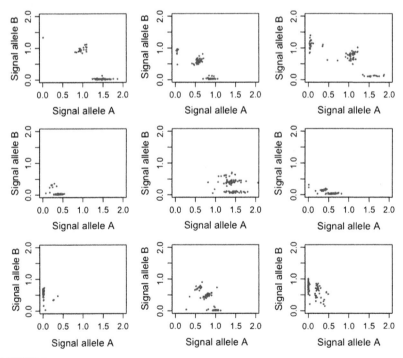

FIGURE 5.1   A sample of SNP signal plots. Each panel shows the pooling of allelic signals for one SNP across 96 arrays. Taken as a whole, the figure illustrates the diversity of the SNP signal plots and the flexibility that an algorithm must have in order to accurately cluster them.

GENOTYPE CALLING FROM A SINGLE ARRAY

The proprietary dynamic model mapping (DM) algorithm by Affymetrix [10] uses a likelihood-based approach for calling the genotypes at each SNP for every microarray individually and independently from both the rest of the SNPs on the same microarray, and from other microarrays. The algorithm makes use of the perfect match (PA, PB) and mismatch intensities (MA, MB) from the probe quartets, and accounts for the level of variation in the pixel fluorescent intensities.

DM assumes that there are two underlying Gaussian distributions for the intensities of every probe quartet for each SNP, belonging to the foreground and background intensities under each of three theoretical models corresponding to the three possible genotypes: {AA, AB, BB}. The method also allows a further NULL model to be considered, which assumes there are no detectable signals from either of the perfect match intensities. Suppose the indices 1, 2, 3 and 4 correspond to the probe cells PA, MA, PB and MB, respectively, within each probe quartet. Let X_{ij} denote the random variable for the intensity for probe index j in the ith probe quartet. Let $\mu_B^{(i)}$ and $\mu_F^{(i)}$ denote the means of the distributions for the background and foreground intensities at the ith probe quartet, respectively, with corresponding standard deviations of $\sigma_B^{(i)}$ and $\sigma_F^{(i)}$. The designations of the foreground and background intensities under each of these four models are:

- NULL model: All the observed intensities are assumed to be distributed as the background intensity:

$$X_{ij} \sim \text{Normal}\left(\mu_B^i, (\sigma_B^i)^2\right) \quad \text{for } j \in \{1, 2, 3, 4\}$$

- AA model: The intensity for the perfect match of allele A (PA) is assumed to follow the distribution for the foreground intensity, while the rest (MA, PB, MB) are assumed to be distributed as the background intensity:

$$X_{ij} \sim \text{Normal}\left(\mu_F^i, (\sigma_F^i)^2\right) \quad \text{for } j = 1$$

$$X_{ij} \sim \text{Normal}\left(\mu_B^i, (\sigma_B^i)^2\right) \quad \text{for } j \in \{2, 3, 4\}$$

- AB model: The intensities for the perfect matches (PA, PB) are assumed to follow the distribution for the foreground intensity while the mismatch intensities (MA, MB) are assumed to be distributed as the background intensity:

$$X_{ij} \sim \text{Normal}\left(\mu_F^i, (\sigma_F^i)^2\right) \quad \text{for } j \in \{1, 3\}$$

$$X_{ij} \sim \text{Normal}\left(\mu_B^i, (\sigma_B^i)^2\right) \quad \text{for } j \in \{2, 4\}$$

- *BB* model: The intensities for the perfect matches (PB) are assumed to follow the distribution for the foreground intensity, while the rest (PA, MA, MB) are assumed to be distributed as the background intensity:

$$X_{ij} \sim \text{Normal}\left(\mu_F^i, (\sigma_F^i)^2\right) \quad \text{for } j = 3$$

$$X_{ij} \sim \text{Normal}\left(\mu_B^i, (\sigma_B^i)^2\right) \quad \text{for } j \in \{1, 2, 4\}$$

In addition, DM makes the assumption that the intensities between probes for each quartet at each SNP are mutually independent.

Let the sample estimates of the parameters under each model for the ith probe quartet be denoted as $\widehat{\mu}_B^{(i)}$, $\widehat{\mu}_F^{(i)}$, $\widehat{\sigma}_B^{(i)}$ and $\widehat{\sigma}_F^{(i)}$. Let x_{ij} and s_{ij} denote the observed mean and standard deviation of the pixel intensities for probe j in the ith probe quartet, with $j \in \{1, 2, 3, 4\}$ corresponding to the probe-cell ordering of $\{PA, MA, PB, MB\}$. Under the NULL model, the maximum likelihood estimates of the mean and the variance for the distribution of the background intensity are:

$$\widehat{\mu}_B^{(i)} = \frac{\sum_{j=1}^{4} x_{ij}}{4}, \quad \left(\widehat{\sigma}_B^{(i)}\right)^2 = \frac{\sum_{j=1}^{4}\left(s_{ij}^2 + x_{ij}^2\right)}{4} - \left(\widehat{\mu}_B^{(i)}\right)^2$$

Under model *AA*, the maximum likelihood estimates of the means and variances for the distributions of the foreground and background intensities are:

$$\widehat{\mu}_F^{(i)} = x_{i1}, \quad \left(\widehat{\sigma}_F^{(i)}\right)^2 = s_{i1}^2$$

and

$$\widehat{\mu}_B^{(i)} = \frac{\sum_{j\neq1}^{3} x_{ij}}{3}, \quad \left(\widehat{\sigma}_B^{(i)}\right)^2 = \frac{\sum_{j\neq1}^{3}\left[s_{ij}^2 + (\widehat{\mu}_B^{(i)} - x_{ij})^2\right]}{3}$$

Under model *AB*, the maximum likelihood estimates of the means and variances for the distributions of the foreground and background intensities are:

$$\widehat{\mu}_F^{(i)} = \frac{x_{i1} + x_{i3}}{2}, \quad \left(\widehat{\sigma}_F^{(i)}\right)^2 = \frac{\left[s_{i1}^2 + (\widehat{\mu}_F^{(i)} - x_{i1})^2\right] + \left[s_{i3}^2 + (\widehat{\mu}_F^{(i)} - x_{i3})^2\right]}{2}$$

and

$$\widehat{\mu}_B^{(i)} = \frac{x_{i2} + x_{i4}}{2}, \quad \left(\widehat{\sigma}_B^{(i)}\right)^2 = \frac{\left[s_{i2}^2 + (\widehat{\mu}_B^{(i)} - x_{i2})^2\right] + \left[s_{i4}^2 + (\widehat{\mu}_B^{(i)} - x_{i4})^2\right]}{2}$$

Under model BB, the maximum likelihood estimates of the means and variances for the distributions of the foreground and background intensities are:

$$\hat{\mu}_F^{(i)} = x_{i3}, \quad \left(\hat{\sigma}_F^{(i)}\right)^2 = s_{i3}^2$$

and

$$\hat{\mu}_B^{(i)} = \frac{\sum_{j\neq 3}^3 x_{ij}}{3}, \quad \left(\hat{\sigma}_B^{(i)}\right)^2 = \frac{\sum_{j\neq 3}^3 \left[s_{ij}^2 + (\hat{\mu}_B^{(i)} - x_{ij})^2\right]}{3}$$

From the assumption of independence between the probes in each quartet, the log-likelihood for probe quartet i under each model can be written as:

$$l_i(NULL) = -\frac{1}{2}\sum_{j=1}^4 \left[\ln\left(2\pi\left(\hat{\sigma}_B^{(i)}\right)^2\right) + \frac{s_{ij}^2 + \left(x_{ij} - \hat{\mu}_B^{(i)}\right)^2}{(\hat{\sigma}_B^{(i)})^2}\right];$$

$$l_i(AA) = -\frac{1}{2}\left\{\sum_{j\neq 1}\left[\ln\left(2\pi\left(\hat{\sigma}_B^{(i)}\right)^2\right) + \frac{s_{ij}^2 + \left(x_{ij} - \hat{\mu}_B^{(i)}\right)^2}{(\hat{\sigma}_B^{(i)})^2}\right] + \ln\left(2\pi\left(\hat{\sigma}_F^{(i)}\right)^2\right)\right.$$
$$\left. + \frac{s_{i1}^2 + \left(x_{i1} - \hat{\mu}_F^{(i)}\right)^2}{(\hat{\sigma}_F^{(i)})^2}\right\};$$

$$l_i(AB) = -\frac{1}{2}\left\{\sum_{j\in\{2,4\}}\left[\ln(2\pi\left(\hat{\sigma}_B^{(i)}\right)^2) + \frac{s_{ij}^2 + \left(x_{ij} - \hat{\mu}_B^{(i)}\right)^2}{(\hat{\sigma}_B^{(i)})^2}\right]\right.$$
$$\left. + \sum_{j\in\{1,3\}}\left[\ln\left(2\pi\left(\hat{\sigma}_F^{(i)}\right)^2\right) + \frac{s_{ij}^2 + \left(x_{ij} - \hat{\mu}_F^{(i)}\right)^2}{(\hat{\sigma}_F^{(i)})^2}\right]\right\};$$

$$l_i(BB) = -\frac{1}{2}\left\{\sum_{j\neq 3}\left[\ln\left(2\pi\left(\hat{\sigma}_B^{(i)}\right)^2\right) + \frac{s_{ij}^2 + \left(x_{ij} - \hat{\mu}_B^{(i)}\right)^2}{(\hat{\sigma}_B^{(i)})^2}\right] + \ln\left(2\pi\left(\hat{\sigma}_F^{(i)}\right)^2\right)\right.$$
$$\left. + \frac{s_{i3}^2 + \left(x_{i3} - \hat{\mu}_F^{(i)}\right)^2}{(\hat{\sigma}_F^{(i)})^2}\right\}$$

A score for each model m for quartet i, $S_i(m)$, is defined as the difference between the log-likelihood from the most likely model and the log-likelihood of the second most likely model:

$$S_i(m) = l_i(m) - \max\{l_i(k), k \in \{NULL, AA, AB, BB\}, k \neq m\}$$

As $S_i(m) > 0$ indicates that model m is most likely for probe quartet i, it thus provides evidence in support for model m. Conversely $S_i(m) < 0$ suggests there exists a better model and that model m may not be the best fitting model. A vector of scores for each model m is obtained by considering the scores from all the probe quartets:

$$V(m) = \{S_1(m), S_2(m), \dots, S_n(m)\}, \quad m \in \{NULL, AA, AB, BB\}$$

with n denoting the number of probe quartets (which is either 6 or 10).

The evidence in support of each model is then robustly aggregated across the probe quartets by performing a Wilcoxon signed rank test on each of these four vectors to test the hypothesis that the median of each vector $V(m)$ is zero, against the one-sided alternative that the median is larger than zero. The intuition behind this formulation is: should the median for $V(m)$ be larger than zero, there is more evidence in favor of model m. Applying this test across the four models yields a p-value under each model, and the genotype corresponding to the model with the lowest p-value is assigned provided the minimum p-value is below a predetermined threshold α. When the minimum p-value is above the threshold or if the smallest p-value corresponds to the NULL model, no genotype call will be made for this sample at this particular SNP. Although the terminology of "NULL genotype" has been used to refer to the situation when no genotype call has been made, this may be due to either a failure to pass the p-value thresholding, or that the data support the NULL model.

GENOTYPE CALLING OF MULTIPLE ARRAYS SIMULTANEOUSLY

Bayesian Robust Linear Model with Mahalanobis Distance (BRLMM)

The BRLMM algorithm is one of the first genotype calling algorithms to adopt a clustering strategy for assigning genotypes to multiple samples assayed on the Affymetrix 500K microarray simultaneously by pooling intensity information across these samples at each SNP [13]. Due to the nature of pooling data across multiple microarrays, there is an additional complication that the intensity profiles of the microarrays should only reflect variations in hybridization according to the genotypes found at each SNP, and not systematic variations introduced by laboratory or non-genetic factors. The process of standardizing the raw hybridization intensities relies on quantile normalization as described above.

BRLMM adopts a clustering approach in integrating the intensity information from multiple samples, where the processed signal intensities θ_i have been transformed to a *Contrast* and *Strength* scale for clustering. Denote (S_A, S_B) as the exponential of (θ_A, θ_B), then:

$$Contrast = \frac{\sinh^{-1}\left(\frac{4(S_A - S_B)}{S_A + S_B}\right)}{\sinh^{-1}(4)}$$

$$Strength = \log(S_A + S_B)$$

BRLMM assumes every SNP on the chip belongs to one of three genotype classes, denoted generically as *AA*, *AB* and *BB*, which are assigned using an ad-hoc Bayesian procedure. Let $Y_i = (C_i, S_i)$ denote the coordinate pair for the *Contrast-Strength* coordinates for individual i at one SNP. Under each model $k \in \{AA, AB, BB\}$, Y_i is assumed to follow a multivariate normal distribution:

$$Y_i \sim \text{MVNorm}(\mu_k, \Sigma_k)$$

The priors on $(\mu_{AA}, \mu_{AB}, \mu_{BB})$ follow a multivariate normal distribution, and each of the Σ_k follows an Inverse-Wishart distribution, of the form:

$$(\mu_{AA}, \mu_{AB}, \mu_{BB}) | m, M \sim \text{MVNorm}(m, M)$$

$$\Sigma | p, S \sim \text{InvWishart}(p, S)$$

with

$$\Sigma = \begin{pmatrix} \Sigma_{AA} & 0 & 0 \\ 0 & \Sigma_{AB} & 0 \\ 0 & 0 & \Sigma_{BB} \end{pmatrix}$$

each of the zeroes in the matrix represents a 2×2 array of zeroes, p fixed at 40 and the rest of the parameters for the priors ($m, M, S, \Sigma_{AA}, \Sigma_{AB}$ and Σ_{BB}) are determined from a set of SNPs where the genotype labels are known. Here, Σ_k is a 2×2 variance-covariance matrix for genotype group k, S is the 6×6 block-diagonal variance-covariance matrix of within-genotype class transformed signals, such that $s_{ij} = 0$ for $|i - j| > 1$ with s_{ij} denoting the entry corresponding to the ith row and jth column in S. In the implementation of BRLMM, a random sample of at least 10,000 SNPs with at least two initial DM calls for each genotype class is used to estimate these parameters.

For each SNP, BRLMM uses DM calls made at a confidence threshold of 0.17 to estimate the attributes of each genotype cluster, before deriving an approximate posterior distribution for the genotype membership conditional on the observed data. Let v and W denote the sample estimates of m and Σ that are derived from the appropriate DM calls, and let N denote a 6×6 diagonal matrix with entries (n_{AA}, n_{AA}, n_{AB}, n_{AB}, n_{BB}, n_{BB}), where n_k denote the number of valid DM calls for genotype k.

Based on the estimates of the cluster centers and variance-covariance matrices, the posterior estimates of μ_k and Σ_k can be obtained by performing point-wise shrinkage towards the prior estimates as:

$$(\mu'_{AA}, \mu'_{AB}, \mu'_{BB}) = (M^{-1} + (N\ S)^{-1})^{-1}(M^{-1}m + (N\ S)^{-1}v)$$

$$\Sigma'_k = \frac{(n_k - 1)W_k + pS_k}{n_k - 1 + p}$$

with W_k and S_k corresponding to the 2×2 block diagonal elements in W and S for genotype class k.

The genotypes for the samples, including those that were previously determined by DM, are reassigned based on the posterior estimates of μ and Σ using the Mahalanobis distance, which for the observed intensity data for the ith individual y_i and under the model for genotype k is given by:

$$d_k(y_i) = \sqrt{(y_i - \mu_k)^T \Sigma_k^{-1}(y_i - \mu_k)}$$

The distances d_{AA}, d_{AB} and d_{BB} are obtained under the three genotype models, and the genotype corresponding to the model with the smallest distance is assigned to the chip. A confidence score, defined as the ratio of the smallest distance to the second-smallest distance, is attached to the assigned genotype. This score is bounded between zero and one, and can be used to assess the quality of the genotype call, where typically only valid calls are made when the confidence scores are less than 0.5, and NULL genotypes are assigned otherwise.

The ILLUMINUS Algorithm

The ILLUMINUS algorithm was specifically designed for assigning genotypes for samples that have been assayed using the Illumina

BeadArray technology [14]. As with any genotype calling algorithm that performs pools information across multiple samples at an SNP to perform concurrent genotype assignment of multiple samples, the process of normalizing the raw hybridization intensities across the samples is particularly important for minimizing non-biological fluctuations that can affect calling accuracy. For Illumina microarrays, this process of normalization adopts a six-degree of freedom affine transformation that consists of the following features [15]:

1. **Outlier removal:** where in each collection of bead intensities, SNPs with extreme hybridization intensities are removed.
2. **Background estimation:** where a random sampling of bead intensities for candidate homozygotes is performed for each allele to estimate a linear model fit, where an empirical origin is defined by the intercept of the two linear model fits for the two alleles.
3. **Rotational estimation:** where all the bead intensities are rotated with respect to the new origin.
4. **Shear estimation:** where all the bead intensities are sheared with respect to the new origin.
5. **Scaling estimation:** where all the bead intensities are scaled according to prior-defined control points with respect to the new origin.

The ILLUMINUS algorithm performs a clustering-based analysis on the contrast-strength scale, defined in a similar fashion with:

$$Contrast = \frac{x_{jl} - y_{jl}}{x_{jl} + y_{jl}}$$

$$Strength = \log(x_{jl} + y_{jl})$$

where x_{jl} and y_{jl} denote the normalized signal intensities for the two alleles for sample j at SNP l, respectively. A three-component bivariate mixture model is used to model the distributions of the contrast-strength data across the samples, where each component assumes a multivariate truncated t-distribution and corresponds to one of the three possible genotype classes AA, AB and BB (Fig. 5.2).

Suppose $f(x, \mu, \Sigma, \nu)$ denote the density function for data x at a t-distribution with location parameter μ, variance-covariance matrix Σ and ν degrees of freedom, the density for the observed data for sample j at SNP l can be expressed as:

$$F(X_{jl}) = \sum_{k=1}^{3} \lambda_k \phi_k(X_{jl} : \mu_k, \Sigma_k, \nu_k)$$

where $\phi_k(X_{jl}: \mu_k, \Sigma_k, \nu_k)$ denote the density function for a truncated t-distribution corresponding to the kth component, such that:

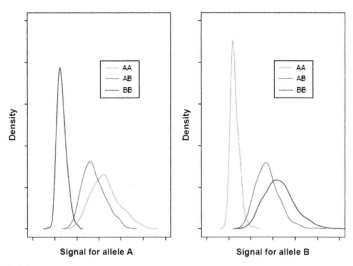

FIGURE 5.2   Allelic signal from each genotype. Density functions of the signals at each allele for the three genotype classes forms the motivation for three-component mixture modeling in concurrent genotype calling across multiple samples. Please refer to color plate section

$$\phi_1(X_{jl} : \mu_1, \Sigma_1, \nu_1) = \frac{f(X_{jl} : \mu_1, \Sigma_1, \nu_1)}{1 - \int_{-\infty}^{-1} f(X_{jl} : \mu_1, \Sigma_1, \nu_1)dc};$$

$$\phi_2(X_{jl} : \mu_2, \Sigma_2, \nu_2) = \frac{f(X_{jl} : \mu_2, \Sigma_2, \nu_2)}{\int_{-1}^{1} f(X_{jl} : \mu_2, \Sigma_2, \nu_2)dc};$$

$$\phi_3(X_{jl} : \mu_3, \Sigma_3, \nu_3) = \frac{f(X_{jl} : \mu_3, \Sigma_3, \nu_3)}{1 - \int_{1}^{\infty} f(X_{jl} : \mu_3, \Sigma_3, \nu_3)dc}$$

where the integrals in the denominators acts on the contrast axis and $(\lambda_1, \lambda_2, \lambda_3)$ denote the mixture proportions according to the principle of Hardy-Weinberg equilibrium.

A fourth component is introduced to act as an outlier class where samples with atypical intensity profiles that clearly are not members of any of the three valid genotype classes are assigned to. This component takes the form of a bivariate Gaussian distribution with zero covariance and considerably large variances such that the density is effectively flat across the possible range of values.

The parameters of the t-distribution are directly estimated from the empirical data, while the degrees of freedom for each of the three classes are predetermined, with $v_1 = v_3 < v_2$ as the Illumina BeadArray technology has the feature where, for most SNPs, homozygotes yield sharp and consistent signals for the homozygotes as compared to the heterozygotes, and this feature is accommodated by the use of t-distributions with heavier tails for the homozygous clusters. For the remaining SNPs where the profile of the contrast-axis is similar across homozygotes and heterozygotes, a constant degree of freedom is fitted for all three components ($v_1 = v_2 = v_3$).

The parameters of the mixture model (μ_k, Σ_k, v_k) are calibrated with an Expectation-Maximization (EM) procedure, which alternates between assigning genotypes conditional on these parameters (the E-step) and inferring these parameters by their maximum-likelihood estimates conditional on the assigned genotypes (the M-step). This iterative EM procedure proceeds until stable genotype calls are obtained for all the samples.

At each E-step, the genotype for a sample is assigned to a specific genotype class if the posterior probability of the corresponding class is above a predetermined threshold (typically set at 0.95), otherwise a NULL genotype is assigned and will not be used to update the parameters of the three genotype classes in the mixture model.

At each M-step, updating the parameters for genotype class k is given by:

$$\mu_k = \left(\bar{c}^k, \bar{s}^k \right) = \left(\frac{1}{n_k} \sum_j^{n_k} c_{jl}^k, \frac{1}{n_k} \sum_j^{n_k} s_{jl}^k \right);$$

$$\Sigma_k = \frac{1}{n_k - 1} \begin{pmatrix} \sum_j^{n_k} \left(c_{jl}^k - \bar{c}^k \right)^2 & \sum_j^{n_k} \left(c_{jl}^k - \bar{c}^k \right) \left(s_{jl}^k - \bar{s}^k \right) \\ \sum_{jl}^{n_k} \left(c_{jl}^k - \bar{c}^k \right) \left(s_{jl}^k - \bar{s}^k \right) & \sum_j^{n_k} \left(s_{jl}^k - \bar{s}^k \right)^2 \end{pmatrix}$$

where n_k denote the number of samples that have been assigned to the kth genotype class, with c_{jl}^k and s_{jl}^k representing the contrast and strength coordinates of the j sample at SNP l, respectively.

OTHER GENOTYPE CALLING ALGORITHMS

We have discussed three genotype calling algorithms that perform considerably differently. The DM algorithm assigns genotypes to each microarray without incorporating information from other samples. While it has been shown that pooling information from multiple samples can improve the calling performance, the ability to assign genotypes to a single microarray is attractive, particularly in the initial process of assessing whether the laboratory genotyping process has been successful where DM still plays an integral role despite being supplanted by other advanced and more accurate algorithms for formal genotype assignment.

The BRLMM and ILLUMINUS algorithms are two examples of clustering-based algorithms that simultaneously assign the genotypes to multiple samples. While both algorithms adopt a similar mathematical framework of modeling mixtures of density functions, BRLMM requires initial genotype calls made by DM to seed the characteristics of the mixture model, whereas ILLUMINUS adopts an Expectation-Maximization framework to calibrate these characteristics from the data empirically. In general, most of the available calling algorithms that perform multi--sample genotype calling adopt similar structures as BRLMM and ILLU-MINUS, except that additional features have been incorporated to increase the sophistication and utility of these algorithms in different settings (see Table 5.1).

Of notable mention is the Markov chain Monte Carlo-based algorithm CHIAMO, the procedure used by the Wellcome Trust Case Control Consortium to assign genotypes on 3,000 controls and 14,000 cases across seven human diseases [3]. CHIAMO adopts a Bayesian hierarchical mixture modeling framework in genotype calling, where priors are assigned to the parameters of the genotype classes instead of assuming them to be fixed. This scheme allows the genotypes of different cohorts of samples to be assigned simultaneously by allowing the parameters of the mixture model to vary across the different cohorts, and allows available information on the allelic spectrum of the SNPs to be incorporated to inform the algorithm.

Another algorithm that should be highlighted is the BIRDSEED algorithm designed for the latest Affymetrix 5.0 and 6.0 microarrays [16]. The BIRDSEED algorithm is another example of a multi-sample genotype calling algorithm, but incorporates SNP-specific information from a set of training data that informs the algorithm on the characteristics of each of the three genotype classes. Due to the different genomic sequence of the oligonucleotide probes, different hybridization performance is expected across different SNPs. By learning about the expected hybridization profiles from well-characterized samples (like those from the International HapMap Project [17, 18]), BIRDSEED anchors the expected positions of the genotype clusters which a subsequent Expectation-Maximization iterative scheme will optimize according to the empirical data. This has been shown to result in significant improvements to the accuracy and call rate of genotyping calling.

Finally, there are algorithms like BEAGLE [19] and that introduced by Yu and colleagues [20] that leverage on the correlation structure between SNPs, or linkage disequilibrium, to improve the accuracy of genotype assignment. These algorithms typically infer genotypes from the hybridization data as well as from the genotypes of neighboring markers

TABLE 5.1

Platform	Algorithm	Type	Reference	Notable features
Affymetrix	DM	Single	Di et al. [12]	Genotype assignment at a single-chip level allows assessment of genotyping performance for QC purposes.
	RLMM	Multiple	Rabbee and Speed [21]	Uses training data to determine expected characteristics of genotype clusters.
	BRLMM	Multiple	Affymetrix Inc. [13]	Bayesian implementation that incorporates prior information on expected characteristics of genotype clusters.
	CHIAMO	Multiple	The Wellcome Trust Case Control Consortium [3]	Bayesian hierarchical model framework allows multiple cohorts to be called simultaneously by modeling the trans-cohort correlation of the characteristics of genotype cluster.
	XTYPING	Multiple	Plagnol et al. [4]	One-dimensional calling scheme which accounts for differential bias between cases and controls in genotype scoring.
	BRLMM-P	Multiple	Affymetrix Inc. [13]	One-dimensional calling scheme and does not rely on secondary calling algorithm to seed the characteristics of the genotype clusters.
	CRLMM	Multiple	Carvalho et al. [22]	Accounts for variation in hybridization intensities due to probe sequence content.
	BIRDSEED	Multiple	Korn et al. [16]	Expectation-Maximization procedure that uses HapMap data to estimate and predict expected genotype cluster characteristics.
Illumina	GenCall	Single	Illumina Inc.	Proprietary calling algorithm for Illumina microarrays.
	ILLUMINUS	Multiple	Teo et al. [14]	Fast and robust clustering scheme that can accommodate whole-genome amplified DNA.
	GenoSNP	Single	Giannoulatou et al. [23]	Highly accurate genotype inference based on within-sample information, allowing for small sample sizes and assessment of genotyping performance for QC purposes.
Generic	BEAGLE	Multiple	Browning and Browning [19]	Integrates phasing and imputation procedures to improve the accuracy of genotype assignment.
			Yu et al. [20]	Incorporate LD information to improve the accuracy of genotype assignment.

that are in LD with the target SNP. This process has been shown to yield more accurate genotypes than simply relying on hybridization intensities for genotype inference [20].

References

[1] D.G. Clayton, N.M. Walker, D.J. Smyth, R. Pask, J.D. Cooper, L.M. Maier, et al., Population structure, differential bias and genomic control in a large-scale, case-control association study, Nat. Genet. 37 (2005) 1243–1246. 10.1038/ng1653.

[2] M.I. McCarthy, G.R. Abecasis, L.R. Cardon, D.B. Goldstein, J. Little, J.P. Ioannidis, et al., Genome-wide association studies for complex traits: consensus, uncertainty and challenges, Nat. Rev. Genet. 9 (2008) 356–369. 10.1038/nrg2344.

[3] Wellcome Trust Case Control Consortium, Genome-wide association study of 14,000 cases of seven common diseases and 3,000 shared controls, Nature 447 (2007) 661–678. 10.1038/nature05911.

[4] V. Plagnol, J.D. Cooper, J.A. Todd, D.G. Clayton, A method to address differential bias in genotyping in large-scale association studies, PLoS Genet. 3 (2007) e74. 10.1371/journal.pgen.0030074.

[5] C.A. Anderson, F.H. Pettersson, J.C. Barrett, J.J. Zhuang, J. Ragoussis, L.R. Cardon, et al., Evaluating the effects of imputation on the power, coverage, and cost efficiency of genome-wide SNP platforms, Am. J. Hum. Genet. 83 (2008) 112–119. 10.1016/j.ajhg.2008.06.008.

[6] C.C. Spencer, Z. Su, P. Donnelly, J. Marchini, Designing genome-wide association studies: sample size, power, imputation, and the choice of genotyping chip, PLoS Genet. 5 (2009) e1000477. 10.1371/journal.pgen.1000477.

[7] S.A. McCarroll, F.G. Kuruvilla, J.M. Korn, S. Cawley, J. Nemesh, A. Wysoker, et al., Integrated detection and population-genetic analysis of SNPs and copy number variation, Nat. Genet. 40 (2008) 1166–1174. 10.1038/ng.238.

[8] F.J. Steemers, W. Chang, G. Lee, D.L. Barker, R. Shen, K.L. Gunderson, Whole-genome genotyping with the single-base extension assay, Nat. Methods 3 (2006) 31–33. 10.1038/nmeth842.

[9] B.M. Bolstad, R.A. Irizarry, M. Astrand, T.P. Speed, A comparison of normalization methods for high density oligonucleotide array data based on variance and bias, Bioinformatics 19 (2003) 185–193.

[10] D. Holder, R.F. Raubertas, B. Pikounis, V. Svetnik, K. Soper, Statistical analysis of high density oligonucleotide arrays: a SAFER approach, The GeneLogic Workshop (2001).

[11] J.W. Tukey, Exploratory Data Analysis, Addison-Wesley (1977).

[12] X. Di, H. Matsuzaki, T.A. Webster, E. Hubbell, G. Liu, S. Dong, et al., Dynamic model based algorithms for screening and genotyping over 100 K SNPs on oligonucleotide microarrays, Bioinformatics 21 (2005) 1958–1963. 10.1093/bioinformatics/bti275.

[13] Affymetrix Inc, BRLMM: an improved genotype calling method for the GeneChip Human Mapping 500K Array Set (2006).

[14] Y.Y. Teo, M. Inouye, K.S. Small, R. Gwilliam, P. Deloukas, D.P. Kwiatkowski, et al., A genotype calling algorithm for the Illumina BeadArray platform, Bioinformatics 23 (2007) 2741–2746. 10.1093/bioinformatics/btm443.

[15] B.G. Kermani, Artificial intelligence and global normalization methods for genotyping. US Patent 20060224529 (2005).

[16] J.M. Korn, F.G. Kuruvilla, S.A. McCarroll, A. Wysoker, J. Nemesh, S. Cawley, et al., Integrated genotype calling and association analysis of SNPs, common copy number polymorphisms and rare CNVs, Nat. Genet. 40 (2008) 1253–1260. 10.1038/ng.237.

[17] International HapMap Consortium, A haplotype map of the human genome, Nature 437 (2005) 1299–1320. 10.1038/nature04226.

[18] International HapMap Consortium, K.A. Frazer, D.G. Ballinger, D.R. Cox, D.A. Hinds, L.L. Stuve, R.A. Gibbs, et al., A second generation human haplotype map of over 3.1 million SNPs, Nature 449 (2007) 851–861. 10.1038/nature06258.

[19] B.L. Browning, S.R. Browning, A unified approach to genotype imputation and haplotype-phase inference for large data sets of trios and unrelated individuals, Am. J. Hum. Genet. 84 (2009) 210–223. 10.1016/j.ajhg.2009.01.005.

[20] Z. Yu, C. Garner, A. Ziogas, H. Anton-Culver, D.J. Schaid, Genotype determination for polymorphisms in linkage disequilibrium, BMC Bioinformatics 10 (63) (2009). 10.1186/1471-2105-10-63.

[21] N. Rabbee, T.P. Speed, A genotype calling algorithm for Affymetrix SNP arrays, Bioinformatics 22 (2006) 7–12. 10.1093/bioinformatics/bti741.

[22] B. Carvalho, H. Bengtsson, T.P. Speed, R.A. Irizarry, Exploration, normalization, and genotype calls of high-density oligonucleotide SNP array data, Biostatistics 8 (2007) 485–499. 10.1093/biostatistics/kxl042.

[23] E. Giannoulatou, C. Yau, S. Colella, J. Ragoussis, C.C. Holmes, GenoSNP: a variational Bayes within-sample SNP genotyping algorithm that does not require a reference population, Bioinformatics 24 (2008) 2209–2214. 10.1093/bioinformatics/btn386.

6

Data Handling

N. William Rayner

Wellcome Trust Centre for Human Genetics,
University of Oxford, UK

Genome-wide association (GWA) studies generate large data files; handling these files can be tricky. This chapter concentrates on how to deal successfully with these data files using widely and generally freely available software.

The first stage of dealing with GWA data is to determine the format. There are many possible genotype formats available; if the data have already been processed, they may already be formatted for such programs as PLINK [1] or SNPtest [2]; however, the format could be that from the genotype calling algorithm which may not be directly usable in your program of choice.

The easiest way to determine the format is usually by looking at the data but the files will likely be too large to open in a text editor; in this case a good way to visualize them is to use the Unix functions, *head, tail, less* and *more*, which will by default show, respectively, the first 10 lines of the file, the last 10 lines and the first page of data — *less* and *more* operate in the same way but *less* allows backward movement in the file as well as forward; if the files are compressed using gzip (.gz file extension) use *zmore* and *zless*. All these Unix commands are also available on the newer Mac operating systems; if you only have a Windows machine available then *Cygwin* [3] can be installed giving a command prompt with all the Unix functions. Further information on any Unix commands mentioned here can be obtained using *man*, e.g., *>man command* will give the full manual page for the command in question.

There are two main types of genotype files. Both are tabular format, the first having one row (line) per SNP and column, or columns, per individual as exemplified by those used by SNPtest and the second having one row per individual and column(s) per SNP as exemplified by PLINK format pedigree (ped) files. It is less common to get files with one row per genotype; these tend to have few columns but many rows and give the largest file size.

When checking the format, apart from looking at the data layout it is advisable to check the allele coding and missing value identifiers, usually a zero, and where possible to look for other potential errors in, for example, gender and affection status. The number of lines present in the file can be checked using the Unix word count *wc* function >*wc -l your-file*; on the tabular formats this will give the number of samples or SNPs in the file. In addition, running an analysis, e.g., *plink — file filename*, will provide summary data that will highlight format or allele coding problems and provide genotype counts.

Alleles in the file can be presented in a number of ways — the simplest coding being A, C, G, T; however, some programs require numeric coding, in this case the alleles can be recoded 1, 2, 3, 4, respectively. Where the alleles are required to be coded as 1 and 2, this can be accomplished in a number of ways. Some programs code allele 1 as the minor allele; this works well on a single study but can pose problems when SNPs have a minor allele frequency close to 50% — great care should be taken when merging datasets to ensure that all alleles are coded the same way. The other way to code alleles is alphabetically. The allele closer to the start of the alphabet is always coded 1; this is more complicated but facilitates dataset merging when the data have been strand standardized. In this scheme allele A is always coded as 1 and T always as 2, the coding of C and G depends on the other allele (see Table 6.1).

Files may contain duplicate SNPs or samples and these may be intentional or unintentional. Intentional duplicates will have the same

TABLE 6.1   Allele Codings using an Alphabetical Coding System

Alleles	Allele coded as 1	Allele coded as 2
A/T	A	T
A/C	A	C
A/G	A	G
C/G	C	G
C/T	C	T
G/T	G	T

identifier and may be used to check genotyping concordance; unintentional duplicates will have a different name. For SNPs this can occur when the build changes and SNP rs numbers are merged; when this occurs the resultant SNP takes the lower rs number and the higher one disappears. Some programs will warn you when finding intentional duplicates. Others will not, so it is advisable to check for these in advance.

A genome-wide study will include phenotype data which in general tends come in many formats, as these may have been collected over a number of years using a number of techniques and programs. The most likely file formats are a text file, either comma, .csv, or tab separated, .txt or Excel files, .xls. As with genotype files these need checking before use. Some checks are similar. What is the missing value identifier? Is the missing value identifier consistent between the different phenotypes and files? For numeric values, age, height, etc. it is worth checking lowest, highest, mean and median values and checking values over three standard deviations from the mean. Other fields, whether mixed numeric and text or just text, should be checked for errors, such as capitalization, spelling mistakes, or other inconsistencies. If merging several phenotype files check that the column order is the same, and the columns are using the same units. Phenotype files should contain a header row which should be clear and give units where appropriate.

When using programs such as Excel or Open Office spreadsheet to manage data, be aware of the auto formatting on import, e.g., a plate well id of 1234E10 will be changed to $1.23e^{10}$ unless you ensure on import that you specify the column type as text; using/or - as a delimiter for numeric data could lead to some values being interpreted as date values.

Editing files and converting between different formats is commonly required. This section covers some techniques to accomplish this. When manipulating files it is a good idea, and essential if using a new technique or tool, to check that the reformatting has been successful; with large files it is very easy to introduce problems and for these to pass unnoticed.

Converting or manipulating standard format files such as ped files can be accomplished using the relevant programs; this reduces the possibility of introducing errors into the file. It also ensures that the edited file is saved as a new file thereby ensuring the original data are not lost.

With genotype files it is usual to create a "clean" file on which all the QC metrics have been applied. As genotype files usually contain information on the SNP's genomic position and chromosome, it is helpful to ensure that this is standardized in the "clean" file by aligning all SNPs to the forward strand of the current NCBI build. This will facilitate merging genome-wide data files, a process which requires care as it can be difficult to spot the small number of differences that can occur between datasets, such as rs number, position and allele differences. Using the allele coding, A, C, G, T or 1, 2, 3, 4, will allow for easier allele difference checking.

Editing large files can be problematic. While Unix text editors can often open and display them they can be awkward to use. TextPad [4], a Windows program, can open files up to 1 Gb in size and allow use of regular expressions for manipulating the data. Excel or Open Office spreadsheet can be useful in processing data files, especially if manipulating columns or sorting data; however, these are also limited by the size of the files they can open. The alternative is to use Unix commands, for example if extracting columns from text files using the *cut* command, e.g., >*cut −f1,4,6-10 file1* >*file2*, will extract columns 1, 4 and 6, 7, 8, 9 and 10 from *file1* and place them in that order in *file2*. By default *cut* will use tab as the column delimiter but by using the −*d* flag this can be changed; −*d* ',' will use commas as the column delimiter. For sorting data files the Unix *sort* command allows a text file to be sorted on a single index column, e.g., >*sort -k4 -o file2 file1* will sort file1 on column 4 and put the data in file2. If column 4 were numeric the −*n* option would be needed.

To merge data, for example a phenotype and genotype file, the *join* function is useful. The data need to be sorted first; >*join -1 2 -2 5 file1 file2* > *file3* will join file1 and file2 using column 2 in file1 and column 5 in file2 as the common field. To simply concatenate files *cat* can be used which will append one file to another; >*cat file1 file2 file3* > *new-file.txt* will append the files in order into new-file.txt. As *cat* does not exist on Windows the *copy* command can be used to concatenate files; >*copy file1 + file2 + file3 new-file.txt* will accomplish the same thing.

If you are not sure of the changes between versions of files the *diff* command allows a summary of these to be generated; >*diff −y −suppress-common-lines file1 file2* > *differences.txt* compares file1 and file2 and prints the differences side by side in the output differences.txt file.

The Unix command *grep* is also a very useful tool. It can be used to extract rows of data matching, say one SNP identifier, e.g., >*grep rs1234 filename.txt* > *file-of-rs1234.txt*, or individual ID or even using a list of identifiers in a file to extract multiple data lines, >*grep -f ids.txt filename.txt* > *file-of-ids.txt*. Grep can also take a regular expression as the extraction string using the -G option. Regular expressions or regex are a very useful way of dealing with the data in the files, as it specifies patterns in the data to match which can then be used to manipulate or extract data.

There are many ways of using regular expressions such as awk, sed and probably the most common, Perl. Perl will not be detailed here as there are many resources [5] on this subject; however, a small amount of installing and running some basic Perl to accomplish data manipulation will be covered.

On Unix and Mac machines Perl is most likely already installed; typing perl -v will display the version and other information if it is installed. On Windows there is the option of using Perl in CygWin+ or to install the Windows version [6].

Running Perl scripts is simple. There are several ways to accomplish this but the most consistent across all the platforms is to type >*perl scriptname*, a .pl file extension is used to indicate Perl scripts.

An example script, space-to-tab.pl, listed in the useful perl scripts section, uses regular expressions to replace multiple spaces found, for example, in PLINK output files, with a single tab, giving a tab delimited file. The relevant line is s∧s+∧t/g, where the \s+ specifies to find spaces, \s, and the + indicates find 1 or more occurrences; the \t specifies what to change it to, in this case a tab. The g at the end is important here as it specifies a global replace otherwise it will only match the first occurrence. Modifying the script for other delimiters is easy; changing \t to \, giving s∧s+∧,/g, will convert the spaces in the file to commas giving a .csv file.

When running commands multiple times or running long commands, it can be easier to set this up as a list of commands in a single file and to use this to perform the manipulations; these are referred to as batch .bat (Windows) or shell .sh (Unix) files (Table 6.2). There are also advantages in that it is easy to repeat any process just by changing the parameters, if needed, and rerunning the script; this, to a certain, extent documents the process being used to create the different files and can even contain programming if needed.

To run these on Windows, the convention is to use the file extension .bat and then this file can be run from a command prompt or even by double clicking on it. On Unix the script file will need to be made executable using the *chmod* command, e.g., >*chmod u+x*.

Once the data have been formatted and analyzed there will be one or more output files to deal with. Programs like SNPtest and PLINK have defined output formats, but others may not and even the defined formats may need manipulation before further steps can take place; however, the output files generated tend to be smaller and therefore easier to work with.

One final thing to note on file formats, and this applies to both input and output files, concerns the carriage returns, or End of Line (EOL) character. These are the hidden characters at the end of the line indicating

TABLE 6.2 Examples of a Unix shell file and Windows batch file, note the #!/bin/sh on the first line of the Unix shell file

Unix shell file (example.sh)	Windows batch file (example.bat)
#!/bin/sh	
cut −f 1,2 file1 > file2	*copy file1 + file2 + file3 + file4 file5*
cut −f 1,2 file3 > file4	*copy file6 + file7 file8*
cut −f 1,2 file5 > file6	
grep rs1234 file 6 > file7	

a new line. With many files and programs being used, and especially if sharing data, it is likely they will be generated on different platforms and Unix, Windows and Macintosh each uses a different character or characters to indicate the end of line. This can cause problems when viewing and manipulating the files, e.g.: Unix format files will not display correctly in Windows programs such as Notepad and even scripting languages can have problems; the Windows version of Perl is unable to recognize Mac end-of-line characters, but will recognize Unix ones; Unix Perl only recognizes Unix end of lines but there are the Unix utilities *dos2unix* and *unix2dos* which will reformat the end-of-line characters in a text file or files into the relevant format. Programs such as TextPad allow you to choose the format the file is saved in so conversion is possible, assuming the file is small enough to open.

One inevitability is that changes will be made to the file, perhaps due to a change of NCBI build or a QC cleaning step. When doing this it is advisable to create and keep a new version of the file, especially if that file has been used to generate results. The best way to manage these requires planning a sensible directory structure layout and making use of long filenames. Although all platforms can cope with spaces in the name, it is recommended not to use them but rather a delimiter such as − or _. An alternative or a complementary approach to prevent filename and directory length getting excessive is to use a readme file in each directory to describe the files and their contents in that directory.

Storage of files is important, given their raw size and the possibility of multiple versions. One option is to compress the files, if they are not already. There are many formats for this. Currently favored is gzip (.gz files), this is likely to be installed by default on Unix-based systems, a command line version is available for Windows [7] and programs such as 7zip [8] can create .gz files. Gzipped files have a number of advantages. Programs like SNPtest can read and write gzipped files directly and there are many Unix tools such as *zmore, zcat, zless, zdiff* and *zgrep* which can operate on gzipped files which eliminates the need to create a large unzipped file before working on it. One limitation of gzip versus other compression programs is that it can only compress one file at a time. If multiple files or directories are to be compressed, the Unix command *tar* is traditionally used (>*tar −cf new-file.tar file1 file2 etc*) and this can then be gzipped.

If using ped files for data storage, PLINK binary ped files can be used to reduce the size of data files; this reduction can be quite considerable − a 9 Gb ped file can be reduced to ~550 Mb binary ped file, and if needed this can be further compressed using a zip program which will achieve a file size of nearly half again (~300 Mb). This can make zipped binary ped files a good solution for archiving data files.

As the time and effort in file formatting and analysis is considerable you will need to consider backing up the data, or at least having a copy of the data in another location. Fortunately, backups are a fairly trivial task nowadays with the advent of backup software included in most operating systems and the availability of large external disks or even backups to the Internet. If the data are critical then it will be worth considering solutions with timed backups and the possibility of off-site backups, although this can be as simple as taking the laptop or backup media home.

Useful Perl scripts:

space-to-tab.pl, usage >*perl space-to-tab.pl filename*

```
use strict;
use warnings;
open IN, "$ARGV[0]" or die $!;
open OUT, ">tab-$ARGV[0]" or die $!;
while (<IN>)
{
s/\s+/\t/g;
print OUT "$_";
}
```

count-columns.pl usage >*perl count-columns.pl filename*

```
use strict;
use warnings;
my $file = $ARGV[0];
my $line = 0;
open INPUT, "$file" or die;
open OUTPUT, ">Count-columns-$file" or die;
while (<INPUT>)
{
chomp;
my $count = s/\s+//g;
print OUTPUT "$line\t$count\n";
$line++;
}
```

find-duplicates.pl usage >*perl find-duplicates.pl filename*

```
use strict;
use warnings;
open INPUT, "$ARGV[0]" or die;
print "Column to check for duplicates\n>";
my $column = <STDIN>;
my %hash;
chomp $column;
while (<INPUT>)
{
my @temp = split/\s+/;
```

```
if ($hash{$temp[$column]})
{
print "$temp[$column]\n";
}
else
{
$hash{$temp[$column]} = 1;
}
}
```

References

1. S. Purcell, B. Neale, K. Todd-Brown, L. Thomas, M.A.R. Ferreira, D. Bender, et al., PLINK: a toolset for whole-genome association and population-based linkage analysis, Am. J. Hum. Genet. 81 (2007) 559–575.
2. The Wellcome Trust Case Control Consortium, Genomewide association study of 14,000 cases of seven common diseases and 3,000 shared controls, Nature 447 (2007) 661–678.
3. http://www.cygwin.com/
4. http://www.textpad.com/
5. http://www.perl.org/
6. http://www.activestate.com/activeperl/
7. http://gnuwin32.sourceforge.net/packages/gzip.htm
8. http://www.7zip.com

Data Quality Control

Carl A. Anderson

Department of Human Genetics, Wellcome Trust Sanger Institute,
Hinxton, Cambridge, UK

Given the large number of single nucleotide polymorphisms (SNPs) and individuals typically included in genome-wide association (GWA) studies, even a very low genotype error rate can result in an unacceptable level of false-negative and false-positive disease associations. Careful and thorough assessment of data quality is therefore a key step that must be undertaken prior to testing for disease associations. Fortunately, due to the large number of markers and individuals, many quality control (QC) procedures can be implemented to reduce the rate of false-positive and false-negative findings (many of which could not be successfully applied to smaller association studies such as candidate-gene designs). Here, in addition to outlining several procedures to identify and remove poorly genotyped markers and individuals, methods for the identification and removal of confounding variables (such as plate-effects and divergent population ancestry) are discussed. QC procedures for both case-control and family-based association studies are highlighted. Brief details of several software packages that can be used for GWAS QC are also given.

INTRODUCTION

The rate of type-1 error (false-positive association) and type-2 error (false-negative association) is central to the success of genome-wide association (GWA) studies. When GWA studies were first envisaged much was written about the need to derive appropriate significance thresholds to limit the number of false-positive and false-negative associations [1]. Much less attention was given to the fact that even a small amount of genotype error or bias can elevate such findings to an unacceptable level. For example, if one million markers are included in a GWA study where the proportion of poorly genotyped markers is 0.001, up to 1000 markers may be unnecessarily taken forward for replication due to false-positive association. It was not until the first wave of GWA studies were being analyzed that the need for extensive quality control (QC) became apparent, and this is now seen as a fundamental step during the analysis of GWA data. The large sample sizes and dense marker panels typically utilized in GWA studies enable thorough QC to be undertaken to identify (for subsequently removal) those individuals and markers introducing the most bias to a study. The basic QC steps that should be undertaken in all GWA studies will be outlined in this chapter.

Even though QC can lower false-positive and false-negative error rates, care should still be taken during the design and genotyping stages to ensure that bias is not unnecessarily introduced [2]. The sample ascertainment and genotyping guidelines outlined in earlier chapters should help minimize the number of individuals and genotypes removed during QC, thus maximizing power to detect association (power is equal to one minus the false-negative rate). Because ascertainment and genotyping protocols can have such a pronounced effect on data quality it is important that these steps are taken into consideration when designing the QC procedure. For example, if controls are being ascertained from two separate cohorts and analyzed as a single control group, SNPs that significantly differ in allele/genotype frequency between the constituent cohorts should be removed.

Given the large samples sizes and marker panels used in GWA studies, a small reduction in either following QC should have little effect on overall power. That said, every marker removed from a study is potentially an unobserved disease association (particularly for rare variants (minor allele frequency (MAF) <0.05) which are less likely to be captured (or tagged) by other markers on the chip [3]). To ensure that a marker is not falsely removed due to inaccurate genotyping in a subset of poorly genotyped individuals, such individuals can be identified and removed prior to undertaking marker QC. The disadvantage of this approach is that individuals may be falsely removed from the study due to inaccurate genotyping across a subset of poorly genotyped markers. An alternative approach would be to complete both sample and marker QC prior to removing any individuals or markers. This approach is conservative in terms of study power but ensures that only data of very high quality remain under analysis.

There are no strictly enforced protocols or quality thresholds for GWA study QC though some genotype/phenotype databases require that data meet certain quality criteria for inclusion (these thresholds can be somewhat lower than those typical adopted in GWA studies) [4]. GWA QC does not lend itself to the use of globally accepted, strictly enforced thresholds because the distribution of several QC metrics varies across genotyping platforms, genotyping centers, ancestral populations and genotype calling algorithms. For example, the threshold at which individuals should be removed due to elevate heterozygosity will vary depending on the set of genotyped SNPs and the ancestral origin of the group of individuals. Given the lack of a standard QC protocol it is important that the QC procedure is fully described in GWAS publications so that the strengths and weaknesses of the protocol can be fully assessed. Indeed, several guidelines have been published outlining the key information that should be reported in a GWA publication, and QC methods and metrics feature prominently [5–7].

Confounding

A key facet of any GWA study, QC protocol should be the identification and removal of confounding variables. Confounding is a well-studied epidemiological phenomenon which is caused by underlying differences between two groups, other than those directly under study, which correlate with the exposure variable. Therefore, in the context of GWA studies of disease, confounding is caused by underlying differences between the cases and controls, other than disease status, that correlate with allele, genotype or haplotype frequency (depending on the statistical test being implemented). In the presence of a confounding variable association can be falsely declared between the exposure variable and disease status and there are several examples of this in the genetic literature. For example, the Wellcome Trust Case Control Consortium (WTCCC) identified an association between type-2 diabetes and a variant in the FTO gene [8, 9]. Other GWA studies of type-2 diabetes had not detected association at this gene, even though they were powered to do so [10, 11]. The WTCCC was the only GWA study published at the time that did not match cases and controls for body mass index (BMI) (due to the use of a common set of controls across seven GWA studies). WTCCC investigators subsequently showed that variants in the FTO gene are associated with BMI/obesity and not type-2 diabetes *per se* [12]. The association with type-2 diabetes was driven by the correlation between the disease and BMI/obesity. This gives evidence of the importance of considering confounding variables when analyzing GWA data. The identification and removal of various confounding variables will be highlighted during the course of the chapter, which can be broadly separated into sample-based and marker-based QC.

SAMPLE-BASED QC

Sample Mix-ups and Plating Errors

Power can be significantly reduced if case/control status is incorrectly assigned to genotypes due to sample mix-ups or plating errors. A good method for detecting such errors is to compare the sex recorded during sample ascertainment with that estimated from X chromosome data. This QC procedure is particularly pertinent when the sex of individuals will be taken into account during analysis or when samples are supposed to be all of a single sex (for example in an ovarian cancer case cohort). In order to complete a check for sex-discordance it is often necessary to call X chromosome genotypes blind to ascertained sex because several genotype-calling algorithms prevent samples denoted as male from having heterozygous genotypes (and instead any marker where the heterozygous genotype is the most likely is fixed to missing). For samples denoted as female a low heterozygosity rate across X chromosome markers is indicative of a sample mix-up or wrongly ascertained sex. Males should be homozygous across all X chromosome SNPs so an elevated heterozygosity rate is indicative of a sample mix up (in reality heterozygosity may not equal zero due to genotyping error). PLINK automates the comparison between ascertained and genotype sex but requires that genotype calling is independent of ascertained sex as inferences are based entirely upon heterozygosity rate [13]. Other software packages that report heterozygosity rate and genotype call rate can also be used.

Low-quality DNA Samples

When designing a genetic study with several thousand participants it is inevitable that there will be some variation in the quality and concentration of the DNA samples supplied. Poor quality or low concentration DNA often fails to amplify or does not produce strong enough fluorescence to be detect above background levels, resulting in an elevated number of missing genotypes [14]. It is important that individuals with a lower than expected genotype call-rate are removed from study because the genotypes that are called for these individuals may be poorly classified (and thus introduce bias to the study). This step is of particular relevance when case and control DNA has been extracted or handled differently because DNA quality or concentration may correlate with disease status and thus introduce a confounding factor. In addition to using genotype call-rate as a measure of DNA quality, genotype heterozygosity rate can also be used. If two DNA samples have mistakenly been placed in the same well this will produce a higher than expected proportion of heterozygous genotypes. A lower than expected number of heterozygous genotypes is indicative of inbreeding. The genotype call-rate and heterozygosity rate can be quickly calculated for each individual in a study using standard GWA study analysis software such as PLINK [13] or GS2 [15]. Indeed, plotting the log10 call-rate against heterozygosity is a good method for easily identifying samples that are introducing error to a study (Fig. 7.1). Careful scrutiny of the plotted data should allow sensible QC thresholds to be devised such that low-quality samples can be identified and removed. Typically, individuals with more than 3–7% missing genotypes are removed from further investigation [8, 16].

Plate-effects

Even when DNA is of high quality and concentration the genotyping procedure can still introduce error. For high-throughput genotyping technologies, such as those used for GWA studies, the individual DNA samples are randomly assigned to plates (usually containing 96 samples) and genotyped in sequence. However, standard GWA study analysis methods treat the case and control cohorts as homogeneous groups and therefore it is important to check that this is a valid assumption — i.e. that differences in genotype frequency do not exist between plates. Furthermore, if cases and controls have been genotyped on different plates the presence of inter-plate differences in allele or genotype frequency (plate-effects) will introduce confounding.

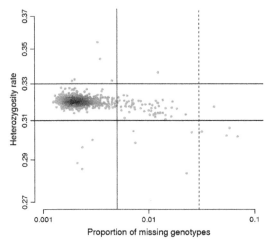

FIGURE 7.1   Proportion of missing genotypes vs. Heterozygosity rate per individual. Shading indicates density of observations (dark indicates a high density). Thresholds for removing individuals due to elevated genotype failure rate and extreme heterozygosity vary between studies. The dashed horizontal line indicates a typical missing genotype QC threshold of 3%. However, plotting the proportion of missing genotypes on the log scale shows that a QC threshold of 0.05% (solid vertical line) is more appropriate for the removal of the extreme tail of the missing genotype distribution. There is little variance in heterozygosity rate and thresholds of 0.31 and 0.33 (solid horizontal lines) are appropriate to remove outliers.

There are two main methods that can be implemented to detect plate-effects. The first involves comparing the frequency of the alleles and/or genotypes on a given plate to those across all the remaining plates and repeating for each plate in turn. A plate effect exists if the number of SNPs showing a significant difference in frequency is greater than expected under the null hypothesis of no association. This method is reasonably straightforward because it uses standard association methodology and software (e.g. chi-square tests and/or Cochran-Armitage trend tests which can both be carried out using PLINK [13]) but can be rather cumbersome and tedious if many plates have been genotyped. The second approach involves using principal components analysis (PCA), a multivariate statistical method which produces a number of uncorrelated variables (or principal components) from a data matrix containing observations across a number of potentially correlated variables (in this case, genetic markers). The first principal component accounts for as much variation within the data as possible in a single component, followed then by the second component and so on. If plate effects are present they will usually present themselves in the first one or two principal components, though this depends on the magnitude of the plate effect and other variation within the data. To thoroughly test for plate effects it is

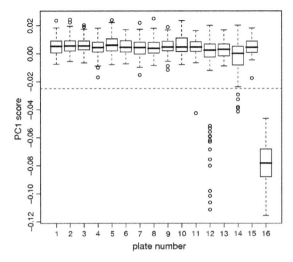

FIGURE 7.2 Detection of plate-effects using principal components analysis. Principal component scores across individuals have been grouped according to plate number and represented in box-whisker plots. The first principal component demonstrates a plate-effect present on plates 11, 12, 14 and 16. Only six individuals are affected across plates 11 and 14, though the whole of plate 16 and 14 individuals on plate 12 appear confounded. These individuals should be removed and the genotypes across remaining individuals re-called.

recommended to calculate around 10 principal components (using, for example, smartpca [17] or R [18]) and for each component in turn plot plate number against PCA score, treating plate number as a factor variable (Fig. 7.2). If a significant plate effect is detected, investigations into the cause of the discordance should take place. For example, is the plate-effect due to a high concentration of individuals with a divergent ancestry or failure rate? When the cause of the plate effect can be identified it is often possible to remove those individuals underlying the plate effect and recall the genotypes of the remaining individuals. An alternative and more conservative approach would be to exclude all the individuals on the discordant plate(s) from further analysis.

Population Stratification

Another form of confounding which has received much attention in the scientific literature is population stratification, which occurs when cases and controls are poorly matched for ancestral population. In this scenario, differences in allele frequency between the ancestral populations can give rise to significant association at alleles with no direct causative effect on disease status [19]. Even a small degree of population stratification can adversely affect a GWA study due to the large sample sizes required to detect common variants underlying most complex diseases [20]. That said, population stratification is more of a problem for GWA studies of ancestrally diverse populations such as the African-Americans [21] than less diverse populations such as the United Kingdom [8]. While a well-designed population case-control study attempts to draw cases and controls from the same population, hidden fine-scale genetic substructure within that single population (or the inadvertent inclusion of individuals from another population) cannot be ruled out [22]. There are several examples of population stratification in the scientific literature. For example, Campbell et al. [23] carried out an association analysis on a panel of American individuals of European ancestry who were discordant for height. They detected significant association to the gene encoding lactase (*LCT*), which has undergone strong selection in certain European populations and, because of this, the frequencies of variants within this gene are very divergent between populations. After matching cases and controls for population ancestry the evidence of association at this locus greatly decreased.

Principal component analysis can also be used to identify individuals of divergent ancestry [17, 24]. To ensure that the PCA produces principal components that differentiate individuals based on ancestry (as opposed to plate effects or any other difference between individuals in the study), genotype data from individuals of known and diverse ancestry should be added to the GWA data during the PCA analysis. Typically, genotype data from the four HapMap phase II and III populations [17, 24] (The Yoruba in Ibadan, Nigeria; Japanese in Tokyo, Japan; Han Chinese in Beijing, China; and the CEPH (U.S. Utah residents with ancestry from northern and western Europe)) are used to assess the ancestry of individuals included in a GWA study [8, 25]. Due to the scale of genetic variation between these diverse populations only two principal components are needed to separately cluster the individuals with regard to ancestry, where one PC loosely separates individuals on a North-South axis and the other on an East-West axis. Non-HapMap datasets containing genotyped individuals of known, ethnically diverse, ancestry can also be used providing they have a sufficiently large number (~100,000) of genotyped SNPs in common with the GWA individuals. Given the large number of markers used in the analysis and the number of individuals typically included in a GWA, PCA is computationally intensive. The analysis can be expedited by building the PCA model using only the reference samples and then applying this to this to GWA study data to predict the PC scores for the GWA individuals (as implemented in smartpca.pl [17]). Following a PCA analysis the two principal component scores can be plotted on a graph and outlying GWA individuals identified and removed from further analysis (Fig. 7.3). The principal component scores for the remaining individuals can be used as covariates in regression models to more fully account for population ancestry. If fine-scale population stratification

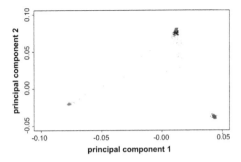

FIGURE 7.3   Identification of individuals with discordant ancestry using principal components analysis. Individuals of known ancestry from the HapMap (CEU (European) — **Red**, YRI (African) — **Green** and CHB&JPT (Asian) — **Purple**) have been used to seed a PCA analysis to detect individuals in of non-European ancestry in a large GWA study of European cases and controls. Crosses indicate GWA-samples removed from further study due to evidence of non-European ancestry and black circles indicate samples remaining under analysis. Please refer to color plate section.

(e.g. within Europe) is a concern then European samples of known country of ancestry can be included to place GWA samples within Europe.

Multi-dimensional scaling (MDS) is an alternative means of identifying individuals of divergent ancestry and is implemented in PLINK [13]. The approach involves calculating a genetic distance (DST) between each pair of individuals based on the number of markers where they share 0, 1 or 2 alleles in common and is given by:

$$1 - \frac{IBS2 + 0.5IBS1}{N}$$

where $IBS2$ is the number of markers where the pair share two alleles identical by state, $IBS1$ is the number of markers where they share one allele identical by state and N is the number of SNPs genotyped in common. Clustering analysis is then carried out on an $N \times N$ matrix of IBS pair-wise distances (where N is the number of individuals in the study). Again, this analysis should be seeded with individuals of known and divergent ancestries to ensure that the first few components explain variation associated with ancestral origin. As with the PCA analysis, individuals with divergent ancestry should be removed and component scores can be used as covariates during analysis to further reduce population stratification.

Duplicate and Related Samples

Another common use of the genetic distance metric is identification of duplicate or related samples. Duplicate samples should be $IBS2$ across all markers and therefore have a DST of zero. In practice, due to errors in genotyping and genotype calling, this is not always the case and a DST of

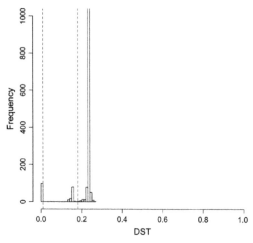

FIGURE 7.4 Identification of duplicate and related individuals. The blue dashed line (DST = 0.01) indicates duplicated individuals and the red dashed line (DST = 0.18) indicates related individuals. These thresholds were chosen based on peaks in the distribution other than the primary peak (plot truncated at a frequency of 1000 to show alternative peaks).

0.01 or above is accepted as indication of a sample duplicate. Identifying related samples based on DST is less straightforward because the distribution of DST varies depending on the marker set and population ancestry of the individuals in the study. If related individuals are present in the sample, it is possible to identify these by scrutinizing the distribution of DST values of each individual and their most related individual. The related samples will have a lower than average DST value and will cause a small peak in the DST distribution (Fig. 7.4). It is then a simple task to remove one individual from each pair of individuals with a diminished DST score (most typically the sample with the most missing data is removed) [26]. It is important that a GWA study does not contain any duplicate or related samples because they will bias the allele/genotype frequency estimates of the case or control groups. A lower than expected DST can also indicate inbreeding and a higher than expected DST can indicate the pair of individuals are of divergent ancestry. For this analysis to be effective both cases and controls should be included.

MARKER-BASED QC

Call-rate and Allele Frequency

The second phase of GWA study QC involves assessing the quality of data on a per-SNP basis. The criterion used to filter out low quality markers differs between studies. Classically, markers with a call rate less than 95% are removed from further study, though cut-offs as low as 90% have previously been implemented. Some studies have used two QC thresholds, one for common SNPs (call rate less than 95%) and a lower threshold (call rate less than 99%) for rare SNPs (MAF < 0.05). This approach is driven by the fact that rare SNPs are more susceptible to genotype calling error due to a smaller number of observations in the rare homozygote and heterozygous genotype clusters. Because of this, many investigators have chosen to remove markers with a small minor allele frequency (typically, less than 1%) prior to analysis. Given that power to detect association to these rare markers is low even in large studies it can be argued that little is being lost through their removal.

Removing SNPs with significant differences in missing genotype rate between cases and controls is another means of reducing confounding and removing poorly genotyped SNPs [27]. Calling case and control genotypes together, or using "fuzzy calls" [28], greatly reduces this confounding but significant differences in genotype failure may still exist in the data and present as false-positive associations. In studies where cases and/or controls have been drawn from several different sources, it is wise to test for significant differences in call rate, allele frequency and genotype frequency between these various groups to ensure that it is fair to treat the combined case or control set as one homogenous group.

Hardy-Weinberg Equilibrium

As samples for GWAS are generally drawn from large, homogeneous, randomly-mating populations, SNPs with allele frequencies that deviate from Hardy-Weinberg equilibrium (HWE) are likely to be subject to high genotype error rates. The vast majority of markers genotyped in GWA studies are diallelic SNP markers with alleles a and A and population frequencies of $1-q$ and q, respectively. When the two alleles that an individual carries are inherited independently, the number of copies of the A allele follows a binomial distribution (Binom(2, q)) [29, 30]. Therefore the probabilities of the aa, aA and AA genotypes are $(1-q)^2$, $2q(1-q)$ and q^2, respectively. The Hardy-Weinberg law states that in large, homogeneous, randomly mating populations, where the two copies of a gene that an individual carries are inherited independently, these probabilities are maintained from one generation to the next. In addition, in non-homogeneous populations these probabilities are established after a single generation of mixing provided that random mating takes place. Deviations from HWE do not necessarily indicate genotyping/calling error and may also indicate selection, so a case sample can show deviations from HWE at loci associated with disease, and it would obviously be counter-productive to remove these loci from further investigation [31]. Therefore, only control samples should be used when testing for deviations for HWE. The significance threshold for declaring SNPs to be in HWE has varied greatly between studies (p-value thresholds between 0.001 and 5×10^{-8} have been reported in the literature). However, those studies which have set very low thresholds for HWE deviations have done so on proviso that all genotype cluster plots for SNPs showing some evidence of deviation from HWE (say, $p < 0.001$) will be examined manually for quality. In practice this means that many SNPs with a HWE p-value less than 0.001 will be removed, though robustly genotyped SNPs with small HWE p-values can remain under study.

FAMILY-BASED STUDIES

While the majority of the GWA studies completed to date have been case vs. control designs, family-based designs can also be used. These designs are not as cost-efficient in terms of power to detect association as case-control designs [32] but they do have advantages, particularly when carrying out QC [33]. An obvious advantage is the increased ability to detect genotyping errors, particularly Mendelian inheritance errors, through the comparison of parental and offspring genotypes. Such comparisons give highly accurate estimates of genotyping error rate and provide a more thorough mechanism for detecting sample mix-ups than is possible with case-control data. Several software packages have implemented checks for Mendelian inheritance error including PLINK [13] and MERLIN [34]. Family-based designs are also more robust to population stratification than case–control studies because typically the cases (the offspring) have two population-matched controls (the parents).

POST-ANALYSIS QC

Even the most stringent QC protocol will not eliminate all type-1 and type-2 error, so care is still needed when interpreting association signals. Intensity data should be manually inspected for genotype clustering errors prior to designing replication studies, which ideally should utilize a different genotyping platform to that used in the GWA study. Cluster plots should be checked on a per cohort basis (for example, at a single SNP the cluster plot for the case genotypes should be compared to that in the controls). Markers that are inconsistently called should be removed from further analysis and not taken forward for replication.

SUMMARY

In summary, QC of GWA data is essential if false-positive and false-negative findings are to be kept to a minimum. Given the number of markers and individuals typically genotyped in GWA studies many sophisticated QC procedures can be undertaken. These procedures include (but are not limited to) the identification of poorly genotyped individuals and SNPs, the detection of individuals with discordant ancestry, the elucidation of duplicate and related pairs of individuals, and the exposure of plate-effects. These QC procedures have been seen to produce high-quality data that has led to the identification of novel susceptibility loci for many common human diseases and complex traits.

References

[1] N. Risch, K. Merikangas, The future of genetic studies of complex human diseases, Science 273 (1996) 1616.

[2] K.T. Zondervan, L.R. Cardon, Designing candidate gene and genome-wide case-control association studies, Nat. Protoc. 2 (2007) 2492–2501.

[3] J.C. Barrett, L.R. Cardon, Evaluating coverage of genome-wide association studies, Nat. Genet. 38 (2006) 659–662.

[4] N.B. Freimer, C. Sabatti, Guidelines for association studies in Human Molecular Genetics, Hum. Mol. Genet. 14 (2005) 2481–2483.

[5] The Wellcome Trust Case Control Consortium, Genome-wide association study of 14,000 cases of seven common diseases and 3,000 shared controls, Nature 447 (2007) 661–678.

[6] T.A. Manolio, L.L. Rodriguez, L. Brooks, G. Abecasis, D. Ballinger, M. Daly, et al., New models of collaboration in genome-wide association studies: the Genetic Association Information Network, Nat. Genet. 39 (2007) 1045–1051.

[7] J. Little, J.P. Higgins, J.P. Ioannidis, D. Moher, F. Gagnon, E. von Elm, et al., Strengthening the reporting of genetic association studies (STREGA): an extension of the strengthening the reporting of observational studies in epidemiology (STROBE) statement, J. Clin. Epidemiol. 62 (2009) 597–608. e4.

[8] S.J. Chanock, T. Manolio, M. Boehnke, E. Boerwinkle, D.J. Hunter, G. Thomas, et al., Replicating genotype-phenotype associations, Nature 447 (2007) 655–660.

[9] E. Zeggini, M.N. Weedon, C.M. Lindgren, T.M. Frayling, K.S. Elliott, H. Lango, et al., Replication of genome-wide association signals in UK samples reveals risk loci for type 2 diabetes, Science 316 (2007) 1336–1341.

[10] R. Saxena, B.F. Voight, V. Lyssenko, N.P. Burtt, P.I.W. de Bakker, et al., Genome-wide association analysis identifies loci for type 2 diabetes and triglyceride levels, Science 316 (2007) 1331–1336.

[11] L.J. Scott, K.L. Mohlke, L.L. Bonnycastle, C.J. Willer, Y. Li, W.L. Duren, et al., A genome-wide association study of type 2 diabetes in Finns detects multiple susceptibility variants, Science 316 (2007) 1341–1345.

[12] T.M. Frayling, N.J. Timpson, M.N. Weedon, E. Zeggini, R.M. Freathy, C.M. Lindgren, et al., A common variant in the FTO gene is associated with body mass index and predisposes to childhood and adult obesity, Science 316 (2007) 889–894.

[13] S. Purcell, B. Neale, K. Todd-Brown, L. Thomas, M.A. Ferreira, D. Bender, et al., PLINK: a tool set for whole-genome association and population-based linkage analyses, Am. J. Hum. Genet. 81 (2007) 559–575.

[14] W. Fu, Y. Wang, Y. Wang, R. Li, R. Lin, L. Jin, Missing call bias in high-throughput genotyping, BMC Genomics 10 (2009) 106.

[15] F. Pettersson, A.P. Morris, M.R. Barnes, L.R. Cardon, Goldsurfer2 (Gs2): a comprehensive tool for the analysis and visualization of genome wide association studies, BMC Bioinformatics 9 (2008) 138.

[16] M.S. Silverberg, J.H. Cho, J.D. Rioux, D.P. McGovern, J. Wu, V. Annese, et al., Ulcerative colitis-risk loci on chromosomes 1p36 and 12q15 found by genome-wide association study, Nat. Genet. 41 (2009) 216–220.

[17] N. Patterson, A.L. Price, D. Reich, Population structure and eigenanalysis, PLoS Genet. 2 (2006) e190.

[18] R Development Core Team (2005). R: a language and environment for statistical computing. Vienna, Austria.

[19] L.R. Cardon, L.J. Palmer, Population stratification and spurious allelic association, Lancet 361 (2003) 598–604.

[20] J. Marchini, L.R. Cardon, M.S. Phillips, P. Donnelly, The effects of human population structure on large genetic association studies, Nature Genetics 36 (2004) 512–517.

[21] R.A. Kittles, W. Chen, R.K. Panguluri, C. Ahaghotu, A. Jackson, C.A. Adebamowo, et al., CYP3A4-V and prostate cancer in African Americans: causal or confounding association because of population stratification? Hum. Genet. 110 (2002) 553–560.

[22] M.L. Freedman, D. Reich, K.L. Penney, G.J. McDonald, A.A. Mignault, N. Patterson, et al., Assessing the impact of population stratification on genetic association studies, Nature Genetics 36 (2004) 388–393.

[23] C.D. Campbell, E.L. Ogburn, K.L. Lunetta, H.N. Lyon, M.L. Freedman, L.C. Groop, et al., Demonstrating stratification in a European American population, Nat. Genet. 37 (2005) 868–872.

[24] A.L. Price, N.J. Patterson, R.M. Plenge, M.E. Weinblatt, N.A. Shadick, D. Reich, Principal components analysis corrects for stratification in genome-wide association studies, Nat. Genet. 38 (2006) 904–909.

[25] M. Jallow, Y.Y. Teo, K.S. Small, K.A. Rockett, P. Deloukas, T.G. Clark, et al., Genome-wide and fine-resolution association analysis of malaria in West Africa, Nat. Genet. 41 (2009) 657–665.

[26] Y.Y. Teo, Common statistical issues in genome-wide association studies: a review on power, data quality control, genotype calling and population structure, Curr. Opin. Lipidol. 19 (2008) 133–143.

[27] V. Moskvina, N. Craddock, P. Holmans, M.J. Owens, M.C. O'Donovan, Effects of differential genotyping error rate on the type1 error probability of case–control studies, Human Hered. 61 (2006) 55–64.

[28] V. Plagnol, J.D. Cooper, J.A. Todd, D.G. Clayton, A method to address different bias in genotyping in large-scale association studies, PLoS Genet. 3 (2007) e74.

[29] G.H. Hardy, Mendelian proportions in a mixed population, Science 28 (1908) 49–50.

[30] W. Weinberg, Über den Nachwels der Vererbung beim Menschen, Jahresh. Wuertt. Ver.vateri.Natkd 64 (1908) 369–382.

[31] D.G. Clayton, N.M. Walker, D.J. Smyth, R. Pask, J.D. Cooper, L.M. Maier, et al., Population structure, differential bias and genomic control in a large-scale, case-control association study, Nat. Genet. 37 (2005) 1243–1246.

[32] J. Teng, N. Risch, The relative power of family-based and case-control designs for linkage disequilibrium studies of complex human diseases. II. Individual genotyping, Genome Res. 9 (1999) 234–241.

[33] B. Benyamin, P.M. Visscher, A.F. McRae, Family-based genome-wide association studies, Pharmacogenomics 10 (2009) 181–190.

[34] G.R. Abecasis, S.S. Cherny, W.O. Cookson, L.R. Cardon, Merlin – rapid analysis of dense genetic maps using sparse gene flow trees, Nat. Gen. 30 (2002) 97–101.

Single-locus Tests of Association for Population-based Studies

Mark M. Iles, Jennifer H. Barrett

Section of Epidemiology and Biostatistics, Cancer Genetics Building,
St James's University Hospital, Leeds, UK

OUTLINE

INTRODUCTION

This chapter discusses the most basic test of genetic association: the single locus test. It is fundamentally very simple — an association between a trait (either categorical or quantitative) and a locus is hypothesized. If such an association exists then we expect that the trait values will differ between genotypes. There are a variety of statistical tests available for this purpose, such as a chi-squared test or regression on the trait value, but all are based around the idea of comparison of trait values between genotypes. Such an approach has several clear advantages:

1. **Ease and reliability:** The analysis usually employs classical statistical methods readily available in any standard statistical package. This means that such tests are easy to apply and unlikely to suffer from errors (either in the fundamental statistical properties of the test or its implementation as a computer package) that might arise in recently developed methods.
2. **Comprehension:** Similarly, the properties of such tests will have been thoroughly investigated and the interpretation of results will be straightforward to most statistically literate scientists.
3. **Flexibility:** The fact that such tests are easily implementable in statistical packages also means that they are more easily adapted to non-standard situations. For example, relatedness between samples may be accounted for by extension of a model to include covariance between samples, or covariates may be included in the model with ease. More complex methods may either not be extendable in this way, or may only be implementable in specially tailored packages that do not offer such options.
4. **Speed:** Perhaps the biggest advantage of single locus tests and a major reason for their current popularity, now that hundreds of thousands of loci are routinely being tested, is their speed. Hundreds of thousands of loci may be tested for association with a trait in a sample of tens of thousands of individuals in a matter of hours on a standard desktop computer.

A simple test like this also has disadvantages:

1. **Non-genetic:** The statistical methods employed were not developed with genetics in mind. Thus more specialized approaches may make better use of the genetic information: e.g., analysis of haplotypes, uncertainty of genotype calls or use of the coalescent to include population history in the model.
2. **Non-independence of loci:** If a single candidate locus is being tested, there is no problem. But if many markers in the same small region are

being tested, the results of testing each one individually will not be independent as a result of linkage disequilibrium (LD) between them. Adjusting for this can be difficult given the complex correlation structure.

3. Too simple: It is very tempting when tests are so simple to repeatedly apply slight variations on the same test, and reporting the most significant. This gives rise to a multiple testing problem.

Regardless of the pros and cons of single-locus tests, very few population-based studies nowadays would not apply such a test given the ease of this. More involved tests are likely to be applied to follow up interesting SNPs/regions.

Chi-squared Test

Population-based studies typically consist of the collection of a number of "cases" and "controls" for a categorical trait, followed by genotyping at a number of markers, either in a candidate region/gene or, as is increasingly common, at high density genome-wide. The terms "case" and "control" may literally apply to those with and without a disease or they may represent some other binary trait, such as relapsers and non-relapsers for a cancer. The results of the study for each genotyped marker could be recorded in a 2×3 table of case/control status against genotype (AA/AB/BB) and an association between the two be tested using the classical chi-squared test with 2 d.f. Here the standard cautions apply, such as the asymptotic assumption requiring an expected value of at least 5 in each cell. This is a problem if one genotype is rare, in which case either Fisher's exact test may be applied or rare genotypes may be merged such that the test becomes one of carriage of the rare allele against non-carriage (AA/AB or BB). This implicitly assumes a dominant model (see Genetic models, below).

Logistic Regression

An alternative to this is the use of logistic regression. Here the trait value, coded as a binary variable, is regressed on the genotype. If the genotype is coded as a categorical variable with three levels (one for each genotype), this is asymptotically equivalent to the chi-squared test. Logistic regression has the advantage over the chi-squared test in that it is easier to test alternative genetic models (see Genetic models, below) or to account for covariates (see Covariates, below).

The analysis of quantitative traits is considered below (Quantitative traits).

GENETIC MODELS

The test statistics described so far have made no assumption about the underlying genetic model so that the relative risks of each genotype are unrelated. However, it may be more powerful to use a specific genetic model. The classical modes of inheritance are: additive (where the relative risk of carrying two copies of the high-risk allele is the square of the risk of carrying one copy, since "additive" refers to the log-scale), dominant (where the risk of carrying two copies of the risk allele is the same as carrying one copy) and recessive (where there is no increased risk associated with carrying one copy of the risk allele, but there is an increased risk associated with carrying two copies). Of course, if the mode of inheritance of the trait is known, then this will be what is modeled, but this is unusual. In reality the mode of inheritance is unlikely to be anything as simple as a purely dominant model, particularly given that there may be further, interacting loci. One of the classical models may be close enough to the true model to give a good fit — but which one?

The more general model, as applied in a chi-squared test, has the advantage that it makes no assumption about the mode of inheritance, but the disadvantage that it requires two degrees of freedom as it is modeling two different levels of risk compared to the baseline, so loses power. Assuming either an additive, dominant or recessive model has the advantage that only one degree of freedom is used, thereby increasing power, but losing power if the model assumed is unrealistic.

It would seem sensible to restrict the model somewhat, for example by assuming that the risk associated with carrying one copy of an allele is between that of carrying no copies and two copies. Examples exist where this is not the case (most famously the heterozygous advantage at the sickle-cell anemia locus due to its effect on malaria resistance [1]) but for most loci heterozygous advantage is biologically rather implausible. This is an assumption not made by the general model, which allows for the possibility of heterozygote advantage. For this reason the model that is most commonly applied is the additive model. This is because it uses only one degree of freedom coupled with a monotonic increase in risk by genotype, but at the same time it assumes a model that is less "extreme" than either the dominant or recessive.

One solution is to conduct three tests of association at each locus assuming a dominant, recessive and additive mode of inheritance and pick the best of these, then adjust the significance level [2]. Alternatively one could make the (reasonable) assumption that the risk associated with the heterozygous genotype lies somewhere between the risk associated with the two homozygous genotypes [3, 4]. Both approaches have the advantage of not being restricted to a single genetic model, but by limiting the genetic model to more realistic models than the range covered by the 2 d.f. chi-squared test may have more power.

But what if the locus being tested is likely not to be the causative locus, but is in fact a marker in LD with it? Despite the fact that increasingly dense sets of markers are being genotyped these still represent only a fraction of the total number of genetic variants in the genome. Recent work suggests that, as LD between the two loci decreases, so the mode of inheritance at the marker (as opposed to the causative locus) looks increasingly additive [5, 6]. This is due to the faster decay of the dominance component with decreasing LD than the additive component. This would suggest that assuming an additive mode of inheritance may be more robust to a range of models than has been suggested by previous consideration of the power of assuming a specific mode of inheritance.

The implementation of tests assuming various modes of inheritance is relatively straightforward. The 2 d.f. chi-squared test makes no assumption about mode of inheritance, but this can be adapted to a dominant or recessive mode of inheritance by merging counts from the heterozygous genotypes with those of one of the homozygous genotypes and conducting a 1 d.f. chi-squared test. Similarly, an additive mode of inheritance is assumed if the Cochran-Armitage trend test is applied. Easier still is to implement these in a logistic regression in which case the genotypes AA/AB/BB are coded as categorical for the "model-free" test, 0/0/1 or 0/1/1 for the dominant or recessive test and 0/1/2 for the additive test.

COVARIATES

One advantage of a regression-based approach to testing for genetic association is the flexibility with which other risk factors can be taken into account in the analysis by including these as covariates in the statistical model. Here we consider several ways in which non-genetic factors might be of importance in understanding a genetic association and different methods of analysis that can be used to address these aspects. It is good practice to consider for each potential covariate whether and how it should be included in the risk modeling, rather than doing this automatically.

Confounding and Population Stratification

A confounder is a factor that is related to the risk factor of interest and, independently of this, to outcome (disease status), but is not an intermediate factor on the causal path between the risk factor and outcome (see, for example, Rothman and Greenland [7]). It is well understood in epidemiology that great care must be taken to control for confounders to avoid making incorrect inferences, and this is of key importance in case–control studies (the usual context of genetic association analysis), since unintended differences between the cases and controls in the sample can give rise to spurious associations. For example, an association between alcohol consumption and disease might be spuriously induced if cases and controls are actually sampled at different frequencies according to socio-economic status. Unless the design includes matching for age and sex, most case–control studies of environmental risk factors must be adjusted for age and sex in the analysis. This is because most diseases vary in frequency according to age and sex, as does exposure to most environmental risk factors. In *genetic* case–control studies there is usually no good reason to consider age and sex as potential confounders, since, although related to disease outcome, they are unlikely to be related to genotype.

The most important potential for confounding in genetic association studies comes from population stratification. Since allele frequencies and disease rates often both vary by ethnic group, ethnicity has long been recognized as a strong potential confounder, and usually distinct ethnic groups are analyzed separately. Hidden population stratification presents a more complex problem and is discussed in more detail in another chapter. If genome-wide data, or information on key ancestrally informative markers, are available, then summaries of these (e.g. from principal components analysis) can be included as covariates. In the absence of such data, adjustment can be made based on detailed self-reported ancestry (such as place of birth of grandparents).

Inclusion of other Known Risk Factors and Subgroup Analyses

Potential confounding is by no means the only reason to include covariates in a model. A further consideration is that power may be improved by allowing for other known important risk factors. For example, smoking is a strong risk factor for cardiovascular disease (CVD), so including smoking information in the model will explain some of the variability in risk and may thus improve the power to detect association between the genetic risk factor and disease. A similar argument may be applicable to age and sex.

For categorical variables, including these in the model, is very similar to carrying out analyses by subgroup and then forming a smoothed estimate of the effect of the genetic risk factor across the subgroups. It may be useful to carry out a Mantel-Haenszel test to do this explicitly, so that the subgroup-specific estimates of risk can be clearly seen. Additionally a test for a difference in effect size across subgroups can be carried out. If the effect of the genetic risk factor differs across subgroups, this is known as effect modification or interaction. This can also be tested for by including an interaction term between the genetic and environmental factors in the regression model.

Another common practice is to carry out separate analyses (and significance tests) for each subgroup. This may uncover disease-gene associations that would otherwise be hidden by the presumption of homogeneity of risk, but the disadvantage of subgroup analyses is the increase in the number of statistical tests and consequent difficulties in interpretation of results. One approach is to only consider separate subgroup analyses if there is evidence of interaction.

Understanding Disease Etiology

A further example of the use of covariates is in the analysis of inter-mediate phenotypes — characteristics of the subject that may be geneti-cally determined precursors of disease. Examples of these include cholesterol levels for CVD and number of nevi for melanoma. Although these may be associated with both the genotype and disease, they are to be distinguished from confounders since they lie on the same causal pathway. Including such phenotypes as covariates in the regression model will remove at least some of the evidence for association between the genotype and disease. It can still be illustrative to consider regression models of the genotype and covariate in relation to disease, where now the aim is to see whether the effect of the genotype on disease risk is mediated through its effect on the phenotype.

Table 8.1 shows an example of the effect on analysis of including an intermediate factor. A particular SNP is genotyped in a set of 906 mela-noma cases and 485 controls from the same population, and the rarer allele is significantly associated with reduced risk of melanoma (per-allele odds

TABLE 8.1 Logistic Regression Analysis of Melanoma Status on Genotype Showing the Effect of Inclusion of an Intermediate Phenotype

Risk factor	Odds ratio	95% CI	P-value
	Model without covariate		
SNP: number of copies of T allele	0.78	0.67, 0.93	0.004
	Model with intermediate factor as covariate		
SNP: number of copies of T allele	0.89	0.74, 1.07	0.23
Nevus count: logarithm of (number of nevi over 2 mm in diameter +1)	2.60	2.28, 2.97	$<10^{-43}$

TABLE 8.2 Linear Regression of Log-transformed Nevus Count on Genotype in Population-based Sample

Genotype	Mean	Standard deviation	N
AA	2.86	1.03	193
AT	2.67	0.97	226
TT	2.62	1.05	65
Linear regression model (AA as baseline)			
	Coefficient	95% CI	P-value
AT	−0.19	−0.39, −0.00	0.05
TT	−0.24	−0.52, 0.04	0.10
	Overall F-statistic from ANOVA F(2, 481) = 2.48, P = 0.09		
Linear regression model on number of T alleles (AA as baseline)			
Per T allele	−0.14	−0.27, −0.01	0.04

ratio (OR) 0.78, 95% confidence interval (0.67, 0.93), $P = 0.004$. However, the SNP is also associated with nevus phenotype, even in subjects without melanoma (see Table 8.2, discussed below). Since number of nevi is a very strong predictor of melanoma risk, this is an example of an intermediate phenotype, and the effect of the SNP on melanoma risk is likely to be through its effect on nevus count. By including both the SNP and the phenotype in a logistic regression model, it can be seen that the effect of the SNP is now no longer statistically significant ($P = 0.23$) and the OR is closer to 1, supporting this interpretation. (Note that intermediate phenotypes are usually measured with error, which could account for some residual effect of the SNP on disease risk.)

Statistical Modeling

The inclusion of covariates usually presents numerous modeling decisions in addition to those discussed above.

Continuous Variable or Factor

Apart from simple examples like sex (male/female), most covariates present us with a choice of how to code them. If we start with a continuous variable, this could be included in the model as continuous, binary, or categorized into three or more groups. The advantage of using the original variable is that no information is being thrown away, but the disadvantages are that a particular model for disease risk is assumed (usually a linear effect of the variable on the log-odds scale) and that interaction is more difficult to interpret. Binary data are simple to interpret, but dichotomizing represents the other extreme and probably discards too much information. A compromise may be to split into n categories ($n = 3$ or 4 say), either on the basis of quantiles of the distribution in controls or on some *a priori* basis. The categorized variable (coded 1, 2, ..., n) could then be included in the model as continuous (assuming a trend in risk from low to high categories, which is often a reasonable approximation and can be tested) or as categorical (with the disadvantage of increasing the number of parameters in the model).

Overfitting

If a large number of covariates are included in a model, this can lead to effect estimates that are both unstable and biased, especially if the sample size is not very large [8]. This can also arise with a smaller number of covariates if interaction terms are included in the model. Generally it is recommended to aim for parsimonious models, where covariates and their interactions with genotype are kept to a minimum. If a large number of terms are to be considered, then the use of methods such as penalized logistic regression will help to avoid some of the problems associated with overfitting [9].

Genetic Covariates

Although the focus of the previous discussion has been on non-genetic covariates, many of the same considerations apply to the inclusion of other genetic information. Where there is a strong known genetic risk factor (such as HLA genotype for many auto-immune diseases), it may be useful to include this in the model to potentially remove heterogeneity and to investigate interaction. The joint analysis of many genetic loci will be considered in more detail in other chapters.

GENOME - WIDE ASSOCIATIONS STUDIES AND GENERAL INTERPRETATION

To have reasonable power to detect association, the genotyped marker must be close enough to the causative locus that they are in strong LD. It is for this reason that, until recently, association studies were only conducted on candidate markers or small regions where a dense set of markers could be genotyped. In such studies, researchers would usually conduct single locus tests initially, with the assumption that more complex methods (such as haplotype-based approaches) would then be applied.

However, in the last few years, chips have become commercially available that allow the simultaneous genotyping of hundreds of thousands of markers: enough to conduct a genome-wide association (GWA) study. It is here that single locus tests have come into their own. In part this is because of their speed; a test that takes 1 second per marker to run, on a 600,000 SNP chip would still take almost a week to run. Thus speed is vital and single locus tests can be run genome-wide on several thousand samples in a matter of hours.

But this is only half the story. If single locus tests had failed to find many replicable associations, attention would be turning to more complex methods, in spite of their slower running time, but this is not the case. Single locus tests have been remarkably successful, having identified hundreds of genuine associations across a wide range of traits [5]. This is true to such an extent that the current emphasis is less on developing new methodology to identify associations and more on genotyping as many samples as possible in a virtual "gold rush" to find as many associations as possible using simple single locus tests.

However, with such large sets of markers comes the obvious problem of multiple testing. A straight Bonferroni correction for 600,000 tests would require a p-value of 8.5×10^{-8} to reach a corrected genome-wide p-value of 0.05. This is clearly an overcorrection, as such densely spaced markers will often be in strong LD and so the tests are far from independent. Further arguments for the appropriate p-value threshold are based on factors such as the effective number of tests [10] or the use of an appropriate prior probability of association in a Bayesian framework [11]. But $p = 0.05$ is already a rather arbitrary value to use as a threshold, so worrying too much about the "correct" threshold is unnecessary as long as there is comparability between studies. Consensus seems to be falling on a p-value of 5×10^{-7}, although with the potential for false positives, due to factors such as chance, genotyping errors, or population stratification, an association is not generally regarded as established until it has been replicated in at least one further independent sample.

The most significant results can of course be displayed in a table, but it is common to display the whole range of results in a so-called "Manhattan plot" (see Fig. 8.1). This is a very simple plot of markers in order along the chromosomes on the x-axis plotted against $-\log_{10}$ of their p-values for association on the y-axis. Points are displayed in two complementary colors on alternative chromosomes to improve legibility and the most significant results are seen as the highest points. This has the advantage that if there are clusters of SNPs showing evidence in a region, these are easily seen.

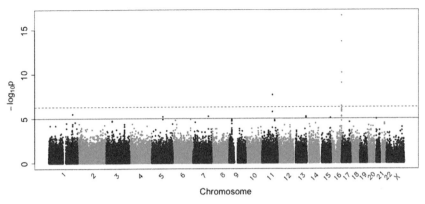

FIGURE 8.1 Example of a Manhattan plot of GWA *p*-values. Taken from Bishop et al. [18]. Please refer to color plate section

The effect size of results is commonly given as an odds ratio. Care should be taken here as the initial finding is likely to be biased towards suggesting stronger effect sizes, in the manner of the "winner's curse" [12, 13]. The odds ratio based on replication data will be less biased, although the original and replication data may be combined to give more reliable results [14].

QUANTITATIVE TRAITS

In this chapter we have thus far focused on case–control studies where the outcome of interest is binary, but sometimes the outcome of interest is a quantitative trait such as blood pressure or body mass index. Many of the above considerations (e.g., the use of covariates and principles of multiple testing in GWA studies) carry over straightforwardly to such applications, but there are some differences in design and analysis.

Population-based Sampling

Quantitative trait association studies are generally based on a sample of unrelated subjects from the population, ascertained independently of phenotype. The standard approach to analysis assumes the following linear model. The phenotype y_{ij} of individual i with genotype j is given by:

$$y_{ij} = \mu_j + e_i$$

where μ_j is the mean for the jth genotype and e_i represents residual environmental and possibly polygenic effects for individual i, assumed to be Normally distributed with mean 0 and variance σ_e^2. The parameters μ_j are estimated in the obvious way by the sample mean for individuals with genotype j. The F-statistic from analysis of variance (ANOVA), the ratio of between- and within-genotype variances, is used to test for association between genotype and phenotype, since under the null hypothesis that all genotypes have the same mean and variance, this ratio should be 1. This is equivalent to a linear regression analysis of phenotype on genotype, the advantage of which is that other factors can be included in the model as covariates. For SNPs, a trend test can be carried out by regressing the phenotype on the number of copies of the minor allele.

Nevus phenotype is under strong genetic control [15]. The total number of nevi above 2 mm in diameter was counted in 484 population-based individuals. The count was log-transformed (after adding 1 to the total to deal with zero counts) and regressed on genotype at a SNP associated with melanoma risk (discussed in "Covariates", above). The mean logged nevus count for each genotype is shown in Table 8.2. Regressing nevus count on genotype according to the above model, there is no overall evidence of an effect ($P = 0.09$ from F-test). However, a simpler model assuming a trend in nevus count with number of T alleles shows some evidence of a reduction in nevi as the number of T alleles increases ($P = 0.04$).

Non-parametric Analysis

The linear regression/ANOVA approach assumes that the residuals are Normally distributed; sometimes a logarithmic or similar transformation might be useful to make this assumption more reasonable. A non-parametric alternative is the Kruskal-Wallis test which is based on ranks rather than the trait values themselves. The complete set of N trait values is ranked from 1 to N, and the average rank in each genotype group is calculated. The test statistic is based on comparing the genotype-specific average ranks with the overall average rank of $(N + 1)/2$. Under the null hypothesis of no genotype–phenotype association, the test statistic follows a chi-squared distribution with 2 degrees of freedom (assuming three genotypes), and a significantly higher value indicates that the distributions differ. Although reasonably powerful the test alone is not very informative, and in general the estimates provided by a model-based approach are useful.

Sampling of Extremes

An alternative approach that also avoids distributional assumptions is to use a sampling scheme that selects individuals on the basis of extreme phenotypes (e.g. [16]). This is most practical when the phenotype is relatively easy to measure, so that large numbers of individuals can easily be screened to select extremes for genotyping. Subjects may be selected conditional on their phenotype being below or above particular thresholds, or the upper and lower n percentiles of a random sample from the population may be included. Having thus created a binary phenotype, the methods discussed earlier for case–control analyses can be applied.

Power

The power of several methods of analysis, variants of those described here, has been compared in a simulation study [17]. Selecting on the basis of extreme phenotype might be a potentially attractive study design, since extremes are more informative. However, under the models considered, linear regression generally performed better than a variant of the comparison of extremes, assuming the same number of genotyped individuals, since most of the information on phenotype is lost in the reduction to categories.

CONCLUSION

Overall, single locus tests of association are fast, efficient and reliable. They are easy to implement whether the trait is categorical or continuous. Single locus tests are easily extendable to a variety of situations, whether this requires adjustment for covariates, consideration of specific genetic models or accounting for potential population stratification. While they may not represent the optimal test (particularly since they are not specifically developed for genetic applications) they have proved hugely successful in genome-wide association studies and are without doubt the first type of test that any large-scale genetic association study would apply.

References

[1] J.S. Haldane, Disease and evolution, La Ricerca Scientifica 19 (1949) 68–76. Supplement.
[2] J.R. González, J.L. Carrasco, F. Dudbridge, L. Armengol, X. Estivill, V. Moreno, Maximizing association statistics over genetic models, Genet. Epidemiol. 32 (3) (2008) 246–254.
[3] K. Wang, V.C. Sheffield, A constrained-likelihood approach to marker-trait association studies, Am. J. Hum. Genet. 77 (5) (2005) 768–780.
[4] R. Yamada, Y. Okada, An optimal dose-effect mode trend test for SNP genotype tables, Genet. Epidemiol. 33 (2) (2009) 114–127.
[5] M.M. Iles, What can genome-wide association studies tell us about the genetics of common disease? PLoS Genet. 4 (2) (2008) e33.

[6] M.M. Iles, The impact of incomplete linkage disequilibrium and genetic model choice on the analysis and interpretation of genome-wide association studies, Ann. Hum. Genet. 74 (4) (2010) 375–379.

[7] K.J. Rothman, S. Greenland, Modern Epidemiology, second ed., Lippincott-Raven, Philadelphia, 1998, pp. 120–125.

[8] P. Peduzzi, J. Concato, E. Kemper, T.R. Holford, A.R. Feinstein, A Simulation study of the number of events per variable in logistic regression analysis, J. Clin. Epidemiol. 49 (1996) 1373–1379.

[9] M.Y. Park, T. Hastie, Penalized logistic regression for detecting gene interactions, Biostatistics 9 (2008) 30–50.

[10] F. Dudbridge, A. Gusnanto, Estimation of significance thresholds for genome-wide association scans, Genet. Epidemiol. 32 (2008) 227–234.

[11] The Wellcome Trust Case Control Consortium, Genome-wide association study of 14,000 cases of seven common diseases and 3,000 shared controls, Nature 447 (2007) 661–678.

[12] H.H. Göring, J.D. Terwilliger, J. Blangero, Large upward bias in estimation of locus-specific effects from genomewide scans, Am. J. Hum. Genet. 69 (6) (2001) 1357–1369.

[13] S. Zöllner, J.K. Pritchard, Overcoming the winner's curse: estimating penetrance parameters from case-control data, Am. J. Hum. Genet. 80 (4) (2007) 605–615.

[14] J. Bowden, F. Dudbridge, Unbiased estimation of odds ratios: combining genomewide association scans with replication studies, Genet. Epidemiol. 33 (5) (2009) 406–418.

[15] R.C. Wachsmuth, R.M. Gaut, J.H. Barrett, C.L. Saunders, J.A. Randerson-Moor, A. Eldridge, et al., Heritability and gene-environment interactions for melanocytic nevus density examined in a UK adolescent twin study, J. Invest. Dermatol. 117 (2001) 348–352.

[16] N.J. Schork, S.K. Nath, D. Fallin, A. Chakravarti, Linkage disequilibrium analysis of biallelic DNA markers, human quantitative trait loci, and threshold-defined case and control subjects, Am. J. Hum. Genet. 67 (2000) 1208–1218.

[17] G.P. Page, C.I. Amos, Comparison of linkage-disequilibrium methods for localization of genes influencing quantitative traits in humans, Am. J. Hum. Genet. 64 (1999) 1194–1205.

[18] D.T. Bishop, F. Demenais, M.M. Iles, M. Harland, J.C. Taylor, E. Corda, et al., Genome-wide association study identifies three loci associated with melanoma risk, Nat. Genet. 41 (8) (2009) 920–925.

Effects of Population Structure in Genome-wide Association Studies

Yurii S. Aulchenko

Erasmus MC Rotterdam, Rotterdam, The Netherlands

OUTLINE

INTRODUCTION

In genetic association studies, we look for association between a genetic polymorphism and the value of a trait of interest. The best scenario — the one we always hope for — is that the observed association results from causation, that is the polymorphism studied is functionally involved in the control of the trait. However, association has no direction, and making causal inference in epidemiology in general and in genetic epidemiology in particular is usually not possible based on statistical analysis only.

In fact, most associations observed in genetic studies are due to a confounder — an (unobserved) factor which is associated with both the genetic polymorphism and the trait analyzed. Presence of such a factor leads to induced, "secondary" correlation between the trait and the polymorphism; if we would have controlled for that factor in the association model, the relation between the polymorphism and the trait would have gone.

There are two major types of confounders leading to induced correlation in genetic association studies. One type is "good" confounding of association by the real, unobserved functional variant, which is, as a rule, not present on the SNP array, but is in linkage disequilibrium (LD) with typed SNP(s). Under this scenario, the functional variant is associated with the trait because of causative relation; at the same time it is associated with a typed polymorphism located nearby because of LD. This confounding induces secondary correlation between the typed polymorphism and the trait, making localization of the true functional polymorphism (LD mapping) possible.

The other major type of confounding observed in genetic association studies is confounding by population (sub)structure. Let us consider a study in which subjects come from two distantly related populations, say Chinese and European. Due to genetic drift, these two populations will have very different frequencies at many loci throughout the genome. At the same time, these two populations are different phenotypically (prevalence of different disease, mean value of quantitative traits) due to accumulated genetic, but also environmental and cultural differences. Therefore any of these traits will show association with multiple genomic loci. While some of these associations may be genuine genetic associations in a sense that either the polymorphisms themselves or the polymorphisms close by are causally involved, most of these associations will be genetically false positives — noise associations generated by strong genetic and phenotypic divergence between the two populations.

The scenario described above is extreme and indeed it is hard to imagine a genetic association study in which two very distinct populations are so bluntly mixed and analyzed not taking this mixture into account. However, a more subtle scenario where several slightly genetically different populations are mixed in the same study is frequently the case and a matter of concern in genome-wide associations (GWA) studies.

In the first section, we will define genetic structure, and will show how it can be quantified; in the next section the effects of genetic structure on the standard association tests are discussed; the final section presents specific association tests which take possible genetic structure into account.

GENETIC STRUCTURE OF POPULATIONS

A major unit of genetic structure is a genetic population. Different definitions of genetic population are available, for example Wikipedia defines population (biol.) as "the collection of inter-breeding organisms of a particular species". The genetics of populations is defined as "the study of the allele frequency distribution and change under the influence of…evolutionary processes". In the framework of population genetics, the main characteristics of interest of a group of individuals are their genotypic distributions frequencies of alleles, and the dynamics of these distributions in time. While the units of interest of population genetics are alleles, the evolutionary processes are acting upon organisms. Therefore a definition of a genetic population should be based on the chance that different alleles, present in the individuals in question, can mix together; if such chance is zero, we may consider such groups as different populations, each described by its own genotypic and allelic frequencies and their dynamic. Based on these considerations, a genetic population may be defined in the following way:

> Two individuals, I_1 and I_2, belong to the same population if (a) the probability that they would have an offspring in common is greater than zero and (b) this probability is much higher than the probability of I_1 and I_2 having an offspring in common with some individual I_3, which is said to belong to other genetic population

Here, to have an offspring in common does not imply having a direct offspring, but rather a common descendant in a number of generations.

However, in gene discovery in general and GWA studies in particular we are usually not interested in future dynamics of allele and genotype distributions. What is a matter of concern in genetic association studies is potential common ancestry — that is, that individuals may share common ancestors and thus share in common the alleles, which are exact copies of the same ancestral allele. Such alleles are called "identical-by-descent", or IBD for short. If the chance of IBD is high, this reflects a high degree of genetic relationship. As a rule, relatives share many features, both environmental and genetic, which may lead to confounding.

Genetic relationship between a pair of individuals is quantified using the "coefficient of kinship", which measures the chance that gametes, sampled at random from these individuals, are IBD.

Thus for the purposes of gene discovery we can define genetic population using retrospective terms based on the concept of IBD:

> Two individuals, I_1 and I_2, belong to the same genetic population if (a) their genetic relationship, measured with the coefficient of kinship, is greater than zero and (b) their kinship is much higher than kinship between them and some individual I_3, which is said to belong to other genetic population

One can see that this definition is quantitative and rather flexible (if not to say arbitrary): what we call a "population" depends on the choice of the threshold for the "much-higher" probability. Actually, what you define as "the same" genetic population depends in large part on the scope and aims of your study. In human genetics literature you may find references to a particular genetically isolated population, population of some country (e.g., "German population", "population of United Kingdom"), European, Caucasoid or even general human population. Defining a population is about deciding on some probability threshold.

In genetic association studies, it is frequently assumed that study participants are "unrelated" and "come from the same genetic population". Here, "unrelated" means that while study participants come from the same population (so there is non-zero kinship between them!), this kinship is so low that it has very little effect on the statistical testing procedures used to study association between genes and phenotypes.

In the following sections we will consider the effects of population structure on the distribution of genotypes in a study population. We will start with an assumption of zero kinship between study participants, which would allow us to formulate the Hardy-Weinberg principle (see "Hardy-Weinberg equilibrium", below). In effect, there is no such thing as zero kinship between any two organisms; however, when kinship is very low, the effects of kinship on genotypic distribution are minimal, as we will see in "Inbreeding", below. The effects of substructure — that is when the study sample consists of several genetic populations — on genotypic distribution will be considered in "Mixture of genetic populations: Wahlund's effect", below.

Hardy-Weinberg Equilibrium

To describe genetic structure of populations we will use a rather simplistic model approximating genetic processes in natural populations. First, we will assume that the population under consideration has an infinitely large size, which implies that we can work in terms of probabilities, and no random process takes place. Second, we accept the non-overlapping:

$$\text{generation} \Rightarrow \text{gametic pool} \Rightarrow \text{generation}$$

model. This model assumes that a set of individuals contributes gametes to a genetic pool, and does not participate in further reproduction. The gametes are sampled randomly from this pool in pairs to form individuals of the next generation. The selection acts on individuals, while mutation occurs in the gametic pool. The key point of this model is the abstract of gametic pool: if you use that, you do not need to consider all pair-wise mating between male and female individuals; rather you consider some abstract infinitely large pool, where gametes are contributed to with the frequency proportional to that in the previous generation. Interestingly, this rather artificial construct has a great potential to describe the phenomena we indeed observe in nature.

Let us consider two alleles: wild-type normal allele (N) and a mutant (D), segregating at some locus in the population and apply the "generation $\Rightarrow$ gametic pool $\Rightarrow$ generation" model. Let us denote the frequency of the D allele in the gametic pool as q, and the frequency of the other allele, N, as $p = 1 - q$. Gametes containing alleles N and D are sampled at random to form diploid individuals of the next generation. The probability to sample an "N" gamete is p, and the probability that the second sampled gamete is also "N" is also p. According to the rule, which states that joint probability of two independent events is a product of their probabilities, the probability to sample "N" and "N" is $p \cdot p = p^2$. In the same manner, the probability to sample "D" and then "D" is $q \cdot q = q^2$. The probability to sample first the mutant and then the normal allele is $q \cdot p$, the same is the probability to sample "D" first and "N" second. In most situations, we do not (and cannot) distinguish heterozygous genotypes DN and ND and refer to both of them as "ND". In this notation, the frequency of ND will be $q \cdot p + p \cdot q = 2 \cdot p \cdot q$. Thus, we have computed the genotypic distribution for a population formed from a gametic pool in which the frequency of the D allele was q.

To obtain the next generation, the next gametic pool is generated. The frequency of D in the next gametic pool is $P(D) = P(DD) + \frac{1}{2} \cdot P(DN) = q^2 + \frac{1}{2} \cdot 2 \cdot p \cdot q$. Here, q^2 is the probability that a gamete-contributing individual has genotype DD; $2 \cdot p \cdot q$ is the probability that a gamete-contributing individual is ND, and $\frac{1}{2}$ is the probability that the ND individual contributes the D allele (only half of the gametes contributed by individuals with the ND genotype are D) (see Fig. 9.1). Thus the freqeuncy of D in the gametic pool is $q^2 + \frac{1}{2} \cdot 2 \cdot p \cdot q = q \cdot (q + p) = q$ — exactly the same as it was in the previous gametic pool.

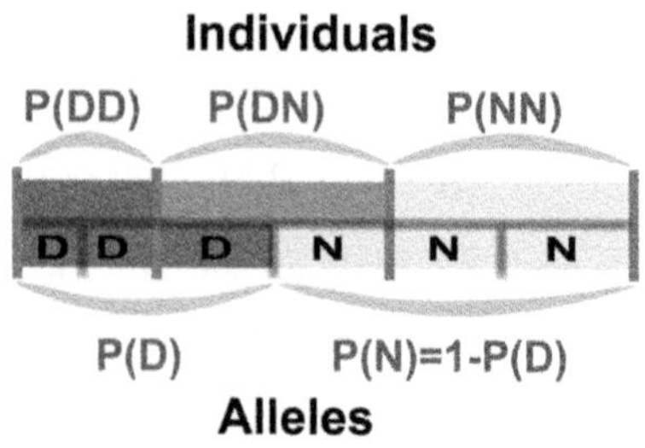

Individuals

P(DD)   P(DN)   P(NN)

P(D)   P(N)=1-P(D)

Alleles

FIGURE 9.1 Genotypic and allelic frequency distribution in a population; $q = P(D) = P(DD) + \frac{1}{2} \cdot P(DN)$. Please refer to color plate section

Thus, if assumptions of random segregation and aggregation hold, the expected frequency of *NN*, *ND* and *DD* genotypes are stable over generations and can be related to the allelic frequencies using the following relation:

$$
\begin{aligned}
P(NN) &= (1-q) \cdot (1-q) & = p^2; \\
P(ND) &= q \cdot (1-q) + (1-q) \cdot q & = 2 \cdot p \cdot q; \\
P(DD) &= q \cdot q & = q^2
\end{aligned}
\tag{1}
$$

which is known as the Hardy-Weinberg equlibrium (HWE) point.

There are many reasons why random segregation and aggregation and, consequently, HWE may be violated. It is very important to realize that, especially if the study participants are believed to come from the same genetic population, most of the time when deviation from HWE is detected, this deviation is due to technical reasons, i.e., genotyping error. Therefore testing for HWE is a part of the genotypic quality control procedure in most studies. Only when the possibility of technical errors is eliminated may other possible explanations be considered. In a case when deviation from HWE cannot be explained by technical reasons, the most frequent explanation would be that the study is composed of representatives of different genetic populations, or other subtle genetic structure is present in the data. However, unless study participants represent a mixture of very distinct genetic populations — the chances of which are low — the effects of genetic structure on HWE are difficult to detect, at least for any given marker, as you will see in the next sections.

Inbreeding

Inbreeding is preferential breeding between (close) relatives. An extreme example of inbreeding is a selfing, a breeding system observed in some plants. Less extreme inbreeding is not uncommon in animal and human populations. Here, the main reasons for inbreeding are usually geographical (e.g., mice live in very small interbred colonies — dems —

which are usually established by a few mice and are quite separated from other dems) or cultural (e.g. royal families of Europe).

Clearly, such preferential breeding between relatives violates the assumption of random aggregation, underlying the Hardy-Weinberg principle. Relatives are likely to share the same alleles, inherited from common ancestors. Therefore their progeny has an increased chance of being *autozygous* — that is, to inherit a copy of exactly the same ancestral allele from both parents. An autozygous genotype is always homozygous, therefore inbreeding should increase the frequency of homozygous, and decrease the frequency of heterozygous, genotypes.

Inbreeding is quantified by the *coefficient of inbreeding*, which is defined as the probability of autozygosity. This coefficient may characterize an individual, or a population in general, in which case it is defined as the expectation that a random individual from the population is autozygous at a random locus. The coefficient of inbreeding is closely related to the coefficient of kinship, defined earlier for a pair of individuals as the probability that two alleles sampled at random from these individuals are IBD. It is easy to see that the coefficient of inbreeding for a person is the same as the kinship between its parents.

Let us compute the inbreeding coefficient for the person J depicted in Figure 9.2. J is a child of G and H, who are cousins. J could be autozygous at, for example, the "red" allele of founder grand-grand-parent A, which could have been transmitted through the meioses A $\Rightarrow$ D, D $\Rightarrow$ G, and G $\Rightarrow$ J, and also through the path A $\Rightarrow$ E, E $\Rightarrow$ H, and H $\Rightarrow$ J (Fig. 9.2B). What is the chance for J being autozygous for the "red" allele? The probability that this particular founder allele is transmitted to D is 1/2, the same as the probability that the allele is transmitted from D to G, and the

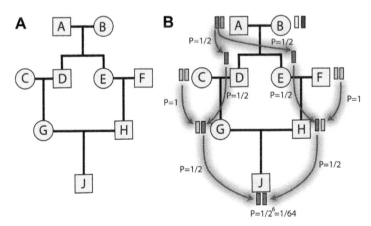

FIGURE 9.2 Inbred family structure (A) and probability of individual "G" being autozygous for the "Red" ancestral allele. Please refer to color plate section

probability that the allele is transmitted from G to J. Thus the probability that the "red" allele is transmitted from A to J is $1/2 \cdot 1/2 \cdot 1/2 = 1/2^3 = 1/8$. Likewise is the chance that that allele is transmitted from A to E to H to J, therefore the probability that J would be autozygous for the red allele is $1/2^3 \cdot 1/2^3 = 1/2^6 = 1/64$. However, we are interested in autozygosity for any founder allele; and there are four such alleles ("red", "green", "yellow" and "blue", Fig. 9.2B). For any of these the probability of autozygosity is the same, thus the total probability of autozygosity for J is $4 \cdot 1/6^4 = 1/2^4 = 1/16$.

Now we shall estimate the expected genotypic probability distribution for a person characterized with some arbitrary coefficient of inbreeding, F — or for a population in which average inbreeding is F. Consider a locus with two alleles, A and B, with frequency of B denoted as q, and frequency of A as $p = 1 - q$. If the person is autozygous for some founder allele, the founder allele may be either A, leading to autozygous genotype AA, or the founder allele may be B, leading to genotype BB. The chance that the founder allele is A is p, and the chance that the founder allele is B is q. If the person is not autozygous, then the expected genotypic frequencies follow HWE. Thus, the probability of genotype AA is $(1 - F) \cdot p^2 + F \cdot p$, where the first term corresponds to the conditional probability that the person is AA given it is not inbred (p^2), multiplied by the prior probability that it is not inbred ($1 - F$), and the second term corresponds to the probability that a person is AA given it is inbred (p), multiplied by the probability that the person is inbred (F). This computation can be easily done for all genotypic classes leading to the expression for HWE under inbreeding:

$$
\begin{aligned}
P(AA) &= (1 - F) \cdot p^2 + F \cdot p &= p^2 + p \cdot q \cdot F \\
P(AB) &= (1 - F) \cdot 2 \cdot p \cdot q + F \cdot 0 &= 2 \cdot p \cdot q \cdot (1 - F) \\
P(BB) &= (1 - f) \cdot q^2 + F \cdot q &= q^2 + p \cdot q \cdot F
\end{aligned}
\tag{2}
$$

How much is inbreeding expected to modify genotypic distribution in human populations and what is the power to detect deviation from HWE due to inbreeding? The levels of inbreeding observed in human genetically isolated populations typically vary between 0.001 (low inbreeding) to 0.05 (relatively high) (see Rudan et al. [1], Pardo et al. [2]). To perform power estimation, we need to estimate the expectation of the χ^2 statistics (the non-centrality parameter, NCP) used to test for HWE under the alternative hypothesis of deviation from HWE because of inbreeding. The test for HWE is performed using standard formula:

$$
T^2 = \sum_i \frac{(O_i - E_i)^2}{E_i}
\tag{3}
$$

where summation is performed over all classes (genotypes); O_i is the count observed in the ith class, and E_i is the count expected under the null hypothesis (HWE). Under the null hypothesis, this test statistic is distributed as χ^2 with number of degrees of freedom equal to the number of genotypes minus the number of alleles.

In our case, when we want to compute the expected value of the test under an alternative hypothesis, we use expectations of counts in different genotypic classes as "observed" values. Thus the expectation of this test statistic for some q, F, and N (sample size) is:

$$
\begin{aligned}
E[T^2] &= \frac{(N(q^2 + pqF) - Nq^2)^2}{Nq^2} + \frac{(N2pq(1 - F) - N2pq)^2}{N2pq} \\
&\quad + \frac{(N(p^2 + pqF) - Np^2)^2}{Np^2} \\
&= \frac{(NpqF)^2}{Nq^2} + \frac{(-2NpqF)^2}{N2pq} + \frac{(NpqF)^2}{Np^2} = Np^2F^2 + 2NpqF^2 + Nq^2F^2 \\
&= NF^2(p^2 + 2pq + q^2) = N \cdot F^2
\end{aligned}
$$

$$(4)$$

Interestingly, the non-centrality parameter does not depend on the allelic frequency. Given the non-centrality parameter, it is easy to compute the power to detect deviation from HWE for any given F. For example, to achieve the power of >0.8 at $\alpha = 0.05$, for a test with one degree of freedom, the non-centrality parameter should be >7.85. Thus, if $F = 0.05$, to have 80% power, $N F^2 > 7.85$, that is the required sample size should be

$$N > \frac{7.85}{F^2} = \frac{7.85}{0.0025} = 3140 \text{ people.}$$

Thus, even in populations with strong inbreeding, rather large sample sizes are required to detect the effects of inbreeding on HWE at a particular locus, even at a relatively weak significance level of 5%.

While the chance that deviation from HWE due to inbreeding will be statistically significant is relatively small, inbreeding may have clear effects on the results of HWE testing in GWA studies. Basically, if testing is performed at a threshold corresponding to nominal significance α, a proportion of markers which show significant deviation will be larger than α. Clearly, how large this proportion will be depends on the inbreeding and on size of the study — expectation of T^2 is a function of both N an F. A proportion of markers showing significant deviation from HWE at different values of inbreeding, sample size, and nominal significance threshold is shown in Table 9.1. While deviation of this proportion from nominal one is minimal at large αs, small sample sizes and coefficients of inbreeding, it may be 10-fold and even 100-fold higher

TABLE 9.1   Expected Proportion of Markers Deviating from HWE in a Sample of N People Coming from a Population with Average Inbreeding F. Proportion of Markers is Shown for Particular Test Statistic Threshold, Corresponding to Nominal Significance α

		α		
N	F	0.05	10^{-4}	$5 \cdot 10^{-8}$
	0.001	0.0501	$1.008 \cdot 10^{-4}$	$5.077 \cdot 10^{-8}$
1,000	0.005	0.0529	$1.205 \cdot 10^{-4}$	$7.025 \cdot 10^{-8}$
	0.010	0.0615	$1.885 \cdot 10^{-4}$	$14.503 \cdot 10^{-8}$
	0.001	0.0511	$1.081 \cdot 10^{-4}$	$5.784 \cdot 10^{-8}$
10,000	0.005	0.0790	$3.544 \cdot 10^{-4}$	$36.991 \cdot 10^{-8}$
	0.010	0.1701	$19.231 \cdot 10^{-4}$	$426.745 \cdot 10^{-8}$

than the nominal level at reasonable values of N and F for smaller thresholds.

Mixture of Genetic Populations: Wahlund's Effect

Consider the following artificial example. Imagine that recruitment of study participants occurs at a hospital, which serves two equally sized villages (V_1 and V_2); however, the villages are very distinct because of cultural reasons, and most marriages occur within a village. Thus these two villages represent two genetically distinct populations. Let us consider a locus with two alleles, A and B. The frequency of A is 0.9 in V_1 and it is 0.2 in V_2. In each population, marriages occur at random, and HWE holds for the locus. What genotypic distribution is expected in a sample ascertained in the hospital, which represents a 1:1 mixture of the two populations?

We can estimate the expected genotypic proportions assuming that HWE holds for each of the populations. If our sample represents a 1:1 mixture of these populations, then the frequency of some genotype is also a 1:1 mixture of the respective frequencies in the individual populations. For example, frequency of AA genotype would be $\dfrac{0.81}{2} + \dfrac{0.04}{2} = 0.425$, and so on (see LINKS, ABEL tutorial chapter[1] "GWA in presence of genetic stratification" for more details of these computations). The frequency of the A allele in the pooled sample will be $0.425 + \dfrac{0.25}{2} = 0.55$. Based on this frequency under HWE we would expect a genotypic frequency

[1] http://mga.bionet.nsc.ru/ryurii/ABEL/abel_beta/ABEL-tutorial.pdf

distribution of 0.3, 0.5 and 0.2, for AA, AB, and BB, respectively; however, the frequencies of genotypes in the admixed population are (0.425, 0.25, 0.325), demonstrating much higher frequencies of homozygous genotypes — excess of homozygotes. It is notable that the differences between the observed homozygote frequencies and those expected under HWE are both 0.125, and, consequently, the observed heterozygosity is less than that expected by $0.125 \cdot 2 = 0.25$.

The phenomenon of deviation from HWE, due to the fact that the considered population consists of two subpopulations, is known as "Wahlund's effect", after the scientist who first considered and quantified genotypic distribution under such a model [3].

Such marked differences between observed and expected under HWE, as we considered in the previous artificial example, are very easily detected; for the above example, a sample of ≈ 35 people is enough to reject the hypothesis of HWE (power >80% at $\alpha = 0.05$). However, the differences we can see in real life are not so marked. For example, the common Pro allele at position 12 of the peroxisome proliferator-activated receptor gamma is associated with increased risk for type 2 diabetes. The frequency of the Pro allele is about 85% in European populations and Caucasian Americans, about 97% in Japan and 99% in African-Americans (see Table 1 from Ruiz-Narvãjez [4]). If we consider a hypothetical study of a sample composed of 50% Caucasians and 50% African-Americans, only a sample as large as 1800 people would allow detection of the deviation from HWE (power >80% at $\alpha = 0.05$) (see LINKS, ABEL tutorial chapter "GWA in presence of genetic stratification" for more details of these computations). One can expect that allelic frequency may be very different in populations which diverged long time ago. However, in populations which diverged not so long ago, the allelic frequency differences are expected to be small and large sample sizes would be required to detect deviation from HWE due to Wahlund's effect.

Let us summarize what genotypic proportions are expected in a sample, which is a mixture of two populations. Let each population be in HWE, and the frequency of the B allele be q_1 in population one and q_2 in population two. Let the proportion of individuals coming from population one be m in the mixed population, and consequently the proportion of individuals from population two be $(1 - m)$. The frequency of the B allele in the mixed population is just the weighted average of the allelic frequencies in the two populations, $\bar{q} = m \cdot q_1 + (1 - m) \cdot q_2$. Let us denote the frequency of the A allele as $\bar{p} = 1 - \bar{q}$. It can be demonstrated that the genotypic frequency distribution in the mixed sample is the function of the frequency of allele B in the sample, $\bar{q}$, and "disequilibrium" parameter F_{st}:

$$
\begin{aligned}
P(AA) &= \bar{p}^2 + \bar{p} \cdot \bar{q} \cdot F_{st} \\
P(AB) &= 2 \cdot \bar{p} \cdot \bar{q} \cdot (1 - F_{st}) \\
P(BB) &= \bar{q}^2 + \bar{p} \cdot \bar{q} \cdot F_{st}
\end{aligned}
\tag{5}
$$

where:

$$F_{st} = \frac{m \cdot (1 - m) \cdot (q_1 - q_2)^2}{\bar{p} \cdot \bar{q}} \qquad (6)$$

You can see that equation (5), expressing the genotypic frequencies distribution under Wahlund's effect, is remarkably similar (actually it is specifically rewritten in a form similar) to equation (2), expressing the genotypic proportions in a population with inbreeding. F_{st} (as well as F of equation (2)) is easily estimated from the data as the ratio between the observed and expected variances of the genotypic distributions. Then the expected non-centrality parameter for the test of HWE is simply $N \cdot F_{st}^2$, where N is the sample size. Therefore our results concerning the proportion of tests expected to pass a particular significance threshold when genome-wide data are analyzed (Table 9.1) hold, with replacement of F with F_{st}.

We can compute that the values of F_{st}, corresponding to the population mixtures considered in PPAR-gamma example (0.067), give us a shortcut to estimate the sample size required to detect deviation from HWE due to Wahlund's effect (at $\alpha = 0.05$ and power 80%): $N > 7.85/0.067^2 \approx$ 1771

A typical value of F_{st} for European populations is about 0.002 (up to 0.023 [5]); very large sample sizes are required to detect deviation from HWE at any given locus at such small F_{st}s. However, the proportion of markers failing to pass the HWE test in GWA may be visibly inflated (Table 9.1).

EFFECTS OF POPULATION STRUCTURE ON STANDARD TESTS FOR ASSOCIATION

Standard Tests for Genetic Association

Standard tests for association between genes and a binary trait are the test for allele frequency difference between cases and controls, and Armitage's trend test for proportions (that the proportion of cases changes across genotypic groups). For quantitative traits, one of the standard tests is the score test for association, which is closely related — even equivalent — to Armitage's trend test.

We will start by presenting the study data as a 2×3 table, where the rows correspond to the case–control status and columns correspond to genotypic groups, and the cells contain counts of events (Table 9.2). For example, r_0 is the number of cases with genotype AA, s_0 is the number of controls with genotype AA and so on.

TABLE 9.2 Counts of Cases and Controls with Different Genotypes

	Genotype			
Status	AA	AB	BB	Total
Case	r_0	r_1	r_2	R
Control	s_0	s_1	s_2	S
Total	n_0	n_1	n_2	N

This table can be rearranged in a 2×2 allelic table, where each cell contains the counts of alleles present in cases and controls, e.g., total number of A alleles in cases is $2 \cdot r_0$ (twice the number of cases that are homozygous for the A allele) plus the number of A alleles present in heterozygous cases (r_1).

Based on such contingency tables, we can test if the allelic frequency is different between the cases and controls, using the standard χ^2 test, formulated as:

$$T^2 = \sum_i \frac{(O_i - E_i)^2}{E_i} \tag{7}$$

where summation is performed over all cells (defined by combination of genotype/allele and phenotype); O_i is the count observed in the ith class, and E_i is the count expected under the null hypothesis (equal frequencies in cases and controls). Under the null hypothesis, this test statistic is distributed as χ^2 with the number of degrees of freedom equal to the number of independent classes.

The null hypothesis assumes that the frequency of the A allele is the same in both cases and controls, and is equal to the frequency for A in the total sample: $\bar{p} = \frac{2 \cdot n_0 + n_1}{2 \cdot N}$. Thus the expected count of A in cases is $2 \cdot R \cdot \bar{p}$, and the expected count of B alleles is $2 \cdot R \cdot (1 - \bar{p})$. Similarly, for cases, the expected count of A is $2 \cdot S \cdot \bar{p}$, and the expected count of B is $2 \cdot S \cdot (1 - \bar{p})$.

Now, we can rewrite the allelic test as:

$$T_A^2 = \frac{((2r_0 + r_1) - 2R\,\bar{p})^2}{2R\,\bar{p}} + \frac{((r_1 + 2r_2) - 2R(1 - \bar{p}))^2}{2R(1 - \bar{p})}$$
$$+ \frac{((2s_0 + s_1) - 2S\,\bar{p})^2}{2S\,\bar{p}} + \frac{((s_1 + 2s_2) - 2S(1 - \bar{p}))^2}{2S(1 - \bar{p})}$$

With some algebra, it can be demonstrated that this expression simplifies to:

$$T_a^2 = \frac{2N\{N(r_1 + 2r_2) - R(n_1 + 2n_2)\}^2}{R(N-R)\{2N(n_1 + 2n_2) - (n_1 + 2n_2)^2\}} \qquad (8)$$

Under the null hypothesis that the frequency of alleles is the same in cases and controls (and, as you will see later, that HWE holds in total sample) the test statistic is distributed as χ^2 with one degree of freedom.

An alternative, Armitage's trend test for proportions, can be used to test the null hypothesis. This test is performed using the 2×3 genotypic Table 9.2. The null hypothesis assumes that the frequency of cases is the same in all genotypic groups; the alternative is that the frequency is not the same, but is, however, not totally arbitrary. As it follows from the name of the test, a trend in proportions is assumed; that is the frequency of cases among people with heterozygous genotypes AB should be exactly between the frequencies of cases in two homozygous classes. This hypothesis may be formalized using parameters β_0, the expected frequency of cases in the AA group, and β_1, the increase in frequency of cases in the AB group. The expected frequency of cases in the AB group is then $(\beta_0 + \beta_1)$; the frequency of cases in the BB group is assumed to be $(\beta_0 + 2 \cdot \beta_1)$. Parameters β_0 and β_1 can be estimated using the maximum likelihood: $\widehat{\beta}_1 = \dfrac{(r_1 + 2r_2) - \bar{p}R}{(n_1 + 4n_2) - \bar{p}N}$, where $\bar{p} = \dfrac{2n_2 + n_1}{2N}$ is the frequency of the B allele in the total sample. Then, β_0 is estimated as $\widehat{\beta}_0 = \dfrac{R}{N} - \widehat{\beta}_1 \bar{p}$. It can be shown that the chi-square test based on these expectations takes the form:

$$T_t^2 = \frac{N\{N(r_1 + 2r_2) - R(n_1 + 2n_2)\}^2}{R(N-R)\{N(n_1 + 4n_2) - (n_1 + 2n_2)^2\}} \qquad (9)$$

You can see that $T_t^2 = T_a^2$ if $n_1 = 2\sqrt{n_0 \cdot n_2}$, which is equivalent to the condition that HWE holds exactly, in which case the counts of heterozygotes should be $2Npq$, with $p = \sqrt{n_0/N}$ and $q = \sqrt{n_2/N}$. When HWE holds for the total sample, the two tests are (at least asymptotically) equivalent. Thus the test T_a^2 is the test for association in the presence of HWE; when HWE does not hold, even in the absence of association, the values of T_a^2 are greater than the values of the T_t^2 test, possibly leading to false positive conclusions about association in the presence of deviations from HWE. Therefore the trend test T_t^2 is to be preferred when testing for genetic association [6].

For study of association between genotype and quantitative traits, linear regression analysis is performed. Let us denote the vector of phenotypes as y, with particular values y_i ($i = 1, \ldots, N$). Let us code the genotypes with a quantitative variable, which reflects the number of

B alleles. Thus, we will code AA as 0, AB as 1, and BB as 2. Let us denote the vector of genotypes as g, with particular values g_i ($i = 1, ..., N$) taking the value of 0, 1 or 2. The linear regression model assumes that the expectation of the trait is:

$$E[y_i] = \mu + \beta_g \cdot g_i$$

where μ is intercept and β_g is the coefficient of regression of the phenotype on the genotype.

The estimate of β_g is provided by a well-known expression:

$$\hat{\beta} = \frac{Cov(y,g)}{Var(g)} = \frac{\sum y_i g_i/N - \sum y_i \sum g_i/N^2}{\sum g_i^2/N - (\sum g_i)^2/N^2} = \frac{N\sum y_i g_i - \sum y_i \sum g_i}{N\sum g_i^2 - (\sum g_i)^2}$$

(10)

Under the null hypothesis, the variance of $\hat{\beta}$ is $Var_0(\hat{\beta}_g) = \dfrac{Var(y)}{N \cdot Var(g)}$, and the score test statistic:

$$T_q^2 = \frac{\hat{\beta}_g^2}{Var_0(\hat{\beta}_g)} = \frac{Cov(y,g)^2}{Var(g)^2} \cdot \frac{N \cdot Var(g)}{Var(y)} = N\frac{Cov(y,g)^2}{Var(y) \cdot Var(g)}$$

(11)

is distributed as χ^2 with one degree of freedom. Here, $Var_0(\hat{\beta}_g)$ denotes the variance under the null hypothesis. Denote the correlation between y and g:

$$\frac{Cov(y,g)}{\sqrt{Var(y) \cdot Var(g)}}$$

as r. Then $T_q^2 = N \cdot r^2$.

It is worth noting that Armitage's test for trend in proportions, used in case–control samples, can be expressed as a linear regression test. Let us code the genotypic groups based on the number of B alleles. Thus, we will code genotype g AA as 0, AB as 1, and BB as 2. Let us also code the phenotypes (y) of cases with "1" and the phenotypes of controls with "0". Then, if we perform linear regression of the case-control status onto the number of B alleles in the genotype, the estimate of the regression coefficient is (following equation (10)):

$$\hat{\beta} = \frac{\sum y_i g_i - \sum y_i \sum g_i/N}{\sum g_i^2 - (\sum g_i)^2/N} = \frac{(r_1 + 2r_2) - \bar{p}\,R}{(n_1 + 4n_2) - \bar{p}\,N}$$

Thus, the expected proportions in Armitage's trend test are provided by the solution of a linear regression equation, in which regression of case–control status, coded as "0" and "1" is performed onto the number of B alleles. It can be further demonstrated that the trend test statistics [9] can be expressed as $T_t^2 = N \cdot r^2 = T_q^2$, where r^2 is the squared coefficient of correlation between y and g.

In the next two sections we will consider the effects of genetic structure on the standard tests for association described above. The extensive treatment of the problem is mainly due to the seminal works of Devlin, Roeder, Bacanu, and Wasserman [7–10]; here, some of their results are repeated.

The genetic structure of the study population (see retrospective definition of genetic population above) will be characterized by the kinship matrix — a square matrix, which, for all pairs of individuals in question, provides their pair-wise kinship coefficients (defined above). The kinship coefficient between persons i and j will be defined as f_{ij}.

Effects of Genetic Structure on Standard Tests

As it was demonstrated above, a test for association (with binary or quantitative trait) can be expressed as $T^2 = \dfrac{\widehat{\beta}^2}{Var(\widehat{\beta})}$. Genetic structure affects both nominator and denominator of this expression. Indeed, if measurements are dependent, the variance of the estimate is likely to be underestimated if dependence is not accounted for, leading to the inflation of the tests statistic. Even in the absence of association, the estimate of the effect may be biased, again leading to increased value of the test statistic.

Let us consider the following artificial example. Study cases are closely related individuals — sibs[2] — coming from one family, and study controls are sibs from another single family. In essence, in an association test we compare the squared frequency difference between the two groups to the variance of this difference. If groups are formed by independent individuals, we can detect arbitrary small frequency differences by increasing the sample size, and consequently, decreasing the variance of the frequency estimate. This definitely is not the case when our study sample consists of a large sibship — whatever the sibship size, only four alleles may be present in the parents, and the precision of the estimate of the allele frequency in the general population is limited — as if we had two people available for our analysis. If we do not take this consideration into account, we are likely to overestimate the precision of the frequency estimate; the denominator of the test statistic becomes small, and the statistic becomes large. The parents of the "case" and "control" sibships have a high chance of being genotypically different; and any genotypic configuration leading to the different number of B alleles will be reflected in the difference in frequencies between sibships, given that the sibships

---

[2]brothers and sisters.

are large. Assuming HWE, the chance that two parental couples will have a different number of B alleles is one minus the chance that they will have the same number of B alleles:

$$
\begin{aligned}
P(\#B_1 \neq \#B_2) &= 1 - (P(\#B_1 = \#B_2 = 0) + P(\#B_1 = \#B_2 = 1) + \ldots,,, \\
&\quad + P(\#B_1 = \#B_2 = 4)) \\
&= 1 - (P(\#B_1 = 0)P(\#B_2 = 0) + \ldots,,, + P(\#B_1 = 4) \\
&\quad P(\#B_2 = 4)) \\
&= 1 - (q^8 + 16p^2q^6 + 36p^4q^4 + 16p^6q^2 + p^8)
\end{aligned}
$$

which, for a common allele with a frequency of 0.2, translates to the probability 0.64.

Let us quantify these two sources of bias - bias in nominator and denominator of the test statistic. Following Devlin et al. [9] let us assume that the number of cases and the number of controls is the same, N. Let us denote the vector of genotypes, (coded as 0, 1, and 2) of cases as X, with X_i ($i = 1, \ldots, N$) being the genotype of the ith case, and the vector of genotypes of controls as Y (with Y_j, $j = 1, \ldots, N$ being the genotype of the jth control). The trend and the allelic test statistics are proportional to the square of:

$$
T = \sum X_i - \sum Y_j \tag{12}
$$

The variance of this statistic, in general form, is:

$$
\begin{aligned}
Var(T) &= \sum_{i=1}^{N} Var(X_i) + \sum_{i=1}^{N} Var(Y_i) \\
&\quad + 2\sum_{i<j} Cov(X_i; X_j) + 2\sum_{i<j} Cov(Y_i; Y_j) \tag{13} \\
&\quad - 2\sum_{i}\sum_{j} Cov(X_i; Y_j)
\end{aligned}
$$

Let us consider a situation in which cases come from one population and controls from the other population; each of the populations is in HWE and the difference between the populations is characterized with F_{st} (see equation (5)). Under this model, $Var(X_i) = Var(Y_i) = 2pq(1 + F_{st})$, and the covariance between any pair of genotypes from the same population is $4pqF_{st}$. Then:

$$
\begin{aligned}
Var(T) &= 2pq(1 + F_{st}) \cdot N + 2pq(1 + F_{st}) \cdot N \\
&\quad + 2 \cdot 4pqF_{st} \cdot N(N-1)/2 + 2 \cdot 4pqF_{st} \cdot N(N-1)/2 + 0 \tag{14} \\
&= 4pqN \cdot (1 + F_{st} + 2F_{st}(N-1)) \approx 4pqN(1 + 2NF_{st})
\end{aligned}
$$

here, $4pqN$ corresponds to the binomial variance of T in the absence of genetic structure (and this is exactly the variance used by allelic test, T_a^2), while $F_{st} + 2 \cdot F_{st} \cdot (N-1)$ reflects the inflation of the variance. As the second term is the function of the sample size, large inflation may be achieved even with small values of F_{st}. Note that here the sample size is 2N, as we assumed N cases and N controls.

Above we have considered an example in which we know the F_{st} between the population of cases and the population of controls. In a practical study, a number of cases and controls are usually sampled from each genetically different population. Let the proportion of individuals sampled from population c among cases be a_c and the proportion of individuals coming from that population among controls be u_c. Then, it can be shown that:

$$Var(T) \approx 4pqN(1 + 2 \cdot N \cdot F_{st} \cdot \sum_c (a_c - u_c)^2) \qquad (15)$$

As it follows from this equation, the variance of the estimated frequency difference depends on the composition of the sample. Maximal inflation of the variance is achieved when the cases and controls are sampled from different populations, while if $\sum(a_c - u_c)^2 = 0$ — which is achieved by sampling an equal number of cases and controls from each subpopulation — the variance inflation is minimal, and corresponds to that used by Armitage's trend test T_t^2.

These results can be generalized to arbitrary relations between cases and controls. Let us denote kinship between cases i and j as f_{ij}^X, kinship between controls as f_{ij}^Y, and the kinship between a case and a control as f_{ij}^{XY}. Then [9]:

$$Var(T) = 4pqN \cdot (1 + F_{st} + \frac{2}{N}\sum_{i<j} f_{ij}^X + \frac{2}{N}\sum_{i<j} f_{ij}^Y - \frac{2}{N}\sum_{i}\sum_{j} f_{ij}^{XY}) \qquad (16)$$

This equation demonstrates that the variance is inflated not only in a situation when cases and controls come from genetically distinct populations, but that also "cryptic relations" among either cases (some $f_{ij}^X > 0$) or controls (some $f_{ij}^Y > 0$) may lead to increased variance of the test. Because summation (e.g., $\sum_{i<j} f_{ij}^X$) is performed over all pairs of cases/controls (which are many!), cryptic relations may have a strong impact on inflation of the variance.

Generally, the variance of the T can be expressed as $Var(T) = 4Npq(1 + F_{st} + D(f^X, f^Y, f^{XY}))$. Here, $4Npq$ corresponds to the binomial variance assuming HWE and independence between cases and controls, the second term $- F_{st}$ — accounts for increase in variance due to deviation from HWE, and the last, which is a function of kinship $- D(f^X, f^Y, f^{XY}) -$ accounts for dependencies between cases and controls. The test:

$$\frac{T^2}{4Npq(1 + F_{st} + D(f^X, f^Y, f^{XY}))},$$

where T is defined by equation (12), is distributed as χ^2 with one degree of freedom.

The allelic and Armitage's trend statistics can be expressed as $T_a^2 = \dfrac{T^2}{4Npq}$ and $T_t^2 = \dfrac{T^2}{4Npq \cdot (1 + F_{st})}$. What is their distribution when there is relatedness/stratification between cases and controls? Clearly, because of overdispersion, both of them are distributed as $\tau^2 \cdot \chi^2$, where $\tau^2 > 1$ is the variance inflation factor. It is easy to show that it is equal to $\tau^2 = 1 + F_{st} + D(f^X, f^Y, f^{XY})$ for the allelic and $\tau^2 = \dfrac{1 + F_{st} + D(f^X, f^Y, f^{XY})}{1 + F_{st}}$ for the trend test.

In the above, we have considered overdispersion of the test statistic due to possible relatedness between the cases and controls, and demonstrated that the allelic and the trend tests are inflated by some constant τ^2, which reflects the inflation of the variance, $\mathrm{Var}(T) = \sigma^2 \cdot \tau^2$, where σ^2 is the variance estimated under the assumption of independence. However, at the beginning of this section we discussed another source of the inflation of the test statistic — the bias in frequencies between cases and controls due to confounding. Let cases and controls be sampled from different populations, whose difference is characterized by F_{st}, and let, for some marker, the allele frequency difference — which occurs entirely due to genetic structure of the samples — between cases and controls be d. The test statistic T is then distributed as Normal with mean $2Nd$ and variance computed in the previous section, $\sigma^2 \cdot \tau^2$, $N(2Nd, \sigma^2\tau^2)$. To address the issue of bias, we need to figure out the distribution of the mean — $2Nd$. In general, the expected frequency difference is zero — indeed, if we consider a large number of subpopulations, or a large number of markers, if allelic frequencies are determined by random effects, the deviation is likely to occur in any direction, resulting in zero on average. The variance of the d is $2pqF_{st}$, thus the variance of $2Nd$ is $(2N)^2 2pqF_{st}$.[3] Thus, finally, T is expected to be distributed as $N(0, \sigma^2 \cdot \tau^2 + 8pqF_{st}N^2)$. If we consider the variance of the allelic test, $\sigma^2 = 4pqN$, then $T \sim N(0, \sigma^2(\tau^2 + 2F_{st}N))$, and, denoting $2F_{st}N$ as v^2, $T \sim N(0, \sigma^2(\tau^2 + v^2))$. A similar expression may be obtained for Armitage's trend test.

Thus, in the presence of genetic structure both allelic and the trend test are distributed as $(\tau^2 + v^2) \cdot \chi^2$. Note that while σ^2 is a function of allelic frequency, the inflation factors τ^2 and v^2 depend on the differentiation between the populations in question, as measured by F_{st}, sample size, N, and the composition of the case-control sample (a_c, u_c), but does not depend on the allele frequency.

---

[3] For the case when some proportion, a_c, among cases comes from population c, and some proportion of controls, u_c, comes from population c $\mathrm{Var}(d) = 2pqF_{st}\sum(a_c - u_c)^2$.

As mentioned earlier, Armitage's trend test can be reformulated as a regression-based score test, in which genotypes are coded as 0, 1, and 2, and phenotypes as 0 and 1. Therefore all the above arguments apply to the analysis of quantitative traits as well. The test statistic $T_q^2 = \dfrac{\widehat{\beta}_g^2}{Var(\widehat{\beta}_g)}$ in which $\widehat{\beta}$ and $Var(\widehat{\beta})$ are estimated using independence assumption is inflated because of overdispersion ($Var(\widehat{\beta})$ is underestimated if the relatedness is not accounted for), and $\widehat{\beta}_g$ may be biased, because different genetic populations may well be characterized by different mean values of the trait (e.g., because of environmental influences). It can also be shown that for quantitative traits T_q^2 is distributed as $(\tau^2 + v^2) \cdot \chi^2$, where τ^2 is inflation due to overdispersion, and v^2 is the inflation because of bias.

Genomic Control

From the previous section it follows that in the presence of genetic structure the standard association statistics may be inflated and the distribution is described as $(\tau^2 + v^2) \cdot \chi^2$. Let us denote $(\tau^2 + v^2)$ as λ for short. λ depends on the genetic structure of the sample, as characterized by pair-wise kinship and sample size, N. For binary trait analysis, it also depends on the composition of the case-control sample, as expressed by the proportion of cases/controls coming from a particular population, c (a_c, u_c). For quantitative traits, it depends on heritability of the trait, and environmentally determined differences co-occurring with the difference in kinship. However, λ does not depend on the allele frequency. Therefore, for any particular study sample, if F_{st} is constant over the genome, λ is also a constant.

Therefore λ can be estimated from the genomic data, using "null" loci — a set of random markers, which are believed not to be associated with the trait. This estimate can then be used to correct the values of the test statistic at the tested loci — a procedure known as "genomic control". The test statistics computed from these loci thus estimate the distribution of the test statistic under the null hypothesis of no association. Let us consider M "null" markers, and denote the test statistic obtained from ith marker as T_i^2. Given genetic structure determines inflation λ, $T^2 \sim \lambda \cdot \chi_1^2$. The mean of a random variable coming from χ_1^2 is equal to 1; if a random variable comes from $\lambda \cdot \chi_1^2$, the mean would be λ. Thus we can estimate λ as the mean of the obtained "null" tests. Bacanu et al. [8] suggested estimating λ as the ratio between the observed median and the one expected for χ_1^2:

$$\widehat{\lambda} = \frac{Median(T_i^2)}{0.4549} \tag{17}$$

Median is used for the genomic control inflation parameter estimation because the median is robust to a few large values at the upper tail of the distribution. This estimator is the one most used in practice.

For the tested markers, the corrected value of the test statistic is obtained by simple division of the original test statistic value on $\hat{\lambda}$, $T^2_{corrected} = T^2_{original}/\lambda$. Note that this procedure is correct only if the same number of study participants was typed for all markers.

As you can see, the genomic control procedure is computationally extremely simple — one needs to compute the test statistic using a simple test (e.g., score test), compute the median to estimate λ, and divide the original test statistics values on $\hat{\lambda}$.

How to choose "null loci" in a GWA study? In a genome, we expect that a small proportion of markers is truly associated with the trait. Therefore, in practice, all loci are used to estimate λ. Of course, if very strong (or multiple weak) true associations are present, true association will increase the average value of the test, and genomic control correction will be conservative. To relax this, it has been suggested [11] to use say, 95% of the least significant associations for estimation of the inflation factor; however, the question of selection of cut-off is not obvious, and all markers are used in practice in most studies.

We have observed that the inflation factor λ is a function of sample size, N — the bigger is the sample size, the larger is inflation. If this is the case, how can we compare inflations between different GWA studies? A good idea is to use a standardized inflation, say, inflation per 1,000 subjects. For a quantitative trait analysis, the inflation factor can be expressed as $\lambda \approx (1 + N * D(sample))$, where $N * D(sample)$ is a term that grows linear with sample size, at some rate determined by sample characteristics.

Therefore, a standardized inflation factor can be estimated as:

$$\hat{\lambda}_{1000} = 1 + \frac{\hat{\lambda} - 1}{N} \cdot 1000,$$

where N is the sample size and $\hat{\lambda}$ is the estimate from the total sample.

What about a case–control sample? For such samples, a standardized λ is computed for a fixed number of cases and controls, say 1000; we will denote such standardized inflation as $\lambda_{1000,1000}$ to distinguish it from the one computed for quantitative traits. If we denote the number of cases as N_a and the number of controls as N_u, then:

$$\hat{\lambda}_{1000,1000} = 1 + \frac{500 \cdot (\hat{\lambda} - 1) \cdot (N_a + N_u)}{N_a \cdot N_u}$$

Clearly, we have assumed that in our GWA study the sample size was equal for all studied markers; this fact allows for a very simple estimation of $\hat{\lambda}$ (equation (17)) and correction of the test statistic. It may happen, however, that the number of participants typed for different marker loci is different; for example, this may happen when all study participants were typed for SNP panel one, and then a part of the study was additionally typed at a different panel. In such a situation, expectation of a T_q^2 test for a quantitative trait can be expressed as:

$$E[T_q^2] = 1 + (\lambda_{1000} - 1) \cdot \frac{N}{1000}$$

It is straightforward to obtain an estimate of λ_{1000} by performing linear regression of the observed test statistic values onto the sample size used; note that the intercept should be fixed to zero in this procedure. Other, more effective procedures may be thought of as well.

For binary traits, if a sample is composed from N_a cases and N_u controls, the expression for the expected value of the test statistic is similar to that obtained for quantitative traits; however, the geometric mean of the number of cases and controls is used [12]:

$$E[T_t^2] = 1 + (\lambda_{1000,1000} - 1) \cdot \frac{\dfrac{1}{1000} + \dfrac{1}{1000}}{\dfrac{1}{N_a} + \dfrac{1}{N_u}}$$

This relation can be used to estimate $\lambda_{1000,1000}$ in a case when sample size varies from SNP to SNP.

We have considered the genomic control procedure for the tests assuming additive effects of the locus onto the phenotype. Under this model, inflation of the tests (λ) does not depend on allelic frequency, which allows a straightforward estimation of λ and further correction of the tests. Does the same apply to other genetic models? Actually, this is not the case, and additive, dominant and overdominant models show very strong dependency of inflation factor on allele frequency (see Zheng et al. [13], and also chapter "GWA in presence of genetic stratification" of ABEL tutorial for details).

Thus, it is important to remember that the genomic control was developed for, and works with, additive genetic models; even in this framework, the behavior of λ depends on allelic frequencies when these are too low or too high — thus application of genome-wide derived inflation factor to such marker loci may be incorrect for loci with low minor allele frequency. This is acceptable statistically because such correction will be conservative, and not liberal — but the same may be worrisome biologically (e.g. missed true positives). For other types of model, lambda can, at least in theory, be estimated taking into account the allelic frequency of the locus in question. However, these methods have not yet been implemented in packages for genome-wide association analyses.

Another important thing to remember about genomic control is that it assumes uniform F_{st} across the genome. This may not be the case for some (e.g., selected) genomic regions. Such regions may still generate a very high test statistic, even after appropriate genomic control correction [14]; basically, the frequency distribution at such loci may vary from population to population much more than the average across the genome. A number of methods taking the genetic structure of the study population into account directly (structured association, EIGENSTRAT, and Mixed Models, reviewed later) can deal with such situations.

Finally, what levels of genomic control inflation parameter are acceptable in GWA studies? As the question is "how much of the total test statistic is affected?", non-standardized inflation may be used to address this question. Despite of lack of clear guidance, the general practice in the field is to consider the values of $\hat{\lambda} < 1.01$ as small and $\hat{\lambda} < 1.05$ as moderate and still acceptable. If $\hat{\lambda} > 1.1$, this suggests a strong influence of genetic structure or other design factors on the test statistic. While GC is a statistically correct method, such analysis lacks power; a good practice is to consider the use of other methods, which take the structure of the sample into account in a direct manner.

ANALYSIS OF STRUCTURED POPULATIONS

A study population may be structured in at least two ways. First, while study participants may be "independent" in that kinship between them is expected to be very low (e.g., random sample from a large outbred population), they may belong to rather different genetic populations. An extreme example of such a study is one aiming to combine the data from a sample obtained by random ascertainment from the population of Beijing and a sample from Amsterdam. While in any of the samples people are only remotely related ("independent"), the two samples in question are characterized by high genetic differences; also the environmental influences may be very different for the two populations. The distribution of the trait under the study may be essentially different between the two populations: not only the mean values, but also variances.

On the other hand, consider a family-based sample from a genetically isolated population. Here, environmental influences are more or less uniform for all study participants; the distribution of the trait may be assumed to be governed by the same set of rules for all study participants. However, they are characterized by high and variable kinship between each other.

These two flavors of stratification may be expressed in terms of kinship. When we talk about different "populations", we assume very low kinship and high genetic differentiation between the members of these populations; when we talk about "relatedness", we assume high kinship and a small degree of differentiation. This is illustrated in Figure 9.3. Roughly speaking, when we expect that study participants shared common ancestors a few (one to four) generations ago, the kinship between study participants is high, and the study can be classified as a family-based one (Study 1 of Fig. 9.3). When common ancestors between study participants are expected say more than five generations ago, the kinship between study participants is low, and such a study may be classified as a population-based sample of "independent" people (Studies 2 and 3 in Fig. 9.3). Now consider two groups of study participants (samples 2 and 3). Any person from group 2 is expected to share a common ancestor with a person from the same study group with much higher probability than with a person from group 3; thus these represent two genetic populations (see above for the retrospective definition of genetic population). It should be kept in mind that if kinship is very low (expected common ancestors dozens of generations ago), this would translate into a high degree of genetic differentiation. Moreover, this reflects a long history of isolation, which usually means geographic separation, and accumulation of cultural and other environmental differences between the populations, which may be crucial in association studies.

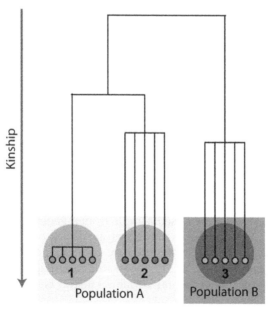

FIGURE 9.3 Three samples from two populations. 1: Family-based sample; 2, 3: random sample of "independent" people. Please refer to color plate section

Of course, a particular study is usually characterized by some mixture of populations and some relatedness between study participants. We will, however, first consider the two extreme scenarios — analysis of samples of "independent" subjects from different populations, and analysis of a family-based study.

Structured Association

If a study population consists of a mixture of several distinct genetic populations, and, within each subpopulation, the study participants are remotely related, structured association analysis may be the method of choice. In such an analysis, effect and its variance are estimated within each strata separately, and then these estimates are pooled to generate global statistic. The strata can be known from design (e.g., place of birth or ethnicity of parents) or estimated from GWA data. By doing this, we allow for arbitrary trait distribution, characterized by stratum-specific mean and variance, in each stratum. Clearly, this may be crucial when different populations characterized by different environments are included in the analysis.

Combining the evidence across strata may be done using a number of methods, e.g., Cochran-Mantel-Haenszel test for binary outcomes. However, one of the simplest ways to combine the evidences coming from multiple strata — or studies — is to use fixed effects inverse variance meta-analysis.

In essence, this method is equivalent to combining likelihoods coming from separate studies, using quadratic approximation. Denote coefficients of regression estimated in S studies/strata as $\widehat{\beta}_i$, and associated squared standard errors of the estimates as $\widehat{s}_i^2$ where $i \in 1, 2, ..., S$. Note that the regression coefficient should be reported on the same scale, e.g., centimeters, meters, or using observations reported on the standard normal scale. Define weights for individual studies as:

$$w_i = \frac{1}{\widehat{s}_i^2}$$

Then the pooled estimate of the regression coefficient is:

$$\widehat{\beta} = \frac{\sum_{i=1}^{S} w_i \widehat{\beta}_i}{\sum_{i=1}^{S} w_i}$$

As you can see, the weights have straightforward interpretation: the bigger the weight of the study (which translates to small standard error), the larger the contribution from this study to the pooled estimate.

The standard error of the pooled estimate is computed as:

$$\widehat{s}^2 = \frac{1}{\sum\limits_{i=1}^{S} w_i}$$

and the χ^2-test for association is computed in standard manner as:

$$T^2 = \frac{\widehat{\beta}^2}{\widehat{s}^2} = \frac{\left(\sum\limits_{i=1}^{S} w_i\widehat{\beta}_i\right)^2}{\sum\limits_{i=1}^{S} w_i}$$

or, alternatively, the Z-test is:

$$Z = \frac{\widehat{\beta}}{\widehat{s}} = \frac{\sum\limits_{i=1}^{S} w_i\widehat{\beta}_i}{\sqrt{\sum\limits_{i=1}^{S} w_i}}$$

When binary traits are studied, and results are expressed as odds ratios with p-values, it is also possible to apply the inverse variance method. For this, you need to transform your odds ratios using natural logarithms, and, on this scale, estimate the standard error. Generic inverse variance pooling may be applied to the data transformed this way; the final results are back-transformed into odds ratio scale using exponentiation.

In the meta-analysis procedure, it is assumed that study populations are composed of genetically homogeneous "unrelated" individuals. Each study is first analyzed separately, and genomic control should either show no inflation of the test statistic, or a small one, which should be corrected with GC prior to combining of strata/studies. If some of the strata demonstrate large residual inflation, this may suggest further substructures present in that strata — it should be dealt with using further subdivision and structure association, or methods described further in this chapter, prior to attempting the pooling of results.

The meta-analytic methodology to perform structured association analysis outlined above has the advantage of simplicity. However, it assumes goodness of quadratic approximation of the likelihood function, which is achieved if numbers are large. If numbers are not large (e.g., rare polymorphisms), a more complicated strategy can be used. A regression model, allowing for study-specific mean, effects of nuisance parameters, and residual variances, but unique (across the studies) effect of interest, may be formulated. Analysis of such a model is equivalent to the above-described meta-analysis procedure, under the large numbers assumption.

Mixed Models-based Approach

In this section we will consider the methods used to analyze samples in which close family relations may be present; these may be family-based samples from outbred populations, or random samples from genetically isolated populations, where, due to limited population size, relatives are likely to be present by chance.

The mixed model-based ideology stems from the classical animal breeding and human heritability analysis methodology, dating back to the works of Fisher [15]. He proposed a model in which a very large number of Mendelian genes contribute small effects to the phenotype (infinitesimal model). We may express this model as:

$$y_i = \mu + G_i + e_i \tag{18}$$

— the value of the trait of the ith person is the sum of the grand population mean, contribution from multiple additively acting genes of small effect (G_i), and residual error (e_i).

Under this assumption, the distribution of the trait in a pedigree is described by a multivariate normal distribution with number of dimensions equal to the number of phenotyped people, with the expectation of the trait value for some individual i equal to:

$$E[y_i] = \mu$$

where μ is grand population mean (intercept). The variance-covariance matrix is defined through its elements V_{ij} — covariance between the phenotypes of person i and person j:

$$V_{ij} = \begin{cases} \sigma_e^2 + \sigma_G^2 & \text{if } i = j \\ 2 \cdot f_{ij} \cdot \sigma_G^2 & \text{if } i \neq j \end{cases} \tag{19}$$

where σ_G^2 is variance due genes, f_{ij} is kinship between persons i and j, and σ_e^2 is the residual variance. The proportion of variance explainable by the additive genetic effects:

$$h^2 = \frac{\sigma_G^2}{\sigma_G^2 + \sigma_e^2}$$

is termed (narrow-sense) "heritability".

Fixed effects of some factor, e.g., SNP, may be included in the model by modifying the expression for the expectation, e.g.:

$$E[y_i] = \mu + \beta_g \cdot g_i$$

where g_i is the genotype of the ith person, leading to the so-called "measured genotypes" model [16].

In essence, this is a linear mixed effects model — the one containing both fixed (e.g., SNP) and random (polygenic) effects having special correlation structure, determined by pedigree. A large body of literature is dedicated to this class of models, and finding the solution for such regression represents no great methodological challenge. Applicability of this model to GWA analysis was first proposed by Yu and colleagues [17]. Two problems occur with application of this model to GWA analysis: first, if the standard Maximum Likelihood (or Restricted Maximum Likelihood) approach is used to estimate this model, the computational time is excessive [18]; second, the pedigree — and thus the kinship matrix — may not be known for study participants.

Let us first consider existing solutions to the first problem. High computational complexity occurs if both the random and fixed effects part of the model are estimated simultaneously, and this procedure is repeated for every SNP analyzed. However, we may assume that the effects of SNP are likely to be small. Thus, inclusion of the SNP into the model is not likely to change the estimates of the components of variance. Therefore a two-step procedure was suggested [19, 20]: first, the mixed model containing all terms but those involving SNPs is estimated (this includes estimation of the components of variance, and possibly the nuisance fixed effects); second, these estimates are then used to compute a test for association for every SNP in analysis. For details on this test, see Chen and Abecasis [19] for original formulation and Aulchenko and Struchalin [20] for extension to multiple variables and interaction effects tested in step 2. The model is implemented in MACH and ProbABEL software packages.

Another method, bearing close resemblance to the method described above, is GRAMMAR — Genome-wide Rapid Analysis using Mixed Models and Regression [18]. Here, in step 1, not only is the mixed model estimated, but also the environmental residuals:

$$\widehat{e}_i = y_i - (\widehat{\mu} + \widehat{G}_i)$$

are estimated. These are free of correlations, and can be used as the new "trait" in further analysis. This procedure allows for very fast GWA analysis in step 2 (for large sample sizes, much faster than the above-described two-step approach); also it has great flexibility as the environmental residuals can be analyzed using a large variety of methods. At the same time, the GRAMMAR test for association is conservative, and the estimates of the effects obtained in step 2 are downward biased [18]. While conservativity of the test may be dealt with using "reverse genomic control" [21], the issue of bias in effect estimates has no solution yet. The GRAMMAR model is implemented in the GenABEL software package [22].

Strictly speaking, the above-described two-step tests are correct if the distributions of covariates in the first and the second parts of the model are independent conditional on the estimated phenotypic variance-covariance matrix. This assumption is most likely to be true when the covariates included in the base model are environmental ones, and thus are not expected to exhibit conditional correlation with SNPs. However, when endogenous risk factors, such as, e.g., body mass index, are included as the covariates in the base model, some SNPs are expected to exhibit covariance with this covariate. In such a situation the two-step test is not strictly correct, but given relatively weak SNP-phenotype correlations normally observed in GWAS it should normally keep good statistical properties in most situations. The properties of the test under such conditions are still awaiting their description.

The methodology described here applies to the study of association of quantitative traits. For binary traits, no formal practical GWA solution is available yet. The best current strategy to analyze binary traits in samples of relatives may be to treat the binary outcome as if it were quantitative. As you have seen with Armitage's trend test, which is equivalent to the score test for a quantitative variable when outcome is coded with "0" and "1" (see above), this approach leads to correct inferences about significance.

Another problem with the application of mixed models to analysis of GWA data is that pedigree is required to estimate the kinship matrix. The expectation of kinship can be estimated from pedigree data using standard methods, for example the kinship for two outbred sibs is $1/4$, for grandchild−grandparent is $1/8$, etc. For an outbred person, the kinship coefficient is $1/2$ − that is two gametes sampled from this person at random are IBD only if the same gamete is sampled. However, in many situations, pedigree information may be absent, incomplete, or not reliable. Moreover, the estimates obtained using pedigree data reflect the expectation of kinship, while the true realization of kinship may vary around this expectation. In the presence of genomic data it may therefore be desirable to estimate the kinship coefficient from these, and not from pedigree. It can be demonstrated that an unbiased and positive semi-definite estimator of the kinship matrix [23] can be obtained by computing the kinship coefficients between individuals i and j with:

$$\widehat{f}_{ij} = \frac{1}{L} \sum_{l=1}^{L} \frac{(g_{l,i} - p_l)(g_{l,j} - p_l)}{p_l(1 - p_l)} \tag{20}$$

where L is the number of loci, p_l is the allelic frequency at the lth locus and $g_{i,j}$ is the genotype of jth person at the lth locus, coded as 0, $1/2$, and 1, corresponding to the homozygous, heterozygous, and other type of homozygous genotype [21, 23, 24]. The frequency is computed for the allele which, when homozygous, corresponds to the genotype coded as "1".

Interestingly, use of the kinship matrix instead of pedigree kinship (when available) may lead to higher power, especially when "dense" pedigrees and traits with high heritability are considered (YSA, unpublished data). This is likely to happen because genomic-based kinship is likely to reflect true genetic relations better than (possibly not completely correct) pedigree expectations.

EIGENSTRAT and Related Methods

The EIGENSTRAT [24] method combines the ideas underlying structured association and the genomic kinship matrix. Basically, it allows for structured association, with strata and more subtle structure identified through the use of genomic data.

The genomic kinship matrix is computed as described earlier (equation (20)). This matrix reflects genetic similarities between study participants. A reverse metric, distance matrix, is computed as $0.5 - f$. Classical multidimensional scaling (CMDS) is applied to identify a number (say, k) of first principal components (PCs) of variation of the distance matrix. The idea behind CMDS is mapping all the pair-wise distances defined by the original matrix to some k-dimensional space. Each study participant is presented as a dot in this space, and these dots are constructed in such way that the distance between these is maximally close to the distances observed in the original distance matrix. Thus, if two genetically distinct populations are present in a sample, the first principal axe of variation ($k = 1$) will identify these. Generally, if n distinct populations are present, these can be identified by $k = n - 1$ first axes.

This principle is demonstrated in the results of CMDS of the genomic distance matrix for the HapMap participants, and a number of people from Europe, presented in Fig. 2 of Nelis et al. [5] (http://www.plosone.org/article/slideshow.action?uri=info:doi/10.1371/journal.pone.0005472&imageURI=info:doi/10.1371/journal.pone.0005472.g002). On panel A, the results of CMDS of the distance matrix to the space of the first two PCs are present. Intercontinental differences are obvious — there are three distinct clusters, corresponding to Yoruba (yellow-brown), Asian (green and dark brown), and European (blue and red) study participants. You can see that mapping to the first two principal components perfectly distinguishes these three populations. The same methodology can be used to distinguish between more subtly divided populations — at panel B of Fig. 2 of Nelis et al. [5] the mapping of European populations is present. Panel C presents even further details of genetic distances — to the level of a single country.

Note that configuration of European cluster in panel A is not exactly the same as in panel B. This happens because the distance matrix, computed based on equation (20), depends on allelic frequencies, as estimated in the total sample. Clearly, these are different when all continents or Europeans only are used. Consecutive analysis of subsamples allows more and more subtle differences between populations to be distinguished.

Interestingly, the genetic distances coincide very well with geographic distances, i.e., the populations close geographically are also close genetically. Thus, most isolation between human populations is explained by geographic distance.

How can these principal components be incorporated in association analysis? Different populations may have different mean values (or prevalences) of the trait under the study, which, coupled with genetic differences, leads to confounding and false positives in association studies. A simple method would be to try to account for these differences in the means by incorporating the principal components as covariates into association analysis [11]. Another related method would be to perform linear adjustment of both the phenotype and the genotype onto PCs (i.e., compute the residuals from linear regression of genotypes/phenotypes onto PCs). Then a simple score test for association (T_q^2, equation (11)) can be used to analyze association. This comprises the original EIGENSTRAT methodology suggested by Price and colleagues [24].

An interesting question is "how many PCs should be included in the analysis?". By default, EIGENSTRAT uses 10 PCs, but this is a rather arbitrary number. In general, "significant" PCs should be included in the analysis. Patterson and colleagues [25] suggested including the leading PCs, which significantly explain progressively more genetic variation. Alternatively, it was argued that PCs significantly associated with the phenotype should be included [26], as inclusion of other PCs is not only unnecessary, but may also reduce power.

Note that methods described here allow for differences in means between the populations, but not for the difference in variances. It would be interesting to address the question "what effects does violation of this assumption have on association studies?", and develop extended methods allowing for differences in variances. This would probably require a study of interpopulation variances.

Summary: What Method to Use?

What methods should be used for analysis of a particular study? A simplified overview is provided in Figure 9.4.

Basically, if study participants come from the same genetic population, are characterized by a low degree of kinship and confounding, as measured with a small genomic control (GC) inflation parameter λ, GC is enough to correct for residual bias.

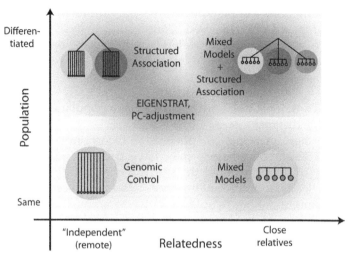

FIGURE 9.4   Applicability of different methods for association analysis. Please refer to color plate section

If study participants come from substantially different populations, and, for each population, the above conditions hold, structured association may be used.

If study participants are characterized by a high degree of kinship, mixed model-based methods should be used.

EIGENSTRAT and PC-based adjustment methods should be used in somewhat intermediate situations when differentiation between the populations and kinship between study participants are not too high: while substantial population differentiation may mean differences in variances, which EIGENSTRAT/PCBA does not account for, it is also known that these methods do not perform very well in pedigrees (YSA, Najaf Amin, unpublished data).

Clearly, a particular study may have its own specifics and should be considered separately. As the difference between outlined scenarios is quantitative rather than qualitative, there is no single recipe for analysis when genetic structure is present in a study sample.

LINKS

Here are some useful links to software which can be used for analysis in structured populations:

EIGENSTRAT:
   http://genepath.med.harvard.edu/~reich/EIGENSTRAT.htm

GenABEL and ABEL tutorial:
  http://mga.bionet.nsc.ru/~yurii/ABEL/abel_beta/
MACH:
  http://www.sph.umich.edu/csg/abecasis/mach/

References

[1] I. Rudan, N. Smolej-Narancic, H. Campbell, A. Carothers, A. Wright, B. Janicijevic, et al., Inbreeding and the genetic complexity of human hypertension, Genetics 163 (3) (2003) 1011–1021.

[2] L.M. Pardo, I. MacKay, B. Oostra, C.M. van Duijn, Y.S. Aulchenko, The effect of genetic drift in a young genetically isolated population, Ann. Hum. Genet. 69 (Pt 3) (2005) 288–295. <http://dx.doi.org/10.1046/j.1529-8817.2005.00162.x>.

[3] S. Wahlund, Zusammensetzung von Population und Korrelationserscheinung vom Standpunkt der Vererbungslehre aus betrachtet, Hereditas 11 (1928) 65–106.

[4] E. Ruiz-Narvājez, Is the Ala12 variant of the PPARG gene an "unthrifty allele"? J. Med. Genet. 42 (7) (2005) 547–550. <http://dx.doi.org/10.1136/jmg. 2004. 026765>.

[5] M. Nelis, T. Esko, R. Mägi, F. Zimprich, A. Zimprich, D. Toncheva, et al., Genetic structure of Europeans: a view from the North-East, PLoS One 4 (5) (2009) e5472. <http://dx.doi.org/10.1371/journal.pone.0005472>.

[6] P.D. Sasieni, From genotypes to genes: doubling the sample size, Biometrics 53 (4) (1997) 1253–1261.

[7] B. Devlin, K. Roeder, Genomic control for association studies, Biometrics 55 (4) (1999) 997–1004.

[8] S.A. Bacanu, B. Devlin, K. Roeder, The power of genomic control, Am.. J. Hum. Genet. 66 (6) (2000) 1933–1944. <http://dx.doi.org/10.1086/302929>.

[9] B. Devlin, K. Roeder, L. Wasserman, Genomic control, a new approach to genetic-based association studies, Theor. Popul. Biol. 60 (3) (2001) 155–166. <http://dx.doi.org/10.1006/tpbi.2001.1542>.

[10] S.A. Bacanu, B. Devlin, K. Roeder, Association studies for quantitative traits in structured populations, Genet. Epidemiol. 22 (2002) 78–93. <http://dx.doi.org/10.1002/gepi.1045>.

[11] R. Sladek, G. Rocheleau, J. Rung, C. Dina, L. Shen, D. Serre, et al., A genome-wide association study identifies novel risk loci for type 2 diabetes, Nature 445 (7130) (2007) 881–885. <http://dx.doi.org/10.1038/nature05616>.

[12] M.L. Freedman, D. Reich, K.L. Penney, G.J. McDonald, A.A. Mignault, N. Patterson, et al., Assessing the impact of population stratification on genetic association studies, Nat. Genet 36 (4) (2004) 388–393. <http://dx.doi.org/10.1038/ng1333>.

[13] G. Zheng, B. Freidlin, Z. Li, J.L. Gastwirth, Genomic control for association studies under various genetic models, Biometrics 61 (2005) 186–192. <http://dx.doi.org/10.1111/j.0006-341X.2005.t01-1-.x>.

[14] C.D. Campbell, E.L. Ogburn, K.L. Lunetta, H.N. Lyon, M.L. Freedman, L.C. Groop, et al., Demonstrating stratification in a European American population, Nat. Genet. 37 (8) (2005) 868–872. <http://dx.doi.org/10.1038/ng1607>.

[15] R.A. Fisher, The correlation between relatives on the supposition of Mendelian inheritance, Trans. R. Soc. Edinb. 52 (1918) 399–433.

[16] E. Boerwinkle, R. Chakraborty, C.F. Sing, The use of measured genotype information in the analysis of quantitative phenotypes in man. I. Models and analytical methods, Ann. Hum. Genet. 50 (1986) 181–194.

[17] J. Yu, G. Pressoir, W.H. Briggs, I.V. Bi, M. Yamasaki, J.F. Doebley, et al., A unified mixed-model method for association mapping that accounts for multiple levels of relatedness, Nat. Genet. 38 (2) (2006) 203–208. <http://dx.doi.org/ 10.1038/ng1702>.

[18] Y.S. Aulchenko, D.J. de Koning, C. Haley, Genomewide rapid association using mixed model and regression: a fast and simple method for genomewide pedigree-based quantitative trait loci association analysis, Genetics 177 (2007) 577–585. <http://dx.doi.org/10.1534/genetics.107.075614>.

[19] W.M. Chen, G.R. Abecasis, Family-based association tests for genomewide association scans, Am. J. Hum. Genet. 81 (5) (2007) 913–926. <http://dx.doi.org/10.1086/521580>.

[20] Aulchenko, Y.S. and Struchalin, M.V. ProbABEL package for genome-wide association analysis of imputed data. Bioinformatics 11 (2010) 134.

[21] N. Amin, C.M. van Duijn, Y.S. Aulchenko, A genomic background based method for association analysis in related individuals, PLoS One 2 (12) (2007) e1274. <http://dx.doi.org/10.1371/journal.pone.0001274>.

[22] Y.S. Aulchenko, S. Ripke, A. Isaacs, C.M. van Duijn, GenABEL: an R library for genome-wide association analysis, Bioinformatics 23 (10) (2007) 1294–1296. <http://dx.doi.org/10.1093/bioinformatics/btm108>.

[23] Astle, W. and Balding, D.J. Population structure and cryptic relatedness in genetic association studies. Statistical Science. 24 (4) (2009) 451–471.

[24] A.L. Price, N.J. Patterson, R.M. Plenge, M.E. Weinblatt, N.A. Shadick, D. Reich, Principal components analysis corrects for stratification in genome-wide association studies, Nat. Genet. 38 (8) (2006) 904–909. <http://dx.doi.org/10.1038/ ng1847>.

[25] N. Patterson, A.L. Price, D. Reich, Population structure and eigenanalysis, PLoS Genet. 2 (12) (2006) e190. <http://dx.doi.org/10.1371/ journal.pgen.0020190>.

[26] J. Novembre, M. Stephens, Interpreting principal component analyses of spatial population genetic variation, Nat. Genet. 40 (5) (2008) 646–649. <http://dx.doi.org/ 10.1038/ng.139>.

Genotype Imputation

Jonathan Marchini

Department of Statistics, University of Oxford, UK

OUTLINE

Genotype imputation is the term used to describe the process of predicting or imputing genotypes that are not directly assayed in a sample of individuals. There are a number of distinct scenarios in which genotype imputation is desirable, but the term now most often refers to the situation in which a reference panel of haplotypes at a dense set of single-nucleotide polymorphisms (SNPs) is used to impute into a study sample of individuals genotyped at a subset of the SNPs. The set of SNPs might cover the whole genome as part of a genome-wide association study (GWAS) or a more focused region as part of a fine-mapping study. The goal is to predict the genotypes at the SNPs not directly genotyped in the study sample. These "in silico" genotypes can then be used to boost the number of SNPs that can be tested for association and so increase the power of the study, the ability to resolve or "fine-map" the causal variant and facilitate meta-analysis. Figure 10.1 illustrates this process.

In samples of unrelated individuals the haplotypes of the individuals over short stretches of sequence will be related to each other identical-by-descent (IBD). The local pattern of IBD can be described by an (unobserved) genealogical tree, which will differ at different loci throughout the genome due to recombination. Imputation methods attempt to identify sharing between the underlying haplotypes of the study individuals and the haplotypes in the reference set and use this sharing to impute the missing alleles in study individuals. For this reason there are strong connections between the models and methods used to infer haplotype phase and those used to perform genotype imputation [1, 2], as well as strong connections to tagging SNP-based approaches [3–5] and methods used in linkage studies [6, 7].

The HapMap 2 haplotypes [8] have been widely used to carry out imputation in studies of samples with ancestry close to those of the HapMap panels. The CEU, YRI and JPT + CHB panels consist of 120, 120 and 180 haplotypes, respectively, at a very dense set of SNPs across the genome. Most studies have used a two-stage procedure that starts by imputing the missing genotypes based on the reference panel without taking the phenotype into account. Imputed genotypes at each SNP, together with their inherent uncertainty, are then tested for association with the phenotype of interest in a second stage. The advantage of the two-stage approach is that different phenotypes can be tested for association without the need to redo the imputation.

The rest of this chapter has sections that cover the main uses of imputation, a description and comparison of the different methods that have been proposed for imputation, a discussion of the factors that affect imputation performance and how imputed genotypes can be used to test for association.

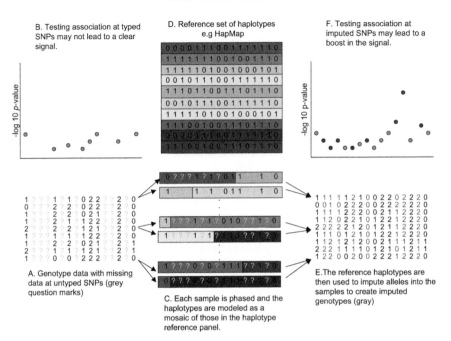

FIGURE 10.1 The figure illustrates the idea of genotype imputation in a sample of unrelated individuals. The raw data consists of a set of genotyped SNPs with a large number of SNPs without any genotype data (A). Testing for association at just these SNPs may not lead to a significant association (B). Imputation attempts to predict these missing genotypes. Algorithms differ in their details but all essentially involve phasing each individual in the study at the typed SNPs. The figure highlights three phased individuals (C). These haplotypes are compared to the dense haplotypes in the reference panel (D). Strand alignment between datasets is an important practical task that must be done before the comparison takes place. In the figure the phased study haplotypes have been colored according to which reference haplotypes they match. This highlights the idea implicit in most phasing and imputation models that the haplotypes of a given individual are modeled as a mosaic of haplotypes of other individuals. Missing genotypes in the study sample are then imputed using those matching haplotypes in the reference set (E). In real examples, the genotypes are imputed with uncertainty and a probability distribution over all three possible genotypes is produced. It is necessary to take account of this uncertainty in any downstream analysis of the imputed data. Testing these imputed SNPs can lead to more significant associations (F) and a more detailed view of associated regions. Please refer to color plate section

USES OF IMPUTATION

Boosting Power

Imputation can lead to a notable boost in the power of a GWAS [9]. For a study with 2000 cases and 2000 controls from a European population, a relative risk at the causal SNP of 1.3 and using a p-value threshold of 5×10^{-7}, the power of imputation compared to testing just genotyped SNPs was estimated to be: Affymetrix 100k (0.242 vs 0.178), Affymetrix 500k (0.450 vs 0.363), Illumina 300k (0.467 vs 0.392), Illumina 610k (0.488 vs 0.439), Illumina 650k (0.492 vs 0.443), Affymetrix 6.0 (0.478 vs 0.420) and Illumina 1M (0.493 vs 0.457). This illustrates that imputation produces a power gain of between 5 and 10% in this scenario and the power of the imputation of the most dense chips is close to that of a hypothetical complete chip consisting of all SNPs in HapMap (power = 0.499). Other simulations have shown that the most benefit occurs for rare SNPs which are harder to tag [10].

Fine-mapping

Imputation provides a much higher resolution view of an associated region than would be seen by just considering genotyped SNPs (Fig. 10.2) and increases the chance that a causal SNP can be directly identified. When imputed SNPs produce larger signals than any of the genotyped SNPs they can become better candidates for replicating in new samples. The limiting factor that determines precision of fine-mapping (in both linkage studies and association studies) is effectively the number of recombination events that have occurred in a region, or the amount of LD in the region. Increasing sample size or mapping in populations with lower levels of LD will both act to increase the chance that a true casual variant can be identified. Imputation methods can also help elucidate when multiple variants or allelic heterogeneity occurs in a region of interest [11, 12].

Meta-analysis

Imputation has been widely used to facilitate meta-analysis of GWAS from different cohorts that may have been genotyped using different genotyping chips, i.e., different sets of SNPs. Imputation effectively enlarges and equates the set of SNPs in each study for which genotypes are available for testing. One of the earliest applications of this approach identified at least six previously unknown loci with robust evidence for association with type 2 diabetes [13]. Other examples include type 1 diabetes [14], colorectal cancer [15], multiple sclerosis [16] and fat mass, weight and risk of obesity [17]. Meta-analysis of GWAS adds a further layer of complexity on top of what is already a sophisticated task. A good practical guide is provided by de Bakker et al. [18]. Results from cohort-specific GWAS are combined using fixed effects models rather than combining the raw data from all studies and then carrying out one association test. This approach allows the conditioning on cohort specific covariates that may be available and accounts for possible population structure effects on a cohort-by-cohort basis.

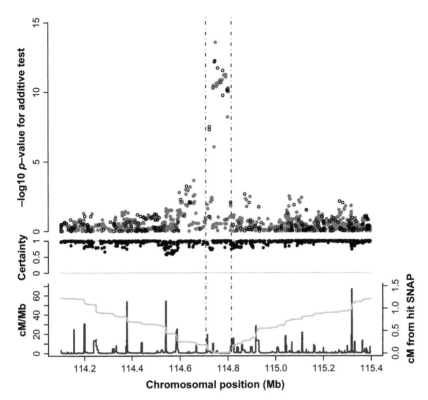

FIGURE 10.2 The results of imputation in and around the TCF7L2 gene in the WTCCC GWAS for type 2 diabetes. The upper part of the plot shows the log10 P-value for the additive model versus a model of no association. The p-values were calculated using called genotypes (black circles) and imputed genotypes (gray circles), called at a threshold of 0.9. The middle panel shows a measure of certainty for each SNP, which is defined as the average maximum posterior genotype call probability. The lower panel shows the fine-scale recombination rate across the region (dark gray) and the cumulative recombination rate measured away from the most highly associated genotyped SNP (horizontal gray). The vertical dashed lines on the plot delineate the main region of association. The largest log10 P value at a genotyped SNP (rs4506565) is 12.25, whereas the largest log10 P-value at an imputed SNP (rs7903146) is 13.57 (taken from Marchini et al. [10]). Please refer to color plate section

GENOTYPE IMPUTATION METHODS

Several different methods have been proposed to carry out genotype imputation. These can be broadly grouped into two main classes of methods: (1) SNP tagging-based approaches that use only a small number of tagSNPs to impute each SNP, and (2) approaches that use more of the flanking genotype data to impute each SNP through the use of various types of hidden Markov models (HMMs). An overview of these two groups of methods is given in the subsequent sections. Table 1 in Marchini and Howie [47] summarizes the properties of each of the methods divided into subsections that deal with properties of the reference panels the methods can handle, properties of the study samples, relevant program options and features, computational performance, error rates and properties and ways of using the output files.

SNP TAGGING-BASED APPROACHES

The idea of SNP tagging in association studies has been established for some time [5, 19, 3] and the methods for genotypes imputation implemented in the programs PLINK [20], SNPMSTAT [21], UNPHASED [22] and TUNA [23] all have strong connections to these approaches. For each SNP to be imputed the reference dataset is used to search for a small set of flanking SNPs that, when phased together with the SNP, leads to a haplotype background that has high LD with the alleles at the SNP. The genotype data from the study and the reference panel are then jointly phased at these SNPs and the missing genotypes in the study are imputed as part of the phasing. The advantage of this approach is that it is simple and quick. The downside is that these approaches generally do not provide as accurate results as other methods because they do not use all the data and the phasing is carried out using a simple multinomial model of haplotype frequencies [1].

HIDDEN MARKOV MODEL-BASED APPROACHES

Several methods have been proposed that use more sophisticated model-based approaches to utilize as much of the genotype data as possible when imputing an SNP, as opposed to just using a small number of tagSNPs. By doing so these methods are able to provide a boost in imputation accuracy, mostly at rarer SNPs and those SNPs which are not well tagged by a small number of flanking SNPs. The method used is the very useful class of statistical model in genetics research called the HMM. This type of model can be used to relate an observed process across the genome to an underlying, unobserved process of interest.

IMPUTE v1

IMPUTE v1 [10] is based on an extension of the HMM models originally developed as part of importance sampling schemes for simulating coalescent trees [24, 25] and for modeling linkage disequilibrium and estimating recombination rates [26]. The method is based on the HMM of each individual's vector of genotypes conditional upon a reference set of haplotypes and an estimate of the fine-scale recombination map across the region. The two underlying haplotypes of each individual are modeled as imperfect mosaics of the haplotypes in the reference panel. Exact marginal probability distributions for the missing genotypes are obtained using the forward-backward algorithm for HMMs [27]. Using a simple modification to the algorithm it is also possible to obtain a marginal distribution for genotypes that are not missing. This provides a useful method of validating observed genotypes and is a valuable feature of the IMPUTE v1 software that allows users to assess the quality of the imputation runs they perform.

IMPUTE v2

IMPUTE v2 [28] takes a different and more flexible approach. SNPs are first divided into two sets: a set T that is typed in both study sample and reference panel, and a set U that is untyped in the study sample but typed in the reference panel. The algorithm involves estimating haplotypes at SNPs in T (using the IMPUTE v1 HMM) and then imputing alleles at SNPs in U conditional upon the current estimated haplotypes. Since the imputation step is haploid imputation it is very fast ($O(N)$) compared to diploid imputation ($O(N^2)$) carried out in IMPUTE v1. Phase uncertainty is accounted for by iterating these steps using an MCMC approach. Since imputation performance is driven by accurate matching of haplotypes the method focuses on accurate haplotype estimation at the SNPs in T using as many individuals as possible.

Alternating between phasing and haploid imputation at a carefully chosen subset of SNPs is particularly suited to study designs where different amounts of genotype data are available in different cohorts of a study. For example, IMPUTE v2 can use both the set of haplotypes from the pilot data of the 1000 Genomes project and haplotype sets from the HapMap 3 dataset as reference panels for imputation. Compared to imputation from HapMap2, this provides a much larger set of imputed SNPs and a notable boost in accuracy at those SNPs included in the HapMap3 SNP set. Other methods can be made to handle this imputation scenario but IMPUTE v2 has been shown to be the most accurate approach [28] and the program makes it straightforward to apply.

When a phenotype is strongly correlated with the genotyping platform Howie et al. [28] found that imputed untyped SNPs in cases from SNPs present in a dense set of genotype data from controls did not lead to elevated false positive rates. However, if cases and controls are typed on different chips then imputing SNPs untyped in both cases and controls from a haplotype panel can lead to false positive associations. SNPs that are imputed accurately from one chip but poorly from the other chip may lead to differences in allele frequency that just reflect allele frequency difference between the haplotype reference panel and the study population, in the manner that population structure can cause problems in GWAS. Ideally, situations like this are best avoided by sensible design. If that isn't possible we recommend QC measures to ensure only the most accurately imputed SNPs are used.

fastPHASE/BIMBAM

The fastPHASE [29] method can be used to estimate haplotypes and carry out imputation and has recently been incorporated into an association testing program called BIMBAM [11, 30]. The method uses the observation that haplotypes tend to cluster into groups of closely related or similar haplotypes. The model specifies a set of K unobserved states or clusters that are meant to represent common haplotypes and each individual's genotype data is then modeled as a hidden Markov model on this state space.

An EM algorithm is used to fit the model and missing genotypes are imputed conditional upon the parameter estimates using the forward-backward algorithm. The authors found that averaging over a set of estimates produced much better results than choosing a single best estimate. Empirical experiments [30] suggest that using $K = 20$ clusters and $E = 10$ start points for the EM algorithm represents a practical compromise between speed and accuracy. The model underlying the GEDI [31] approach is very similar to that in fastPHASE.

The models underlying fastPHASE and IMPUTE are quite similar. IMPUTE has the advantage of not needing to estimate any parameters by using real haplotypes as the model's underlying states. In contrast, fastPHASE uses a much smaller set of states, which greatly speeds up the required HMM calculations, but has a large number of parameters that need to be estimated first.

MACH

MACH uses an HMM model very similar to that used by HOTSPOTTER [26] and IMPUTE. The method can carry out phasing and as a consequence it can be used for imputation. The method works by successively updating the phase of each individual's genotype data conditional upon the current haplotype estimates of all the other samples. The model involves "crossover" and "error" parameters that are updated as the algorithm progresses. Any missing genotype data is naturally imputed by this process and marginal genotype probabilities can be reported as a summary. An alternative two-step approach is also recommended that starts by estimating the parameters of the model using a subset of the full dataset and then carries out maximum likelihood genotype imputation based on the estimated parameters. In contrast, IMPUTE v1 uses fixed estimates of its mutation rates and recombination maps. Estimating the parameters allows more flexibility to adapt to the dataset being analyzed. However, it is likely that some parameters will not be estimated well and this will reduce imputation accuracy.

BEAGLE

The BEAGLE method [32–34] is based upon a graphical model of a set of haplotypes. The method works iteratively by fitting the model to the current set of estimated haplotypes and then resampling new estimated haplotypes for each individual. The model is empirical in the sense that it has no parameters that need to be estimated and is applied to a given set of haplotypes in two steps. Possibly the best way to understand the model is by looking at the small example given in Figure 2 and Table 1 of Browning [32]. Browning [35] provides a useful review that contrasts various methods for phasing and imputation.

The model has the property that, in regions where the levels of LD are high, the graph will tend to consist of many edges that capture the distinct common haplotypes that occur there. In contrast, when levels of LD are low, for example across a recombination hotspot, the graph will tend to collapse to 1 or a few nodes. In this way, the model has the attractive property that it can adapt to the local haplotype diversity that occurs in the data. In some sense it can be thought of as a local haplotype clustering model, similar to fastPHASE, but with a variable number of clusters across a region. Another advantage of the model is that, as it involves no parameter estimation steps, it is very fast to fit to a dataset.

Imputation in Related Samples

UNPHASED allows for genotype imputation in nuclear families and this approach has been used to impute sporadic missing SNP genotype data in a study with a mixture of HLA and SNP genotypes in the study of nuclear families and unrelated individuals [36]. A more focused method in which genotypes in founders are imputed down into descendants has also been proposed [37]. Since close relatives will share long stretches of haplotypes the descendants need only be typed at a relatively sparse set of markers for this to work well. Kong et al. [38] proposed a related approach in which surrogate parents are used instead of real parents. For each individual surrogate parents are identified as those who share long stretches of sequence with at least one allele identical-by-state (IBS). Regions where this occurs are assumed to be IBD and this estimated relatedness is used to help phase the individuals accurately over long stretches. This approach only works when a sufficient proportion of the population (>1% as a rule of thumb) has been genotyped but may have useful applications when carrying out imputation if large, densely typed or sequenced cohorts become available. A related idea is that used in IMPUTE v2 [28], in which a "surrogate family" of individuals is used when updating the phase of a given individual over reasonably long stretches of sequence (typically 5 Mb in practice).

Imputation of Untyped Variation

Imputation of SNPs which have not been typed in either the haplotype reference panel or the study sample is also possible. Some methods do this via inference of the genealogy between study sample haplotypes [39, 40, 10, 12] while others aim to identify haplotype effects more directly [41, 41]. These methods can lead to a boost in power, especially when the causal variant is rare, or where there is local heterogeneity in the signal of association [12].

Imputation of Non-SNP Variation

The general idea of imputation is readily extended to other types of genetic variation such as copy number variants (CNVs) [42] and classical HLA alleles [43]. Looking ahead, the imputation of large numbers of small insertions and deletions (indels) that will be discovered from sequencing based projects such as the 1000 Genome project are likely to be widely adopted in GWAS studies.

Sporadic Missing Data Imputation and Correction of Genotyping Errors

Many of the widely used imputation programs allow imputation of sporadic missing genotypes that can occur when calling genotypes from genotyping chips. Genotyping error rates are often very low (0.2% in the WTCCC study [14]) so this type of imputation will not greatly boost power but can help control false positives at SNPs where genotype calling is challenging. Recently, the BEAGLE model [44] has been extended to handle genotype intensity data so that genotypes can be called by utilizing LD information between SNPs and this offers a small improvement in genotyping error rates.

Factors that Affect Imputation Accuracy

Most imputation methods produce a probabilistic prediction of each imputed genotype of the form:

$$p_{ijk} = P(G_{ij} = k | H, G), \quad k \in \{0, 1, 2\}, \quad \sum_k p_{ijk} = 1 \quad (1)$$

where $G_{ij} \in \{0, 1, 2\}$ denotes the genotype of the ith individual at the jth SNP.

To assess the quality of predictions and compare method genotypes can be masked or hidden and then predicted using an imputation algorithm. The most likely predicted genotype above some threshold can be compared to the true genotype, and a plot of the percentage discordance versus the percentage missingness can be constructed for a range of thresholds to illustrate performance. This method was recently used to compare methods using 1377 UK individuals genotyped on both the Affymetrix 500k SNP chip and the Illumina 550k chip (see Fig. X of [28]). Genotypes on the Affymetrix chip were combined with the 120 CEU haplotypes to predict the 22,270 HapMap SNPs on chromosome 10 that were on the Illumina chip but not the Affymetrix chip. The error rate of the *best guess genotype* for various methods was: BEAGLE (default), 6.33%; BEAGLE (50 iterations), 6.24%; fastPHASE ($k = 20$), 6.07%; fastPHASE ($k = 30$), 5.92%; IMPUTE v1, 5.42%; IMPUTE v2 ($k = 40$), 5.23%; IMPUTE v2 ($k = 80$), 5.16%; MACH, 5.46% and these results are consistent with other comparisons in the literature [45, 46]. For the best methods an error rate of between 2 and 3% can be achieved but at the expense of 10% of missing genotypes. Another option involves measuring the squared correlation between the best guess genotype and the truth [34] which can be averaged across SNPs to give a single measure. Another desirable property of imputation methods is that the predicted probabilities they produce should be well calibrated. Most methods in common use have been shown to produce well-calibrated probabilities [29, 10].

The imputation accuracy results from Howie et al. [28] are specific to a UK population using the CEU HapMap and the Affymetrix 500k chip. The study population, properties of the reference panel and genotyping chip will all influence performance, and performance may vary between rare and common alleles. Marchini and Howie [47] carried out some experiments to examine the performance of various genotyping chips and in the three HapMap 2 panels. The conclusions were that: (1) across all imputation panels and genotyping chips, imputation error rate increases as the minor allele frequency decreases, which is inline with previous observations [8] that have shown that rare SNPs are more difficult to tag than common SNPs and thus essentially harder to predict, (2) using a reference panel phased using trio information boosts imputation performance, compared to using a reference panel phased without trio information, (3) the Illumina chips outperform the two Affymetrix chips in the CEU population but in the YRI performance of all chips decreases, but more so for the Illumina chips, thus the use of tagging methods for chip design can influence the imputation performance of a chip, (4) imputation accuracy increases as reference panel size increases [28], and (5) using a combination of CEU, YRI and JPT + CHB haplotype panels can boost the performance of imputation, especially at rare SNPs, compared to using a single haplotype panel.

It is also important to consider the performance of imputation in individuals from other populations other than the three main HapMap panels. Huang et al. [48] examined the portability of the HapMap reference panels for imputation using genome-wide SNP data collected on samples from 29 worldwide populations. When a single HapMap panel was used as the basis for imputation they found that European populations had the lowest imputation error rates, followed by populations from East Asia, Central and South Asia, the Americas, Oceania, the Middle East, and Africa. Within Africa, which has high levels of genetic diversity, imputation accuracy using the YRI panel varied considerably. These results indicate that the difference in genetic diversity between the study population and the reference panel also plays a role in imputation accuracy.

Huang et al. [48] also found that imputation-based mixtures of at least two HapMap panels reduced imputation error rates in 25 of the populations. In 11 of the populations the optimal choice of mixtures was to combine all three HapMap populations together as a reference panel. Seven of these 11 groups (Bedouin, Mozabite, Druze, Basque, Burusho, Daur and Yi) were from Eurasia with some degree of dissimilarity from the HapMap CEU and JPT + CHB panels. The remaining four groups (Melanisian, Papuan, Pima and Colombian) were from Oceania and the Americas. These results can guide the choice of HapMap panels to use, with the caveat that they are specific to the HumanHap550 chip. A related point concerns imputation of admixed individuals. Pasaniuc et al. [49] have shown that imputation conditional upon a local ancestry estimate can be more accurate than unconditional imputation but the biggest gains in accuracy will occur in admixed individuals from genetically dissimilar populations.

More recently, sets of haplotypes from the HapMap3 project and from the pilot phase of the 1000 Genomes project (1KGP) have been made available. The HapMap3 has 10 distinct sets of haplotypes and larger numbers of haplotypes in each set. For example, there are 330 CEU haplotypes. This allows more accurate imputation of rarer SNPs but HapMap3 has a smaller set of SNPs (1481135 QC + SNPs) than HapMap2.

The 1KGP haplotype sets consist of 120 CEU and 120 YRI at 8.5 million SNPs. This large boost in the number of SNPs allows finer resolution of signals in associated regions. When the 1KGP data are complete it is likely that this will become the reference set of choice for imputation into GWAS datasets. The large increase in both the number of SNPs and samples will allow more accurate imputation of the majority of SNPs, indels and other structural variants above 1% frequency.

Post-imputation Information Measures

Once imputation has been carried out it is useful to assess the quality of imputed genotypes at SNPs in the absence of any true set of genotypes to compare them to. If the imputation quality is low at an SNP it may be wise to filter out such SNPs before association testing is performed [13]. MACH, BEAGLE, IMPUTE and SNPTEST all have metrics to assess quality that are designed to lie in the range [0, 1]. A value of 1 indicates that there is no uncertainty in the imputed genotypes, whereas a value of 0 means that there is complete uncertainty about the genotypes. All of these measures can be interpreted in the following way: an information measure of α on a sample of N individuals indicates that the amount of data at the imputed SNP is roughly equivalent to a set of perfectly observed genotype data in a sample size of αN. These metrics were recently compared and described in detail [47] and shown to be very highly correlated in general despite differences in their formulation.

Association Testing using Imputed Data

The probabilistic nature of imputed SNPs means that testing for association at these SNPs requires some care. Using just those imputed genotypes with posterior probability above some threshold (or using the *best guess genotype*) is a reasonable method of comparing the accuracy across methods but it is not recommended when carrying out association tests at imputed SNPs. Removing genotypes in this way can lead to both false positives and loss of power.

To fully account for the uncertainty in imputed genotypes well-established statistical theory for missing data problems can be used [10, 47]. An observed data likelihood is used in which the contribution of each possible genotype is weighted by its imputation probability. This approach is implemented in the SNPTEST program. A Score test is then the most efficient way to use this likelihood to carry out a test but can break down if the sample size is small, the allele frequency is low or when there is considerable genotype uncertainty. The alternative is to use an iterative scheme to maximize the likelihood and carry out a maximum likelihood ratio or test or a Wald test. SNPTEST allows both quantitative and binary traits and can condition on user-specified covariates.

A simple approach for testing imputed SNPs involves using the expected genotype count $e_{ij} = p_{ij1} + 2_{pij2}$ (also called posterior mean [30] or allele dosage). These expected counts can be used to test for association with a binary or quantitative phenotype, using a standard logistic or linear regression model, respectively. This method has been shown to provide a good approximation to methods that take the genotype uncertainty into account when the effect size of the risk allele is small [30], which is the case for most of the common variants found in recent GWAS. This approach is implemented in the programs MACH2DAT/ MACH2QTL, SNPTEST, PLINK and the R package ProbABEL. The ProbABEL package also allows time to event phenotypes to be considered using Cox proportional hazards models.

The use of Bayesian methods for analyzing SNP associations has recently been proposed [10, 11, 30, 50, 51] and has advantages over the use of p-values in power and interpretation. Within the Bayesian framework focus centers on calculation of a Bayes factor (BF), which is a ratio of marginal likelihoods between a model of association (M1) and a null model of no association (M0). These marginal likelihoods can be approximated using a Laplace approximation and a straight-forward modification of the likelihood maximization used by frequentist methods [47]. This approach (implemented in SNPTEST) is much more stable than when maximizing the likelihood, since the prior acts to regularize the parameter estimation. The expected geno-type count can also be used in Bayesian approaches [30] and is implemented in both BIMBAM and SNPTEST. Stephens and Balding [51] provide an excellent review of the use of Bayes factors and includes a good discussion on the choice of priors. Currently, only SNPTEST can calculate Bayes factors for binary traits conditional upon a set of covariates.

In the context of fine-mapping where multiple SNPs in a gene or region may play a role in the underlying causal model it is desirable to consider models that allow multiple SNPs. The BIMBAM approach [11] combines imputation with such an approach and can produce posterior probabilities of association for each SNP and also on the number of associated SNPs in the region.

Bayes Factors versus p-values

At directly genotyped SNPs Bayes factors and p-values can be made equivalent in the sense that they give the same ranking of SNPs [50] but this occurs for a particular choice of prior in which the prior variance of the effect size increases as minor allele frequency decreases (or as the information at the SNP about the effect size parameter decreases). This prior assumes larger effects at rarer SNPs which may be biologically reasonable. At imputed SNPs the level of uncertainty also influences the amount of information there is about the effect size parameter. To make Bayes factors give the same ranking of SNPs as p-values we would need to allow prior variance to increase as the amount of imputation uncertainty increases which makes no sense [51]. So even when adopting a prior that depends upon allele frequency, Bayes factors and p-values will not give the same ranking at imputed SNPs. In practice, studies have tended to filter out SNPs with low information so it seems unlikely that a re-analysis of studies using a Bayes factors will result in very different outcomes, but as we probe rarer and rarer SNPs based on imputation from the 1000 Genomes data it may become more important to take care of these details.

Joint Imputation and Testing

The two-stage process of imputation followed by association testing may underestimate effect sizes since genotypes are effectively imputed under the null model. The SNPMSTAT [21] and UNPHASED approaches allow joint imputation and testing within a single model, but currently only using a simple multinomial for haplotype frequencies at a small number of tagSNPs flanking the locus of interest. A comparison of the joint approach (using SNPMSTAT [21]) to a two stage approach (using IMPUTE and then SNPTEST) on three different datasets suggested that the improved performance of imputation gained by using a method that uses as much flanking genotype data as possible (like IMPUTE, MACH, fastPHASE and BEAGLE) outweighs the advantage of joint imputation and testing [52].

PERSPECTIVES AND FUTURE DIRECTIONS

It seems likely that genotype imputation will continue to play an important role in the analysis of GWAS over the next few years as researchers apply the approach to an increasing set of diseases and traits and work together to combine cohorts through meta-analysis. The main factor that will influence the precise way in which imputation is used will be the increasing availability of next-generation sequencing data. Such data will allow researchers to assess SNPs as well as structural variants such as short indels and CNVs. One public resource of such data will be the 1000 Genomes project. Imputation is being used both *within* the data to reconstruct genotypes from the low-coverage sequencing reads and to impute *from* the data into other cohorts.

Compared to HapMap2 the number of SNPs, the number of haplotypes and the number of populations will increase notably. This resource will include many more SNPs between a 1 and 5% frequency that can be imputed. This may be important if rarer variation is an important part of the etiology of a given trait. The availability of haplotypes on a larger set of populations should lead to an improvement in imputation in populations not well matched to the CEU, YRI and CHB + JPT haplotype sets in HapMap2.

The challenges for imputation methods will be in utilizing the larger more diverse set of haplotypes available for imputation. Since any haplotype estimates produced from the 1000 Genomes data may have more inherent uncertainty than the HapMap2 haplotypes, due to the low-coverage sequencing used and the larger number of rare SNPs, it may be important to take this into account when imputing *from* this data. Along these lines both IMPUTE v2 and BEAGLE already offer the ability to accept genotypes estimated with uncertainty when carrying out imputation and phasing. Care may also be needed when analyzing rare imputed SNPs. It is well known that the asymptotic theory used by frequentist association tests breaks down at rarer SNPs which means that p-values may not be well calibrated and should be interpreted with care.

Software

IMPUTE v1, IMPUTE v2, SNPTEST http://www.stats.ox.ac.uk/~marchini/#software

fastPHASE, BIMBAM http://stephenslab.uchicago.edu/software.html

MACH, MACH2DAT, MACH2QTL http://www.sph.umich.edu/csg/abecasis/MACH/

BEAGLE http://www.stat.auckland.ac.nz/bbrowning/beagle/beagle.html

SNPMSTAT http://www.bios.unc.edu/lin/software/SNPMStat/

GEDI http://dna.engr.uconn.edu/

PLINK http://pngu.mgh.harvard.edu/purcell/plink/

UNPHASED http://homepages.lshtm.ac.uk/frankdudbridge/software/unphased/

TUNA http://www.stat.uchicago.edu/wen/tuna/

ProbABEL http://mga.bionet.nsc.ru/yurii/ABEL/

References

[1] L. Excoffier, M. Slatkin, Maximum-likelihood estimation of molecular haplotype frequencies in a diploid population, Mol. Biol. Evol. 12 (5) (1995) 921–927.

[2] M. Stephens, N. Smith, P. Donnelly, A new statistical method for haplotype reconstruction from population data, Am. J. Hum. Genet. 68 (4) (2001) 978–989.

[3] P. de Bakker, R. Yelensky, I. Pe'er, S. Gabriel, M. Daly, D. Altshuler, Efficiency and power in genetic association studies, Nat. Genet. 37 (11) (2005) 1217–1223.

[4] C. Carlson, M. Eberle, M. Rieder, Q. Yi, L. Kruglyak, D. Nickerson, Selecting a maximally informative set of single-nucleotide polymorphisms for association analyses using linkage disequilibrium, Am. J. Hum. Genet. 74 (1) (2004) 106–120.

[5] G. Johnson, L. Esposito, B. Barratt, A. Smith, J. Heward, G.D. Genova, et al., Haplotype tagging for the identification of common disease genes, Nat. Genet. 29 (2) (2001) 233–237.

[6] R. Elston, J. Stewart, A general model for the genetic analysis of pedigree data, Hum. Hered. 21 (6) (1971) 523–542.

[7] E. Lander, P. Green, Construction of multilocus genetic linkage maps in humans, Proc. Natl. Acad. Sci. USA 84 (8) (1987) 2363–2367.

[8] K. Frazer, D. Ballinger, D. Cox, D. Hinds, L. Stuve, R. Gibbs, et al., A second generation human haplotype map of over 3.1 million SNPs, Nature 449 (7164) (2007) 851–861 (2007).

[9] C.C.A. Spencer, Z. Su, P. Donnelly, J. Marchini, Designing genomewide association studies: sample size, power, imputation, and the choice of genotyping chip, PLoS Genet. 5 (5) (2009) e1000477.

[10] J. Marchini, B. Howie, S. Myers, G. McVean, P. Donnelly, A new multipoint method for genome-wide association studies by imputation of genotypes, Nat. Genet. 39 (7) (2007) 906–913.

[11] B. Servin, M. Stephens, Imputation-based analysis of association studies: candidate regions and quantitative traits, PLoS Genet. 3 (7) (2007) e114.

[12] Z. Su, N. Cardin, Wellcome Trust Case Control Consortium, P. Donnelly, J. Marchini, A Bayesian method for detecting and characterizing allelic heterogeneity and boosting signals in genome-wide association studies, Stat. Sci. 24 (2009) 430–450.

[13] E. Zeggini, L.J. Scott, R. Saxena, B.F. Voight, J.L. Marchini, T. Hu, et al., Meta-analysis of genome-wide association data and large-scale replication identifies additional susceptibility loci for type 2 diabetes, Nat. Genet. 40 (5) (2008) 638–645.

[14] J. Cooper, D. Smyth, A. Smiles, V. Plagnol, N. Walker, J. Allen, et al., Metaanalysis of genome-wide association study data identifies additional type 1 diabetes risk loci, Nat. Genet. 40 (12) (2008) 1399–1401.

[15] R. Houlston, E. Webb, P. Broderick, A. Pittman, M.D. Bernardo, S. Lubbe, et al., Meta-analysis of genome-wide association data identifies four new susceptibility loci for colorectal cancer, Nat. Genet. 40 (12) (2008) 1426–1435.

[16] P. De Jager, X. Jia, J. Wang, P. de Bakker, L. Ottoboni, N. Aggarwal, et al., Meta-analysis of genome scans and replication identify CD6, IRF8 and TNFRSF1A as new multiple sclerosis susceptibility loci, Nat. Genet. 41 (7) (2009) 776–782.

[17] R.J.F. Loos, C.M. Lindgren, S. Li, E. Wheeler, J.H. Zhao, I. Prokopenko, et al., Common variants near MC4R are associated with fat mass, weight and risk of obesity, Nat. Genet. 40 (6) (2008) 768–775.

[18] P. de Bakker, M. Ferreira, X. Jia, B. Neale, S. Raychaudhuri, B. Voight. Practical aspects of imputation-driven meta-analysis of genome-wide association studies, Hum. Mol. Genet. 17(R2) R122-R128.

[19] D. Evans, L. Cardon, A. Morris, Genotype prediction using a dense map of SNPs, Genet. Epidemiol. 27 (4) (2004) 375–384.

[20] S. Purcell, B. Neale, K. Todd-Brown, L. Thomas, M. Ferreira, D. Bender, et al., PLINK: a tool set for whole-genome association and population-based linkage analyses, Am. J. Hum. Genet. 81 (3) (2007) 559–575.

[21] D. Lin, Y. Hu, B. Huang, Simple and efficient analysis of disease association with missing genotype data, Am. J. Hum. Genet. 82 (2) (2008) 444–452.

[22] F. Dudbridge, Likelihood-based association analysis for nuclear families and unrelated subjects with missing genotype data, Hum. Hered. 66 (2) (2008) 87–98.

[23] D. Nicolae, Testing untyped alleles (TUNA)-applications to genome-wide association studies, Genet. Epidemiol. 30 (8) (2006) 718–727.

[24] M. Stephens, P. Donnelly, Inference in molecular population genetics, J.R. Statist. Soc. B. 62 (2000) 605–635.

[25] P. Fearnhead, P. Donnelly, Estimating recombination rates from population genetic data, Genetics 159 (3) (2001) 1299–1318.

[26] N. Li, M. Stephens, Modeling linkage disequilibrium and identifying recombination hotspots using single-nucleotide polymorphism data, Genetics 165 (4) (2003) 2213–2233.

[27] L.R. Rabiner, A tutorial on hidden Markov models and selected applications in speech recognition, Proceedings of the IEEE 77 (2) (1989) 257–286.

[28] B.N. Howie, P. Donnelly, J. Marchini, A flexible and accurate genotype imputation method for the next generation of genome-wide association studies, PLoS Genet. 5 (6) (2009) e1000529.

[29] P. Scheet, M. Stephens, A fast and flexible statistical model for largescale population genotype data: applications to inferring missing genotypes and haplotypic phase, Am. J. Hum. Genet. 78 (4) (2006) 629–644.

[30] Y. Guan, M. Stephens, Practical issues in imputation-based association mapping, PLoS Genet. 4 (12) (2008) e1000279.

[31] J. Kennedy, I. Mandoiu, B. Pasaniuc, Genotype error detection using hidden Markov models of haplotype diversity, J. Comput. Biol. 15 (9) (2008) 1155–1171.

[32] S. Browning, Multilocus association mapping using variable-length Markov chains, Am. J. Hum. Genet. 78 (6) (2006) 903–913.

[33] S. Browning, B. Browning, Rapid and accurate haplotype phasing and missing-data inference for whole-genome association studies by use of localized haplotype clustering, Am. J. Hum. Genet. 81 (5) (2007) 1084–1097.

[34] B. Browning, S. Browning, A unified approach to genotype imputation and haplotype-phase inference for large data sets of trios and unrelated individuals, Am. J. Hum. Genet. 84 (2) (2009) 210–223.

[35] S. Browning, Missing data imputation and haplotype phase inference for genome-wide association studies, Hum. Genet. 124 (5) (2008) 439–450.

[36] R. Pastorino, C. Menni, M. Barca, L. Foco, V. Saddi, G. Gazzaniga, et al., Association between protective and deleterious HLA alleles with multiple sclerosis in central east Sardinia, PLoS ONE 4 (8) (2009) e6526.

[37] J. Burdick, W. Chen, G. Abecasis, V. Cheung, In silico method for inferring genotypes in pedigrees, Nat. Genet. 38 (9) (2006) 1002–1004.

[38] A. Kong, G. Masson, M. Frigge, A. Gylfason, P. Zusmanovich, G. Thorleifsson, et al., Detection of sharing by descent, long-range phasing and haplotype imputation, Nat. Genet. (2008).

[39] S. Zollner, J. Pritchard, Coalescent-based association mapping and fine mapping of complex trait loci, Genetics 169 (2) (2005) 1071–1092.

[40] M. Minichiello, R. Durbin, Mapping trait loci by use of inferred ancestral recombination graphs, Am. J. Hum. Genet. 79 (5) (2006) 910–922.

[41] B. Browning, S. Browning, Efficient multilocus association testing for whole genome association studies using localized haplotype clustering, Genet. Epidemiol. 31 (5) (2007) 365–375.

[42] N. Cardin, C. Barnes, V. Plagnol, D. Clayton, M. Hurles, P. Donnelly, et al., Using lD to predict CNVS and test for disease associations, Presented at the annual meeting of the American Society of Human Genetics, San Diego, California, 2007. 25 October 2007.

[43] S. Leslie, P. Donnelly, G. McVean, A statistical method for predicting classical HLA alleles from SNP data, Am. J. Hum. Genet. 82 (1) (2008) 48–56.

[44] B.L. Browning, Z. Yu, Simultaneous genotype calling and haplotype phasing improves genotype accuracy and reduces false-positive associations for genome-wide association studies, Am. J. Hum. Genet. 85 (6) (2009) 847–861.

[45] Y. Pei, J. Li, L. Zhang, C. Papasian, H. Deng, Analyses and comparison of accuracy of different genotype imputation methods, PLoS ONE 3 (10) (2008) e3551.

[46] K. Hao, E. Chudin, J. McElwee, E.E. Schadt, Accuracy of genomewide imputation of untyped markers and impacts on statistical power for association studies, BMC Genet. 10 (2009) 27.

[47] J. Marchini, B. Howie, Genotype imputation for genome-wide association studies, Nature Rev. Genet. 11 (2010) 499–511.

[48] L. Huang, Y. Li, A. Singleton, J. Hardy, G. Abecasis, N. Rosenberg, et al., Genotype-imputation accuracy across worldwide human populations, Am. J. Hum. Genet. 84 (2) (2009) 235–250.

[49] B. Pasaniuc, S. Sankararaman, G. Kimmel, E. Halperin, Inference of locus-specific ancestry in closely related populations, Bioinformatics 25 (12) (2009) i213–i221.

[50] J. Wakefield, Bayes factors for genome-wide association studies: comparison with p-values, Genet. Epidemiol. 33 (2009) 79–86.

[51] M. Stephens, D. Balding, Bayesian statistical methods for genetic association studies, Nat. Rev. Genet. 10 (10) (2009) 681–690.

[52] J. Marchini, B. Howie, Comparing algorithms for genotype imputation, Am. J. Hum. Genet. 83 (4) (2008) 535–539. author reply 539–540.

11

Haplotype Methods for Population-based Association Studies

Andrew Morris

Wellcome Trust Centre for Human Genetics, University of Oxford, UK

The traditional approach to the analysis of population-based genetic association studies is to test for the presence of a statistical relationship between the phenotype and each genotyped polymorphism, one by one, utilizing the techniques described in detail in Chapter 8. This approach has the advantage of minimal computational burden, and has been successfully applied to the analysis of genome-wide association (GWA) studies, such as those undertaken as part of the main experiment of the Wellcome Trust Case Control Consortium (WTCCC), resulting in the identification of many novel loci, now replicated and confirmed as contributing effects to a wide range of complex traits [1].

However, perhaps the most notable feature of common human genetic variation is the strong statistical correlation between proximal single-nucleotide polymorphisms (SNPs), with specific combinations of alleles, referred to as *haplotypes,* occurring together, in *cis,* more often than would be expected by chance. This so-called *linkage disequilibrium* (LD) between SNPs occurs as a result of the shared ancestry of a population of haplotypes, as described in detail in Chapter 2. Patterns of LD across the genome can be thought of as discrete, with *blocks* of highly correlated SNPs occurring over extended regions of limited recombination which are bounded by highly localized hotspots of meiotic crossover activity. As a result, common genetic variation can be represented by haplotypes within blocks of LD that are, more often than not, transmitted from parents to their offspring undisturbed by recombination.

The block-like structure of common human genetic variation is one of the key factors motivating the use of haplotypes to identify loci contributing effects to complex traits. However, from a statistical perspective, haplotype-based analyses within blocks of LD also reduce the burden of multiple testing over traditional single-locus methods, leading to an expected increase in power, particularly for GWA studies of hundreds of thousands of SNPs. From a biological viewpoint, Clark [2] highlights that the functional properties of a protein are determined by a linear sequence of amino acids which, by definition, correspond to genetic variation on a haplotype. In particular, there is evidence that specific combinations of mutations, occurring together, in *cis,* may interact to form a *super-allele* that has a greater effect on phenotypic variation than any one of the individual causal variants alone. Examples of such super-allelic effects include a specific haplotype in *HPC2/ELAC2* which has been demonstrated to be associated with increased risk of prostate cancer [3], and 2-SNP haplotypes in the region flanking *CDKN2A/B* which have been established as associated with altered risk of type 2 diabetes [4, 5].

Unfortunately, despite the compelling arguments to analyze haplotypes in association studies, current SNP typing technology is typically limited to the generation of *unphased* genotype data. As a result, we cannot generally recover the necessary phase information to reconstruct, unequivocally, the *diplotype* (i.e., pair of phased haplotypes) carried by an individual. Although, in the laboratory, there are a variety of techniques available to determine diplotypes experimentally, these methods are low throughput and extremely costly in comparison to genotyping, and have thus been restricted to studies of candidate genes in samples that are too small to be of practical use in association mapping [6]. Parental genotype data can help to recover haplotypes by fixing phase in the maternal and paternal diplotypes for SNPs at which we can determine the alleles transmitted from each parent to their child (i.e., if the parents and child are not all heterozygous). However, unless parents are genotyped routinely as part of a family-based association study, it is not cost effective to collect this additional information purely on the grounds of haplotype reconstruction.

In this chapter we focus on two key statistical questions: (1) how can we best reconstruct haplotypes from unphased genotype data in samples of unrelated individuals?, and (2) how can we make use of reconstructed haplotypes in testing for association with a phenotype of interest? We begin by comparing and contrasting a range of techniques that have been commonly utilized to recover phase information from the genotype data of unrelated individuals. We also describe how these methods can be used to reconstruct haplotypes across large genomic regions, even entire chromosomes, and review their relative performance in terms of accuracy and computation time. We then go on to describe a general framework for the analysis of reconstructed haplotypes that takes account of the uncertainty attached to the process of phase inference. Finally, we describe a range of techniques that aim to increase the power of haplotype-based association studies by attempting to allow for their expected evolution within blocks of strong LD.

HAPLOTYPE RECONSTRUCTION IN POPULATION-BASED ASSOCIATION STUDIES

Consider a sample of unrelated individuals genotyped at three SNPs, with alleles denoted A and a at the first locus, B and b at the second, and C and c at the third. For an individual with the three-locus genotype $AABBCc$, for example, we can resolve their diplotype without ambiguity, namely ABC/ABc, although without parental data we cannot determine which have been maternally or paternally derived. However, for an individual with genotype $AaBbCc$, there are four possible diplotypes: (1) ABC/abc, (2) ABc/abC, (3) AbC/aBc, and (4) Abc/abC. In fact, there are $2^{(k-1)}$ possible diplotypes consistent with a multi-locus genotype which is heterozygous at k SNPs. It is this problem which has prompted the development of statistical methods to resolve diplotypes directly using unphased SNP genotype data.

Likelihood-based Methods

Consider a sample of N unrelated individuals with unphased genotype data, denoted $\mathbf{G}$, at M proximal SNPs. In this setting, there are a total of

$T = 2^M$ possible distinct haplotypes, occurring with unknown haplotype frequencies denoted $\mathbf{h} = \{h_1, h_2, ..., h_T\}$. Our goal is to make inferences about $\mathbf{h}$ and the unobserved diplotypes, $\mathbf{H}$, of sampled individuals. We can express the likelihood of the observed multi-locus genotype data, given the unobserved haplotype frequencies as:

$$f(\mathbf{G}|\mathbf{h}) = \prod_i f(G_i|\mathbf{h}) \tag{1}$$

where G_i denotes the multi-Locus genotype of the ith individual. The contribution of this individual to the likelihood is given by a summation over all possible diplotypes, H_i, consistent with their genotype, namely:

$$f(G_i|\mathbf{h}) = \sum_{H_i \in G_i} f(H_i|\mathbf{h}) \tag{2}$$

Under the assumption of Hardy-Weinberg equilibrium (HWE):

$$f(H_i|\mathbf{h}) = \begin{cases} h^2_{H_{i1}} & \text{if } H_{i1} = H_{i2} \\ 2h_{H_{i1}}h_{H_{i2}} & \text{if } H_{i1} \neq H_{i2} \end{cases} \tag{3}$$

where H_{i1} and H_{i2} denote the unordered pair of haplotypes forming diplotype H_i. We can obtain *maximum likelihood estimates* of the haplotype frequencies, denoted $\hat{\mathbf{h}}$, by maximizing the likelihood in equation (1) over $\mathbf{h}$.

Expectation-Maximization Algorithm

One numerical solution to the problem of obtaining maximum likelihood estimates of haplotype frequencies is to make use of the expectation-maximization (EM) algorithm [7–9]. The algorithm begins by specifying initial guesses at the haplotype frequencies, $\mathbf{h}^{(0)}$, typically $1/T$ without other prior information. In each subsequent iteration, $t + 1$, of the algorithm, the haplotype frequencies, $\mathbf{h}^{(t)}$, are updated using expectation and maximization steps until the likelihood in equation (1) converges (i.e., the change in the likelihood between iterations is very small).

Expectation Step. Use the current haplotype frequencies, $\mathbf{h}^{(t)}$, to calculate the probability of resolving the genotype, G_i, into each possible consistent unordered diplotype, given by:

$$f(H_i|G_i, \mathbf{h}^{(t)}) = \frac{f(H_i|\mathbf{h}^{(t)})}{f(G_i|\mathbf{h}^{(t)})} \tag{4}$$

for the ith individual, where the terms in the numerator and denominator are given by equations (3) and (2), respectively.

Maximization Step. Update the haplotype frequencies, $\mathbf{h}^{(t+1)}$, using the expected diplotype frequencies from equation (4), such that:

$$h_j^{(t+1)} = \frac{1}{2N} \sum_i \sum_{H_i \in G_i} z_{H_{ij}} f(H_i | G_i, h^{(t)})$$

where z_H is the number of copies (0, 1 or 2) of the jth haplotype in diplotype H_i.

On convergence, we can generate a distribution of diplotypes consistent with the multi-locus genotype of each individual using the expectations in equation (4). Furthermore, we can make a "best guess" of the diplotype for the ith individual, made up of the pair of haplotypes that maximizes $f(H_i | G_i, \hat{\mathbf{h}})$.

One of the key advantages of the EM algorithm is that we can take account of missing genotype data. In the likelihood formulation, we can consider all diplotypes consistent with the observed genotype data, and at the same time, all possible genotypes at untyped SNPs. One possible limitation of the EM algorithm is the assumption of HWE, where diplotypes are formed by random pairs of haplotypes from the population (i.e., *random mating*). However, simulations have demonstrated that haplotype frequency estimates obtained from the EM algorithm are relatively robust to departures from HWE [10].

There are two substantial computational limitations of the EM algorithm. First, the algorithm may fail to converge to a global maximum likelihood solution. This problem can be solved by: (1) choosing sensible initial haplotype frequencies (such as those given by the product of the constituent allele frequencies, rather than uniform probabilities), and (2) carrying out several runs of the algorithm with different initial haplotype frequencies to check for consistency of estimates. However, more importantly, with large numbers of SNPs, there are too many possible distinct haplotypes that the computational burden of the algorithm is likely to be too great to be of practical use in many settings. In the comparison of Haplotype Reconstruction Methods Section, we discuss a number of possible solutions to this problem, and highlight important adaptations to the standard EM algorithm that have been implemented in publicly available software packages.

Bayesian Methods

Consider again the sample of unrelated individuals with which we wish to make inferences about their unobserved diplotypes, **H**, and haplotype frequencies, **h**, given their observed multi-locus genotype data, **G**. Within a Bayesian paradigm, we consider the joint posterior distribution of the unknowns, given by:

$$f(\mathbf{H}, \mathbf{h}|\mathbf{G}) \propto f(\mathbf{G}, \mathbf{H}|\mathbf{h})f(\mathbf{h})$$

In this expression, $f(\mathbf{G}, \mathbf{H}|\mathbf{h})$ denotes the likelihood of diplotypes, **H**, consistent with the observed genotype data, and $f(\mathbf{h})$ corresponds to the prior distribution of haplotype frequencies representing our initial beliefs about haplotype diversity in the population. Note that we can express the likelihood as:

$$f(\mathbf{G}, \mathbf{H}|\mathbf{h}) = \prod_i I(H_i \in G_i)f(H_i|\mathbf{h})$$

where $I(H_i \in G_i)$ is an indicator variable denoting that the diplotype H_i is consistent with genotype G_i, and $f(H_i|\mathbf{h})$ is given by equation (3).

Gibbs Sampling Algorithm

It is not possible to calculate the posterior distribution, $f(\mathbf{H},\mathbf{h}|\mathbf{G})$, directly. However, one approach to generate an approximation to this distribution is to make use of *Gibbs sampling* techniques via implementation of *Markov Chain Monte Carlo* (MCMC) methods [11]. The algorithm begins by assigning initial random values to all the unknowns (i.e. haplotype frequencies and diplotype for each individual). In each iteration, the algorithm updates these unknowns by sampling from their marginal posterior distributions, described in greater detail below. The algorithm is run for an initial burn-in period to allow convergence, assessed using standard MCMC diagnostics [11]. Subsequently, each iteration of the Gibbs sampler generates a random draw from the posterior distribution $f(\mathbf{H},\mathbf{h}|\mathbf{G})$, and thus output from the MCMC algorithm can be used to make inferences about the joint and marginal posterior distributions of the unknowns. For example, we can approximate the posterior probability of a specific diplotype for the ith individual by the proportion of outputs for which that unordered pair of haplotypes was sampled for H_i. Furthermore, we can obtain a posterior mean estimate of the jth haplotype frequency by averaging the values of h_j sampled over all iterations. In principle, output from the algorithm can also be used to approximate the uncertainty in these estimated haplotype frequencies by considering the distribution of sampled values from the MCMC output.

Prior Distribution of Population Haplotype Frequencies

The subjective component of the Bayesian paradigm is the choice of prior distribution for the unknown model parameters, here the population haplotype frequencies. Niu et al. [10] assume a *Dirichlet* prior distribution, Dir(β), commonly used to describe random probabilities that are constrained to sum to one. They have developed a Gibbs sampling algorithm, implemented in the HAPLOTYPER software, that initially assigns random population haplotype frequencies, $\mathbf{h}^{(0)}$, from the $\mathrm{Dir}(\boldsymbol{\beta})$ distribution. In each subsequent iteration, $t+1$, of the algorithm, population haplotype frequencies and individual diplotypes are updated with the following two steps:

1. Update the diplotype, $H_i^{(t+1)}$, for each individual in turn, by sampling from $f(H_i|G_i,\mathbf{h}^{(t)})$, defined in equation (4), for the current set of population haplotype frequencies, $\mathbf{h}^{(t)}$.
2. Update population haplotype frequencies, $\mathbf{h}^{(t+1)}$, by sampling from a $\mathrm{Dir}(\boldsymbol{\beta} + \mathbf{z}^{(t+1)})$ distribution, where $\mathbf{z}^{(t+1)}$ is a vector of counts of each possible distinct haplotype in the current configuration of diplotypes, $\mathbf{H}^{(t+1)}$.

By sampling from the distribution $f(H_i|G_i,\mathbf{h}^{(t)})$ under a Dirichlet prior model, greater weight is given to diplotypes containing common haplotypes than those containing rare haplotypes. The output from the HAPLOTYPER software provides a predicted "best guess" diplotype for each individual, given by the pair of haplotypes that is most often sampled in step 1 of the algorithm, in other words with maximum posterior probability $f(H_i|G_i,\mathbf{h})$. The posterior mean and standard deviation of each population haplotype frequency is also reported, given by counting the number of times it appears in the predicted "best guess" diplotype across the entire sample of individuals.

Within blocks of strong LD, we do not expect haplotypes to occur "at random". Instead, haplotypes tend to group together in similar clusters, generated as a result of the correlation between SNPs. This clustering property can be modeled by means of a *coalescent* prior probability distribution for population haplotype frequencies. Stephens et al. [12] incorporated an approximation to this prior in their PHASEv1 algorithm, which begins by assigning a random diplotype to each individual, given their multi-locus genotype data, denoted $\mathbf{H}^{(0)}$. In each subsequent iteration, $t+1$, the algorithm makes a draw from the posterior distribution of diplotypes for a randomly selected individual, assuming the diplotypes of all other individuals have been assigned correctly, as follows:

1. Select an individual, i, at random.
2. Sample a new diplotype for this individual, $H_i^{(t+1)}$, from the posterior distribution $f(H_i|G_i,H_{-i}^{(t)})$, where $H_{-i}^{(t)}$ denotes the current set of diplotypes for all other individuals.
3. Keep the diplotype fixed for all other individuals so that $H_j^{(t+1)} = H_j^{(t)}$ for all $j \neq i$.

The exact form of the conditional posterior distribution, $f(H_i|G_i,H_{-i}^{(t)})$, cannot be written down explicitly. However, Stephens and Donnelly [13] derived an approximation for the conditional probability of a future sampled haplotype, given a set of previously sampled haplotypes, as a function of SNP mutation rates. In particular:

$$f(H_i | G_i, H_{-i}^{(t)}) \propto f(H_{i1} | H_{-i}^{(t)}) f(H_{i2} | H_{i1}, H_{-i}^{(t)})$$

where H_{i1} and H_{i2} denote the ordered constituent pair of haplotypes in the diplotype H_i, consistent with the observed genotype G_i. Under this model, the diplotype is most likely to be made up of haplotypes that are the same, or similar, to those in the remainder of the sample, provided that the sample size is not too small and the SNP mutation rate is not too high.

The output of the PHASEv1 software includes a predicted "best guess" diplotype, given by the pair of haplotypes that are most often sampled, and thus maximize the posterior probability $f(H_i | G_i, \mathbf{h})$. To account for uncertainty in the predicted diplotype, the output also provides posterior probabilities that the phase call is correct at each SNP. The posterior mean and standard deviation of each population haplotype frequency is also reported, given by counting the number of times it appears in the predicted "best guess" diplotype across the entire sample of individuals, in the same way as HAPLOTYPER.

The coalescent prior model is more consistent than the Dirichlet distribution with patterns of haplotype diversity within strong blocks of LD. We would thus expect that PHASEv1 would produce more accurate estimates of haplotype frequencies and diplotype reconstruction than would HAPLOTYPER. To illustrate the differences between the two prior models, consider individuals genotyped at three proximal SNPs, with alleles coded as above, and assume that there are two common haplotypes, aBC and abc, in the population. An individual with genotype aaBbCc has two possible diplotypes: (1) aBC/abc, and (2) aBc/abC. Since the first diplotype contains the two common haplotypes, it is assigned higher conditional posterior probability than the second under either the Dirichlet or the approximate coalescent prior model. An individual with genotype AABbCc also has two possible diplotypes: (1) ABC/Abc, and (2) ABc/AbC. However, neither of these diplotypes contains common haplotypes, so HAPLOTYPER would assign equal probability to sampling either pair of haplotypes according to the Dirichlet prior model. On the other hand, the approximate coalescent prior model takes account of the fact that the first diplotype contains haplotypes that are more similar to the common haplotypes than those in the second. Specifically, two mutations are required to obtain haplotypes ABC and Abc from the common haplotypes (both at SNP A), while four are required to obtain haplotypes ABc and AbC. As a consequence, the approximate coalescent prior utilized in PHASEv1 would assign greater probability to sampling diplotype (1) than diplotype (2).

Haplotype Reconstruction over Large Genomic Regions

The likelihood-based and Bayesian methods described above are expected to generate accurate estimates of population haplotype frequencies and diplotype reconstruction over regions of limited haplotype diversity, such as blocks of strong LD. However, over large genomic regions, where samples are genotyped at large numbers of SNPs, there are likely to be too many possible haplotype configurations to make the algorithms computationally practical. Furthermore, the approximate coalescent prior distribution of haplotype frequencies utilized in the PHASEv1 software makes an implicit assumption of no recombination, and thus is inappropriate for crossing the boundaries between blocks of strong LD.

Partition Ligation

One successful solution to the problem of phasing long-range haplotypes with many SNPs is partition ligation (PL). With this approach, the region under investigation is first partitioned into short, non-overlapping segments, typically consisting of less than ten adjacent SNPs. Although the specific partitioning of the region is thought to be relatively unimportant, there may be some improvement in the accuracy of the reconstruction process by considering segments that correspond precisely to blocks of strong LD. Haplotype reconstruction can then be applied within each segment to obtain accurate population frequency estimates and diplotype resolution. In the subsequent ligation step, adjacent segments are successively combined, at each stage repeating haplotype reconstruction to the newly formed segment. Typically, to reduce the computational burden of the ligation step, only those diplotypes that can be formed from the most common haplotypes in each of the constituent segments are considered. Simulation studies have demonstrated that the more haplotypes considered in the ligation step, the more accurate the diplotype reconstruction will be, primarily because the "global" diplotype for an individual may not consist of the most likely haplotype configurations within each segment [14].

PL was first implemented in the context of haplotype reconstruction by Niu et al. [10] in the HAPLOTYPER software. Qin et al. [14] developed at PL-EM algorithm, capitalizing on the advantages of PL to overcome the computational limitations of the standard EM algorithm for haplotype reconstruction. Stephens and Donnelly [15] incorporated PL with their approximate coalescent prior model for haplotype frequencies in their PHASEv2 software. The most recent version, PHASEv2.1, incorporates recombination in the prior model [16], allowing more flexible representation of LD over larger genomic regions [17].

The SNPHAP software (www-gene.cimr.cam.ac.uk/clayton/software/snphap.txt) uses similar techniques to reconstruct haplotypes, one SNP at a time, using an EM algorithm, at each stage removing those diplotypes for each individual with low probability $f(H_i|G_i,\mathbf{h})$. By *phase culling* in this way, we remove rare haplotypes with low probabilities of appearing in the sample, although we retain rare haplotypes with high probabilities of appearing in just a few individuals in the sample, and hence reduce the computational burden of the algorithm. However, the process of culling diplotypes at the first few SNPs can fail to provide maximum-likelihood estimates of the population haplotype frequencies [14]. To overcome this issue, we could cull diplotypes only after every jth SNP, perhaps reflecting boundaries between blocks of LD, although there will be a trade-off with the computational efficiency of the algorithm as culling occurs less often.

Haplotype Clustering and Hidden Markov Models

The block-like structure of common genetic variation across the human genome results in small regions of LD within which haplotypes tend to cluster together according to their similarity. Each group can be thought of as having evolved from some common ancestral haplotype, with variation occurring as a result of rare SNP mutation events. Over larger genomic regions, this clustering of haplotypes breaks down because of ancestral recombination events which occur, more often than not, at the boundaries between LD blocks. Each observed haplotype can thus be thought of as a member of a specific cluster within a block, but that cluster membership will vary across blocks as a result of recombination. This process can be represented by means of a *hidden Markov model* (HMM), where each cluster can be thought of as a local group of closely related haplotypes, and each observed haplotype can be considered as a *mosaic* of these unobserved, or "hidden", clusters across the whole region.

According to the hidden Markov model, the allele carried by an observed haplotype, at any given SNP, depends on the "hidden" cluster to which it belongs at that specific position in the genome. Moving across the region, the likelihood of changing cluster membership is modeled by means of transition probabilities, determined by rates of recombination, in a similar way to the modeling of identity by descent in linkage analysis [18]. Scheet and Stephens [19] developed an EM algorithm, implemented in the fastPHASE software, to obtain estimates of the hidden Markov model parameters. Their model assumes either fixed, or variable, numbers of clusters, K, within each block, where large K tend to produce more accurate results, but at the expense of computational time. Conditional on these model parameter estimates, fastPHASE generates an approximation to the posterior distribution of diplotypes consistent with observed genotype data, from which a point estimate of the "best guess" haplotype reconstruction can be obtained. Browning and Browning [20] make use of a similar *localized haplotype-cluster model* in their BEAGLE software, where the number of clusters within a block adapts to the local LD structure.

Comparison of Haplotype Reconstruction Methods

Marchini et al. [21] undertook a detailed simulation study to compare a wide range of likelihood-based and Bayesian methods for haplotype reconstruction in samples of unrelated individuals. Simulations were performed using coalescent-based population genetics models [22] and high-density SNP genotype data made available through the international HapMap project [23]. The study encompassed haplotype reconstruction in samples of 90 unrelated individuals, typed for SNPs across 1 Mb regions, allowing for variation in recombination rates and missing genotype data. The most accurate software for haplotype reconstruction was PHASEv2.1, for which the proportion of individuals who were phased incorrectly was less than 6%. The other methods considered had slightly worse error rates, but required less computational effort. However, despite the computational burden of PHASEv2.1, compared to other software, it has been successfully applied in reconstructing haplotypes for more than three million SNPs across the entire genome in samples of 90 unrelated individuals as part of the international HapMap project [24].

Scheet and Stephens [19] compared fastPHASE with PHASEv2.1 (and a range of other reconstruction methods) using the same simulated data as Marchini et al. [21], as well as real autosomal genotype data from the international HapMap project [23] and X-chromosome data [25]. Their results highlighted that fastPHASE performed slightly less well than PHASEv2.1 for haplotype reconstruction, but at a fraction of the computation time. Browning and Browning [20] compared BEAGLE and fastPHASE using low- and high-density simulated data (1 SNP per ~10 kb and ~3 kb, respectively), and real data from the WTCCC [1]. Their results suggested lower error rates in haplotype reconstruction, in general, for BEAGLE than for fastPHASE, even with a large number of clusters of haplotypes ($K = 20$). Furthermore, BEAGLE requires less computation time than fastPHASE. Browning and Browning [20] demonstrated that BEAGLE could be used to phase genome-wide data at ~500,000 SNPs present on the Affymetrix SNPArray 5.0 in 3002 samples from the WTCCC in just 3.1 days using a Linux server with eight dual-core AMD Opteron 8220 SE processors and 32 GB RAM.

POPULATION-BASED HAPLOTYPE ASSOCIATION ANALYSIS

In order to undertake haplotype-based association with disease, the reconstruction of diplotypes from the unphased genotype data is just the first step in the analysis. The simplest approach to make use of reconstructed haplotypes to test for disease association is to compare frequencies between cases and controls, which would be expected to be equal under the null hypothesis of independence. However, with this simple test, we cannot take account of potential non-genetic risk factors, such as age, diet, or exposure to cigarette smoke, which may be extremely important for complex diseases. We can allow for these covariate effects in a logistic regression modeling framework, parameterized in terms of the odds of disease for each haplotype.

Logistic Regression Modeling Framework

Consider a sample of unrelated cases and controls, genotyped for M SNPs in a small genomic region. The disease phenotype of the ith individual is denoted $y_i = 1$ if affected, and $y_i = 0$ if unaffected, with additional covariate measurements denoted x_i. Let us begin by considering a scenario where the diplotypes, $\mathbf{H}$, of the sample are *known*. It is common to assume that the pair of haplotypes, H_{i1} and H_{i2}, forming the diplotype, H_i, of the ith individual, contribute independent effects to the risk of disease. Under this assumption, the logistic regression model is parameterized in terms of the *log-odds* of disease for each haplotype, so that the likelihood of the observed phenotype and diplotype data is given by:

$$f(\mathbf{y}|\mathbf{H}, \mathbf{x}, \mu, \boldsymbol{\beta}, \boldsymbol{\gamma}) = \prod_i f(y_i|H_i, \mathbf{x}_i, \mu, \boldsymbol{\beta}, \boldsymbol{\gamma})$$

where:

$$f(y_i|H_i, \mathbf{x}_i, \mu, \boldsymbol{\beta}, \boldsymbol{\gamma}) = \frac{(\exp[\mu + \beta_{H_{i1}} + \beta_{H_{i2}} + \sum_k \gamma_k x_{ik}])^{y_i}}{1 + \exp[\mu + \beta_{H_{i1}} + \beta_{H_{i2}} + \sum_k \gamma_k x_{ik}]} \tag{5}$$

In this expression, β_j denotes the log-odds ratio of the jth most frequent haplotype, relative to the baseline haplotype, usually taken to be the most common, so that we impose the constraint $\beta_1 = 0$. Furthermore, μ is the baseline log-odds of disease, and γ_k denotes the regression effect of the kth covariate. Assuming the disease to be rare, an individual carrying diplotype H_i is $\exp[\beta_{H_{i1}} + \beta_{H_{i2}}]$ times more likely to be affected than one carrying two copies of the baseline haplotype.

Under the null hypothesis of no association, the log-odds of disease for each haplotype will be the same so that $\beta = 0$. Therefore, we can construct as likelihood ratio test of association by considering the *deviance*:

$$\Lambda = 2\log\left[f\left(\mathbf{y}\middle|\mathbf{H}, \mathbf{x}, \widehat{\mu}, \widehat{\beta}, \widehat{\gamma}\right) - f\left(\mathbf{y}\middle|\mathbf{H}, \mathbf{x}, \widehat{\mu}^{(0)}, \beta = 0, \widehat{\gamma}^{(0)}\right)\right] \quad (6)$$

where $f(\mathbf{y}|\mathbf{H}, \mathbf{x}, \widehat{\mu}, \widehat{\beta}, \widehat{\gamma})$ and $f(\mathbf{y}|\mathbf{H}, \mathbf{x}, \widehat{\mu}^{(0)}, \beta = 0, \widehat{\gamma}^{(0)})$ are obtained by maximizing the likelihood in equation (6) over the sets of parameters $\{\mu, \beta, \gamma\}$ and $\{\mu, \gamma\}$ subject to the constraint $\beta = 0$, respectively. Under the null hypothesis of no association, the deviance has an approximate chi-squared distribution with $d - 1$ degrees of freedom, where d is the number of distinct haplotypes across the sample diplotypes, $\mathbf{H}$.

Analysis of Reconstructed Haplotypes

The obvious problem with the logistic regression model, above, is that the diplotypes, $\mathbf{H}$, are not generally known, but will have been estimated from SNP genotype data using the haplotype reconstruction software described above. One solution to the problem would be to take the "best guess" diplotype for each individual, and to treat it in the logistic regression model as if it were known. However, this does not take account of the uncertainty in the process of haplotype reconstruction, and effectively assumes that we have more information about the "best guess" diplotypes than we have, in reality. As a consequence, the variances of the estimated haplotype log-odds ratios will be too small, and the false positive error rate of the resulting tests of association will be inflated.

The correct approach to allow for uncertainty in the haplotype reconstruction process is to consider the distribution of diplotypes for the ith individual, $f(H_i|G_i, \mathbf{h})$, consistent with their unphased SNP genotype, and given population haplotype frequencies $\mathbf{h}$. The likelihood of the observed phenotype data, given the unphased genotype data, can then be expressed as a weighted likelihood over all possible consistent diplotypes, such that:

$$f(\mathbf{y}|\mathbf{G}, \mathbf{x}, \mu, \beta, \gamma, \mathbf{h}) = \prod_i \sum_{H_i \in G_i} f(y_i|H_i, \mathbf{x}_i, \mu, \beta, \gamma)f(H_i|G_i, \mathbf{h}) \quad (7)$$

To test the null hypothesis of no association between haplotypes and disease, we can construct a likelihood ratio test in the same way as above. In this setting, the deviance test statistic has an approximate chi-squared distribution with $d - 1$ degrees of freedom, where d is the number of distinct haplotypes consistent with the unphased SNP genotype data. However, calculation of the deviance requires maximizing the likelihood over population haplotype frequencies, in addition the logistic regression model parameters, $\{\mu,\beta,\gamma\}$, which adds an additional level of complexity to the problem.

Zaykin et al. [26] propose a two stage strategy of first obtaining the distribution of diplotypes, $f(\mathbf{H}|\mathbf{G},\mathbf{h})$, consistent with the unphased SNP genotypes by applying the standard EM algorithm to the combined sample of cases and controls, although any of the haplotype reconstruction methods described above could be employed instead. The resulting probabilities, $f(H_i|G_i,\mathbf{h})$, are then treated as fixed in the logistic regression model, and the likelihood in equation (7) is maximized only over the parameters $\{\mu,\beta,\gamma\}$. Their method has been implemented in the HTR (haplotype trend regression) software, which produces empirical p-values through permutation of case and control labels in addition to the asymptotic p-values obtained from the deviance as described above.

It is important to note that the prospective likelihood in equation (7) does not take into account ascertainment; in other words, the fact that affected individuals will be represented with greater frequency in the case–control sample than in the population overall. This over-representation of affected individuals in the case–control sample will lead to overestimation of high-risk population haplotype frequencies, and hence will introduce bias in the estimated haplotype log-odds ratios. One possible solution to the ascertainment problem would be to restrict estimation of the population haplotype frequencies to the control sample. However, this may exclude rare high-risk haplotypes that appear only among cases. The correct approach, implemented by Stram et al. [27], is to include an ascertainment correction in the likelihood in equation (7), and to maximize the likelihood over all logistic regression model parameters, and population haplotype frequencies, simultaneously. However, Stram et al. [27] have demonstrated that haplotype log-odds ratios are generally only slightly biased if we ignore ascertainment and employ a two stage strategy of diplotype reconstruction followed by association testing.

Haplotype Clustering Techniques

One potential problem with haplotype-based analyses, particularly with many SNPs over large genomic regions, is lack of parsimony. Specifically, there may be many haplotypes consistent with the unphased genotype data, some of which may be vary rare. Nevertheless, we require a parameter (i.e., log-odds ratio) for each haplotype, other than the baseline, in the logistic regression model. The odds ratios of rare haplotypes will be difficult to estimate and the resulting test of disease association may require many degrees of freedom, potentially leading to a lack of power. One possible solution to this problem would be to group together all rare haplotypes in one category, and assign the same log-odds ratio to each of these, thus reducing the number of parameters in the logistic regression model. However, the issue here is that we may combine high- and low-risk rare haplotypes together, effectively "canceling out" their effect on disease association, leading to a reduction in power overall.

A more appropriate approach to tackle the dimensionality problem is to take advantage of the expectation that similar SNP haplotypes in a small genomic region tend to share more recent common ancestry with each other, and hence are likely to share the same allele(s) at the causal variant(s) for the disease in the region. As a result, we would expect that similar haplotypes, in terms of their allelic make-up, will have compa-rable disease risks, and thus could naturally be combined in the logistic regression analysis. A number of methods have been proposed that group haplotypes according to some similarity metric, and then assign the same log-odds ratio to all haplotypes within the same cluster [28–36]. In this way, the logistic regression model can be parameterized in terms of the log-odds ratio for each *cluster*, which will be less than the number of haplotypes, and thus will require less degrees of freedom in the corre-sponding association test, without the expected loss of power arising from the grouping of haplotypes with very different disease risks.

Molitor et al. [33] and Morris [35, 36] cluster haplotypes according to a *Bayesian partition model* [37, 38]. The model is defined by specifying the number of clusters of haplotypes, κ, and the center of each cluster, C, taken from the set of distinct haplotypes consistent with the unphased SNP genotype data, without replacement. All remaining haplotypes are then assigned to the nearest cluster center according to an appropriate metric of similarity between pairs of haplotypes. Morris [35] then considers the posterior distribution of logistic regression and Bayesian partition model parameters, given by:

$$f(\mu, \beta, \gamma, \kappa, C | y, G, x, h) \propto f(y | G, x, \mu, \beta, \gamma, \kappa, C, h) f(\mu, \beta, \gamma, \kappa, C)$$

In this expression, $f(\mu, \beta, \gamma, \kappa, C)$ corresponds to the joint prior distribu-tion of model parameters, while $f(y | G, x, \mu, \beta, \gamma, \kappa, C, h)$ denotes the likeli-hood of the observed phenotype data under the logistic regression model, as given by equation (7), but with individual haplotype log-odds ratios replaced by cluster effects.

The posterior density $f(\mu, \beta, \gamma, \kappa, C | y, G, x, h)$ cannot be determined explicitly, but can be approximated by application of a *Metropolis-Hastings* MCMC algorithm [11]. The GENEBPM software begins by reconstructing haplotypes using a modified EM algorithm, which can be applied across blocks of LD using the same phase culling principles as SNPHAP, as describe above. After the subsequent application of the MCMC algorithm, the evidence in favor of association is evaluated by means of an approximation to the *Bayes' factor* [39], given by the ratio of the marginal likelihood under the null hypothesis of independence between SNP haplotypes and disease (i.e., one cluster of haplotypes, $\kappa = 1$, since all have the same disease risk) to that under the alternative (i.e., more than one cluster of haplotypes, $\kappa > 1$). Output from the MCMC algorithm can also be used to estimate posterior mean log-odds ratios for each haplotype, as well as to identify clusters of haplotypes with similar risk of disease. The results of simulations suggest that GENEBPM has greater power to detect association than single SNP tests and other haplotype-based tests, such as HTR [26], that do not allow for clustering, particularly when there are multiple causal variants in the same region.

SUMMARY

Studies of common human genetic variation have demonstrated that the genome can be broadly thought of as arranged into blocks of strong LD containing SNPs that are highly correlated with each other. Within these blocks, much of the common genetic variation can be arranged on a limited number of haplotypes, reducing the multiple testing burden compared with single-SNP tests. From a biological perspective, specific combinations of alleles occurring together on the same haplotype may directly alter the properties of a protein, and thus may have a greater effect on the phenotype than any of the individual variants, alone. However, with current genotyping technology, we do not have the necessary phase information to recover the haplotypes carried by unrelated individuals in population-based association studies.

The first step in any haplotype-based association analysis is to attempt to reconstruct the diplotypes carried by individuals using their unphased genotype data and one of a range of different software packages, which each have their own advantages and disadvantages. The PHASEv2.1 software is perhaps thought of as the "state-of-the-art" technology, since it incorporates the most realistic model of haplotype evolution, and has been demonstrated to have the smallest error rates over a variety of simulated data settings. The latest version of the software incorporates PL, allowing haplotype reconstruction over large genomic regions. Other software packages, such as fastPHASE and BEAGLE, incorporate more simplistic models of haplotype structure, and are considerably more computationally efficient, but at the expense of accuracy in the reconstructed diplotypes.

Having reconstructed haplotypes within a block of strong LD or a small genomic region, we can use simple logistic regression modeling techniques to construct multi-locus tests of association. The most appropriate tests take account of the fact that reconstructed haplotypes are just estimates of the underlying diplotypes by weighting all possible pairs consistent with the unphased genotype data, weighted by their probability, generally given by the phasing software. Even within strong blocks of strong LD, there may be considerable uncertainty in the diplotypes, leading to large numbers of consistent haplotype, many of which will be rare. To develop a parsimonious test of association, we can cluster haplotypes according to their similarity. In this way, we assume that each cluster contains closely related haplotypes which are likely to carry the same alleles at causal variants, and thus have similar disease risks. The GENEBPM software implements clustering according to a Bayesian partition model, and can be used to assess the evidence in favor of haplotype association by means of a Bayes' factor. Haplotype-based association tests, including those incorporated in the GENEBPM software, have been demonstrated to have greater power than single-SNP analyses, particularly when there are multiple causal variants within the same region, or the causal variant is rare.

Haplotype-based association tests have not generally been undertaken as part of the primary analysis of GWA studies, primarily because of computational limitations. Instead, they are being utilized, with some success, in gene-based association studies [3, 40], or in follow-up regions demonstrating "moderate" evidence of association in GWA studies in an attempt to boost the signal, or gain a further understanding of the underlying genetic architecture from the perspective of fine-mapping. However, with improvements in the computational efficiency of phasing and haplotype analysis software, genome-wide haplotype-based association studies are becoming increasingly realistic [41] and show great promise for discovering novel loci contributing effects to complex human traits.

References

[1] The Wellcome Trust Case Control Consortium, Genome-wide association study of 14,000 cases of seven common diseases and 3,000 shared controls, Nature 447 (2007) 661–678.

[2] A.G. Clark, The role of haplotypes in candidate gene studies, Genet. Epidemiol. 27 (2004) 321–333.

[3] S.V. Tavtigian, J. Simard, D.H. Teng, V. Abtin, M. Baumgard, A. Beck, et al., A candidate prostate cancer susceptibility gene at chromosome 17p, Nat. Genet. 27 (2001) 172–180.

[4] E. Zeggini, L.J. Scott, R. Saxena, B.F. Voight, J.L. Marchini, T. Hu, et al., Meta-analysis of genome-wide association data and large-scale replication identifies additional susceptibility loci for type 2 diabetes, Nat. Genet. 40 (2008) 638–645.

[5] E. Zeggini, M.N. Weedon, C.M. Lindgren, T.M. Frayling, K.S. Elliott, H. Lango, et al., Replication of genome-wide association signals in UK samples reveals risk loci for type 2 diabetes, Science 317 (2007) 1035–1036.

[6] B. Hoppe, T. Haupl, R. Gruber, H. Kiesewetter, G.R. Burmester, A. Salama, et al., Detailed analysis of the variability of peptidylarginine deiminase type 4 in German patients with rheumatoid arthritis: a case-control study, Arth. Res. Ther. 8 (2006) R34.

[7] L. Excoffier, M. Slatkin, Maximum-likelihood estimation of molecular haplotype frequencies in a diploid population, Mol. Biol. Evol. 12 (1995) 921–927.

[8] M. Hawley, K. Kidd, HAPLO: a program using the EM algorithm to estimate the frequencies of multi-site haplotypes, J. Hered. 86 (1995) 409–411.

[9] J.C. Long, R.C. Williams, M. Urbanek, An EM algorithm and testing strategy for multiple locus haplotypes, Am. J. Hum. Genet. 71 (1995) 1129–1137.

[10] T. Niu, Z.S. Qin, X. Xu, J.S. Liu, Bayesian haplotype inference for multiple linked single-nucleotide polymorphisms, Am. J. Hum. Genet. 70 (2002) 157–169.

[11] W.R. Gilks, S. Richardson, D.J. Spiegelhalter, in: Markov Chain Monte Carlo in Practice, Chapman and Hall, London, 1996.

[12] M. Stephens, N.J. Smith, P. Donnelly, A new statistical method for haplotype reconstruction from population data, Am. J.Hum. Genet. 68 (2001) 978–989.

[13] M. Stephens, P. Donnelly, Inference in molecular population genetics, J.R. Stat. Soc. Series B. 62 (2000) 605–655.

[14] Z.S. Qin, T. Niu, J.S. Liu, Partition-ligation expectation-maximisation for haplotype inference with single nucleotide polymorphisms, Am. J. Hum. Genet. 71 (2002) 1242–1247.

[15] M. Stephens, P. Donnelly, A comparison of Bayesian methods for haplotype reconstruction from population genetic data, Am. J. Hum. Genet. 73 (2003) 1162–1169.

[16] N. Li, M. Stephens, Modeling linkage disequilibrium and identifying recombination hotspots using single nucleotide polymorphism data, Genetics 165 (2003) 2213–2233.

[17] M. Stephens, P. Scheet, Accounting for decay of linkage disequilibrium in haplotype inference and missing-data imputation, Am. J. Hum. Genet. 76 (2005) 449–462.

[18] E.S. Lander, P. Green, Construction of multilocus genetic maps in humans, Proc. Nat. Acad. Sci. 84 (1987) 2363–2367.

[19] P. Scheet, M. Stephens, A fast and flexible statistical model for large-scale population genotype data: applications to inferring missing genotypes and haplotypic phase, Am. J. Hum. Genet. 78 (2006) 629–644.

[20] S.R. Browning, B.L. Browning, Rapid and accurate phasing and missing data inference for whole genome-association studies by use of localised haplotype clustering, Am. J. Hum. Genet. 81 (2007) 1084–1097.

[21] J. Marchini, D. Culter, N. Patterson, M. Stephens, E. Eskin, E. Halperin, et al., A comparison of phasing algorithms for trios and unrelated individuals, Am. J. Hum. Genet. 78 (2006) 437–450.

[22] S.F. Schaffner, C. Foo, S. Gabriel, D. Reich, M. Daly, D. Altshuler, Calibrating a coalescent simulation of human genome sequence variation, Genet. Res. 15 (2005) 1576–1583.

[23] The International HapMap Consortium, A haplotype map of the human genome, Nature 437 (2005) 1299–1320.

[24] The International HapMap Consortium, A second generation human haplotype map of over 3.1 million SNPs, Nature 449 (2007) 851–861.

[25] S. Lin, D. Cutler, M. Zwick, A. Chakravarti, Haplotype inference in random population samples, Am. J. Hum. Genet. 71 (2002) 1129–1137.

[26] D.V. Zaykin, P.H. Westfall, S.S. Young, M.A. Karnoub, M.J. Wagner, M.G. Ehm, Testing association of statistically inferred haplotypes with discrete and continuous traits in samples of unrelated individuals, Hum. Hered. 53 (2002) 79–91.

[27] D.O. Stram, C.L. Pearce, P. Bretsky, M. Freedman, J.N. Hirschhorn, D. Altshuler, et al., Modelling and EM estimation of haplotype-specific relative risks from genotype data for a case-control study of unrelated individuals, Hum. Hered. 55 (2003) 179–190.

[28] A.R. Templeton, E. Boerwinkle, C.F. Sing, A cladistic analysis of phenotypic associations with haplotypes inferred from restriction endonuclease mapping I. Basic theory and analysis of alcohol dehydrogenase activity in Drosophilia, Genetics 117 (1987) 343–351.

[29] A.R. Templeton, C.F. Sing, A. Kessling, S. Humphries, A cladistic analysis of phenotypic associations with haplotypes inferred from restriction endonuclease mapping II. The analysis of natural populations, Genetics 120 (1988) 1145–1154.

[30] A.R. Templeton, K.A. Crandall, C.F. Sing, A cladistic analysis of phenotypic associations with haplotypes inferred from restriction endonuclease mapping and DNA sequence data III. Cladogram estimation, Genetics 132 (1992) 619–633.

[31] A.R. Templeton, C.F. Sing, A cladistic analysis of phenotypic associations with haplotypes inferred from restriction endonuclease mapping IV. Nested analyses with cladogram uncertainty and recombination, Genetics 134 (1993) 659–669.

[32] J. Molitor, P. Marjoram, D. Thomas, Application of Bayesian spatial statistical methods to the analysis of haplotype effects and gene mapping, Genet. Epidemiol. 25 (2003a) 95–105.

[33] J. Molitor, P. Marjoram, D. Thomas, Fine scale mapping of disease genes with multiple mutations via spatial clustering techniques, Am. J. Hum. Genet. 73 (2003b) 1368–1384.

[34] C. Durrant, K.T. Zondervan, L.R. Cardon, S. Hunt, P. Deloukas, A.P. Morris, Linkage disequilibrium mapping via cladistic analysis of SNP haplotypes, Am. J. Hum. Genet. 75 (2004) 35–43.

[35] A.P. Morris, Direct analysis of unphased SNP genotype data in population-based association studies via Bayesian partition modelling of haplotypes, Genet. Epidemiol. 29 (2005) 91–107.

[36] A.P. Morris, A flexible Bayesian framework for modelling haplotype association with disease allowing for dominance effects of the underlying causative variants, Am. J. Hum. Genet. 79 (2006) 679–694.

[37] L. Knorr-Held, G. Rasser, Bayesian detection of clusters and discontinuities in disease maps, Biometrics 46 (2000) 13–21.

[38] D.G.T. Denison, C.C. Holmes, Bayesian partition for estimating disease risk, Biometrics 57 (2001) 143–149.

[39] R.E. Kass, A.E. Raftery, Bayes' factors and model uncertainty, J. Am. Stat. Assoc. 90 (1995) 773–795.

[40] C.M. Drysdale, Complex promoter and coding region b2-andrenic receptor haplotypes alter receptor expression and predict in vivo responsiveness, Proc. Nat. Acad. Sci. 97 (2000) 10483–10488.

[41] B.L. Browning, S.R. Browning, Haplotypic analysis of Wellcome Trust Case Control Consortium data, Hum. Genet. 123 (2008) 273–280.

Gene-Gene Interaction and Epistasis

David M. Evans

MRC Centre for Causal Analyses in Translational Epidemiology,
Department of Social Medicine, University of Bristol, UK

OUTLINE

INTRODUCTION

There is mounting evidence suggesting that gene-gene or "epistatic" interactions play an important role in the etiology of complex traits and diseases in humans. Studies in experimental organisms such as *Saccharomyces cerevisae* and *Drosophilla melanogaster* suggest that gene-gene interactions occur commonly, may involve several loci, and may be responsible for appreciable portions of the total phenotypic variance [1−5]. With the advent of genome-wide association (GWA) studies, the field has progressed quickly from a handful of discoveries at the turn of 2006, to identifying hundreds of common variants reliably associated with complex traits and disease [6].

While GWA studies have undoubtedly contributed spectacularly to identifying genetic variants involved in complex disease risk, the vast majority of genome scans have only employed single-locus tests of association that do not explicitly consider the possibility of gene-gene interactions. Attention is now turning to whether fitting statistical models which include interaction components may prove profitable in gene finding. Indeed, the existence of undetected epistasis has been offered as one reason why, for many common diseases, even though many risk disposing loci have been identified, the majority of the heritability still remains to be explained [7, 8] (for a contrary view see Hill et al. [9]). In this chapter, we define the differing and sometimes confusing meanings of the words gene-gene interaction and "epistasis". We show how the genetic variance in any system can be decomposed into single-locus and epistatic components of variance. We explain why including interaction effects in models of association might be important. Finally, we describe some of the simpler methods, challenges and issues in fitting models that contain interactions to GWA data.

WHAT IS "EPISTASIS"?

There are few terms in the scientific literature that have generated as much widespread confusion as the word "epistasis". As we shall see, this is partly due to the history of the word, but has certainly not been helped by the many conflicting and often unstated definitions of the term found in the literature since it first appeared in 1909. Epistasis can be loosely defined as the interaction between two or more genes. However, what is meant by the word "interaction" is not always clear, and can vary between biologists, statisticians, statistical geneticists and epidemiologists [10, 11].

"BIOLOGICAL" EPISTASIS

The term "epistatic" was first coined in the genetics literature by William Bateson [12] to describe a masking effect, whereby a variant or allele at one locus masks the expression of a phenotype at *another* locus. This sort of definition is analogous to the traditional meaning of the word "dominance" from Mendelian genetics, which refers to a situation where an allele masks the expression of other alleles at the *same* locus. A classic example of a biological epistatic interaction is the coat color in a Labrador retriever dog [13]. The coat color of Labradors can be black, brown or golden. Coat color is primarily controlled by two different loci — a black/brown locus (B/b) and a gold locus (E/e). The black allele is dominant to the brown allele, so dogs that are heterozygous at this locus will preferentially have a black coat color. However, expression at this locus also depends upon the dog's genotype at the gold locus. Dogs homozygous for the recessive "e" allele at the gold locus have a golden coat color regardless of their genotype at the black/brown locus. In other words, the gold locus "masks" or is epistatic to the effect at the black/brown locus. In fact, the expected ratio of a mating between two dogs that are heterozygous at both loci is 9:4:3 (black:golden:brown).

STATISTICAL EPISTASIS

In the case of quantitative traits, epistasis refers to a deviation from additivity in the effect of alleles at different loci with respect to their contribution to the quantitative phenotype. In other words, it describes the situation where the combined effect of two or more loci cannot be predicted from the sum of their individual single-locus effects. This definition of epistasis, which is often employed by statistical geneticists, was first used by Ronald Fisher in his classic 1918 paper "The correlation between relatives on the supposition of Mendelian inheritance", which unified the disparate Galtonian and Mendelian schools of genetics [14].

The total genetic variance of a quantitative trait can be partitioned into components of variance due to single-locus effects and gene-gene ("epistatic") interactions. Consider a single bi-allelic locus with alleles denoted X and x. Let the mean trait value of individuals with genotypes XX, Xx and xx be μ_{XX}, μ_{Xx} and μ_{xx}, respectively. Without loss of generality we assume that these genotypic means are scaled as deviations from the trait grand mean (i.e., the weighted average of μ_{XX}, μ_{Xx} and μ_{xx}). The degree to which μ_{Xx} deviates from the average trait value of two homozygous means (μ_{XX}, μ_{xx}) quantifies the amount of genetic dominance (i.e., interaction between alleles at the same) at the locus. It is worth pointing out that the concept of dominance is scale dependent in that systems which exhibit dominance on one scale may not necessarily exhibit dominance when transferred to a different scale of measurement. For example, the mean body mass index of individuals with the heterozygous genotype may be exactly midway between the mean of individuals who are homozygous at the same locus when measured in $\log(\mathrm{kg}/\mathrm{m}^2)$, but not when measured in "raw" kg/m^2 units.

Let the allele frequencies of X and x be p_X and p_x, respectively. Assuming Hardy-Weinberg equilibrium, the total genetic variance (V_G) at the locus can be quantified as:

$$V_G = p_X^2\mu_{XX}^2 + 2p_Xp_x\mu_{Xx}^2 + p_x^2\mu_{xx}^2 \tag{1}$$

The genetic variance at this locus can be decomposed further (see, e.g., Sham, [15]) into additive and dominance components of variance:

$$V_A = 2p_Xp_x(p_X(\mu_{XX} - \mu_{Xx}) + p_x(\mu_{Xx} - \mu_{xx}))^2 \tag{2}$$

$$V_D = p_X^2p_x^2(\mu_{XX} - 2\mu_{Xx} + \mu_{xx})^2 \tag{3}$$

The additive genetic variance is the proportion of variance explained by regressing the mean trait value for each genotype on the number of copies of the X allele (i.e., the "gene dosage"). It is equal to (twice) the amount of variance in the offspring's phenotype that can be predicted from knowledge of a single parent's trait value. In contrast, the dominance variance quantifies the residual variance that cannot be explained by the regression. Since additive and dominance components of variance are defined to be orthogonal, the dominance variance equals the total genetic variance at the locus minus the additive component. Notice also in equations (2) and (3) how the amount of variance in each component depends not only on the genotypic means, but also on the allele frequencies at each locus implying that additive and dominance components of variance are population specific.

Extending the partitioning to two loci, consider another bi-allelic locus, this time with alleles Y and y. This time there are nine genotypic means indexing the nine possible genotype combinations at the two loci (μ_{XXYY}, μ_{XXYy}, μ_{XXyy}, μ_{XxYY}, μ_{XxYy}, μ_{Xxyy}, μ_{xxYY}, μ_{xxYy}, μ_{xxyy}). Assuming again that each genotypic mean is scaled as a deviation from the overall trait mean, Hardy-Weinberg equilibrium within each locus and linkage equilibrium between the alleles at the different loci, the total genetic variance at the loci can be quantified:

$$V_G = p_X^2 p_Y^2 \mu_{XXYY}^2 + 2p_X p_x p_Y^2 \mu_{XxYY}^2 + p_x^2 p_Y^2 \mu_{xxYY}^2 + p_X^2 2p_Y p_y \mu_{XXYy}^2$$
$$+ 2p_X p_x 2p_Y p_y \mu_{XxYy}^2 + p_x^2 2p_Y p_y \mu_{xxYy}^2 + p_X^2 p_y^2 \mu_{XXyy}^2$$
$$+ 2p_X p_x p_y^2 \mu_{Xxyy}^2 + p_x^2 p_y^2 \mu_{xxyy}^2 \tag{4}$$

where p_Y and p_y are the allele frequencies at the second locus. Similar to before, the genetic variance can be partitioned into single-locus variance components specific to the first (V_X) and second loci (V_Y) by considering the marginal trait distributions at each locus. The difference between the total genetic variance and the sum of the two single-locus components is the variance due to interaction or the total epistatic variance (V_I):

$$V_I = V_G - V_X - V_Y \tag{5}$$

The epistatic variance is the residual genetic variance that cannot be accounted for by the additive and dominance variances at the individual loci. It arises because of non-linear interaction between the alleles at different loci. Like the single-locus components of variance, the epistatic variance depends upon allele frequencies at each locus, and is therefore population specific. It also depends on the scale of measurement. In fact, the epistatic variance component can itself be decomposed into components arising from interaction between additive x additive (V_{AA}), additive x dominant (V_{AD}), dominant x additive (V_{DA}) and dominant by dominant (V_{DD}) components. However, the derivation of these components is complicated and not particularly enlightening for our purposes so we do not include it here. In fact, a similar partitioning can be extended to include any number of loci, although as the number of loci increases, so does the number of terms describing the possible epistatic interactions between the different components. The interested reader is referred to any of the classic texts on quantitative genetics for a derivation of these components [16].

TWO-LOCUS QUANTITATIVE TRAIT MODELS INCORPORATING EPISTATIC INTERACTIONS

Any two locus bi-allelic system can be fully specified by $3 \times 3 = 9$ genotypic means, and if we are willing to assume Hardy-Weinberg equilibrium within loci and linkage disequilibrium between loci, the allele frequencies at both loci. There are of course an infinite number of possible two-locus models — some which involve epistatic interactions and some which do not. However, if we are prepared to restrict our attention to two-locus models which only include genotypic means of 0 or 1, it is feasible to completely enumerate all possible 50 two-locus models as has been done in Figure 12.1 [17]. Each row represents the genotype at the first locus (i.e., XX, Xx or xx), and each column represents the genotype at the second locus (i.e., YY, Yy or yy) so that each cell in the 3 × 3 box represents the trait mean for individuals who have the corresponding row and column genotypes. So, for example, the mean for all individuals in model M1 who

```
  M1        M2        M3        M5        M7        M10       M11
0 0 0     0 0 0     0 0 0     0 0 0     0 0 0     0 0 0     0 0 0
0 0 0     0 0 0     0 0 0     0 0 0     0 0 0     0 0 1     0 0 1
0 0 1     0 1 0     0 1 1     1 0 1     1 1 1     0 1 0     0 1 1

  M12       M13       M14       M15       M16       M17       M18
0 0 0     0 0 0     0 0 0     0 0 0     0 0 0     0 0 0     0 0 0
0 0 1     0 0 1     0 0 1     0 0 1     0 1 0     0 1 0     0 1 0
1 0 0     1 0 1     1 1 0     1 1 1     0 0 0     0 0 1     0 1 0

  M19       M21       M23       M26       M27       M28       M29
0 0 0     0 0 0     0 0 0     0 0 0     0 0 0     0 0 0     0 0 0
0 1 0     0 1 0     0 1 0     0 1 1     0 1 1     0 1 1     0 1 1
0 1 1     1 0 1     1 1 1     0 1 0     0 1 1     1 0 0     1 0 1

  M30       M40       M41       M42       M43       M45       M56
0 0 0     0 0 0     0 0 0     0 0 0     0 0 0     0 0 0     0 0 0
0 1 1     1 0 1     1 0 1     1 0 1     1 0 1     1 0 1     1 1 1
1 1 0     0 0 0     0 0 1     0 1 0     0 1 1     1 0 1     0 0 0

  M57       M58       M59       M61       M68       M69       M70
0 0 0     0 0 0     0 0 0     0 0 0     0 0 1     0 0 1     0 0 1
1 1 1     1 1 1     1 1 1     1 1 1     0 0 0     0 0 0     0 0 0
0 0 1     0 1 0     0 1 1     1 0 1     1 0 0     1 0 1     1 1 0

  M78       M84       M85       M86       M94       M97       M98
0 0 1     0 0 1     0 0 1     0 0 1     0 0 1     0 0 1     0 0 1
0 0 1     0 1 0     0 1 0     0 1 0     0 1 1     1 0 0     1 0 0
1 1 0     1 0 0     1 0 1     1 1 0     1 1 0     0 0 1     0 1 0

 M99      M101      M106      M108      M113      M114      M170      M186
0 0 1     0 0 1     0 0 1     0 0 1     0 0 1     0 0 1     0 1 0     0 1 0
1 0 0     1 0 0     1 0 1     1 0 1     1 1 0     1 1 0     1 0 1     1 1 1
0 1 1     1 0 1     0 1 0     1 0 0     0 0 1     0 1 0     0 1 0     0 1 0
```

FIGURE 12.1 Fifty bi-allelic two-locus models that incorporate varying degrees of epistasis. Each row indexes the genotype at the first locus and each column refers to the genotype at the second locus. The number within each cell is the trait mean for that two-locus genotype combination. Models range from the very simple (e.g., M1) to the more exotic (e.g., M170) Figure taken from D.M. Evans, J. Marchini, A.P. Morris, L.R. Cardon, Two-stage two-locus models in genome-wide association, PLoS. Genet. 2 (2006) e157

have the genotype "*xxyy*" is one, whereas everyone else has a trait mean of zero.

Since each genotype can have a trait mean of 0 or 1 there are $2^9 = 512$ possible models; however, many of these are redundant because they are simple permutations of one of the others (hence 50 models in total). For example, model M1, in which all individuals have a genotypic mean of zero except for the individual with the genotype *xxyy*, is essentially the same as any model in which all individuals have a genotypic mean of zero except for individuals in one of the single double-homozygote classes. The models range from situations where there is an effect at one locus only (M7, M56), to models which incorporate quite simple interactions (e.g., M1), to exotic-looking models where there are complicated interactions between the different loci (e.g., M170).

Using equations (1), (4) and (5) it is possible to calculate the single-locus and epistatic interaction components of variance for each model across the complete range of possible allele frequencies. This has been done in Figure 12.2 for the simple-looking model M1. The horizontal axes on the graphs represent allele frequencies at the first and second loci, whereas the vertical axes give the total genetic variance due to each component. The figure shows that even for simple-looking models, there are regions of the parameter space where appreciable proportions of the genetic variance reside in the epistatic component. Figure 12.3 displays the proportion of the total genetic variance in the epistatic variance component for eight models from Figure 12.1, ranging from simple-looking models to more exotic-looking ones. It is interesting to note that for many models, there exist sizeable regions of the possible parameter space of allelic frequencies where the majority of variation resides in the epistatic variance component. In these situations, the contributing loci will not be able to be detected by single-locus tests of association. Rather a model which explicitly includes epistatic interactions will need to be fitted to the data.

TEST FOR ASSOCIATION INCORPORATING INTERACTIONS

In the case of two bi-allelic loci, single-locus effects and the (epistatic) interactions between them can be represented mathematically using the following linear model:

$$y = \mu + a_1 x_1 + a_2 x_2 + d_1 z_1 + d_2 z_2 + i_{aa} x_1 x_2 + i_{ad} x_1 z_2 + i_{da} x_2 z_1 + i_{dd} z_1 z_2 \quad (6)$$

where y is the quantitative trait of interest, μ is the intercept and x_1, x_2, z_2, and z_2 are dummy variables coded so as to reflect the underlying additive (x_1, x_2) and dominance effects (z_1, z_2) at the first and second loci,

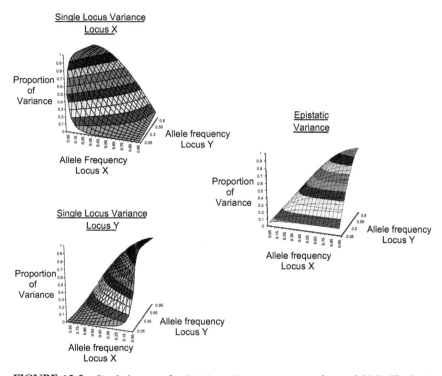

FIGURE 12.2 Single-locus and epistatic variance components for model M1. The horizontal axes on the graphs represent allele frequencies at the first and second loci, whereas the vertical axes give the proportion of the total genetic variance due to each component. Please refer to color plate section

respectively. For example, if the alleles at the first locus are denoted "X" and "x", then we might code the genotype "XX" by $x_1 = -1$ and $z_1 = -0.5$, the genotype "Xx" by $x_1 = 0$ and $z_1 = 0.5$, and the genotype "xx" by $x_1 = 1$ and $z_1 = -0.5$. The terms and x_1x_2, x_1z_2, x_2z_1 and z_1z_2 are the products of these terms, representing interaction between the additive and dominance effects at these loci. The terms a_1, a_2, d_1, d_2, i_{aa}, i_{ad}, i_{da} and i_{dd} are regression coefficients that can be estimated by, e.g., least squares, maximum likelihood, etc.

As there are nine genotypic means in a two-locus bi-allelic system, it is possible to estimate nine different parameters (the additive and dominance terms, the interaction terms and the intercept). If the focus of the investigator is solely on gene discovery, then the full model could be compared to a null model in which a single overall mean is estimated (i.e., an 8-degree-of-freedom test). This test has the advantage that both models can be fitted quickly in closed form (i.e., the maximum likelihood estimates of the parameters are simply the two-locus genotypic means in the

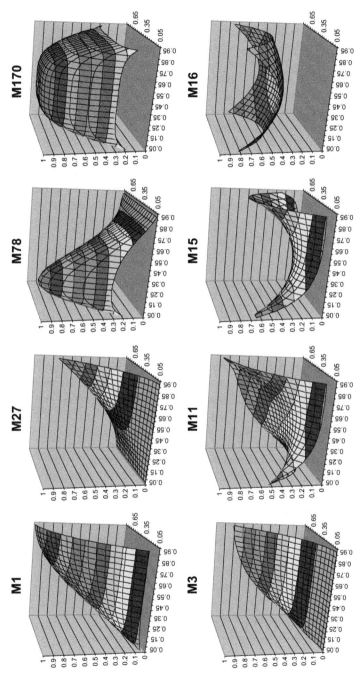

FIGURE 12.3 The proportion of the total genetic variance within the epistatic variance component for eight models from Figure 12.1. The vertical axis denotes the proportion of variance, whereas the horizontal axes denote allele frequencies at each locus. For each model, there are portions of the parameter space where much of the variance is contained within the epistatic component. Please refer to color plate section

case of the full model and the overall mean in the case of the null model) without the need for optimization and/or matrix inversion. A potential problem, however, is that if one of the SNPs already exhibits a large single-locus association, then many combinations of SNPs including this marker will exhibit low p-values solely because of the large effect at the first locus complicating interpretation. If the focus is on detecting statistical interaction between the loci, then the full model could be compared with a model that just contains single-locus effects (i.e., a 4-degree-of-freedom test). The difficulty is that the four parameter model cannot be solved in closed form and so will be more computationally demanding to compute (particularly if many of these models are fitted to the data as in a GWA study). There are many other variations on these two basic tests that are possible. For example, an investigator might be interested in testing the existence of a particular interaction term in the model (e.g., the additive $\times$ additive term). For a useful summary of some of the many options available see Cordell [18].

TWO-LOCUS BINARY MODELS INCORPORATING INTERACTION

In the case of a binary affection status from a case-control study of unrelated individuals, a similar model to equation (6) can be fitted to the data. In this case, the expected log-odds of disease are modeled as a linear function of the genotypes at the two loci using logistic regression:

$$y = \mu + a_1 x_1 + a_2 x_2 + d_1 z_1 + d_2 z_2 + i_{aa} x_1 x_2 + i_{ad} x_1 z_2 + i_{da} x_2 z_1 + i_{dd} z_1 z_2 \quad (7)$$

where y is the log-odds of disease, and the other terms in the equation are the same as in equation (6). This model implicitly assumes that the log-odds scale is the scale of interest. In fact, models with interaction effects on one scale (e.g., log-odds scale) may correspond to models with no interaction effects on another scale (e.g., a penetrance scale) illustrating again that statistical interactions are scale dependent. For example, a model that is additive on the log-odds scale will be multiplicative on the odds scale.

WHY MODEL EPISTASIS?

There are several important reasons why considering genetic interactions might be important in gene finding studies. One reason is that fitting a full model which includes epistatic effects may increase the power to detect individual loci [19, 20]. This is particularly relevant for situations where the majority of the genetic variance resides in the epistatic variance component, in which case fitting single-locus models risks missing trait loci completely (Fig. 12.3). This possibility has been touted as one reason why, despite finding many disease associated loci, for most conditions we have yet to explain the majority of the underlying heritability [7, 8]. It is unclear whether epistatic interactions contribute to an appreciable portion of this missing heritability, although it is worth noting that there is still no replicated example of epistasis underlying a complex disease/trait in humans [9]. It is also worth pointing out that in the case of individual complex disease loci, our initial inability to resolve them was not a result of undetected epistasis, but rather that the sample sizes required to detect single loci of small effect had been vastly underestimated.

Another reason commonly cited is that the discovery of a statistical interaction may have some utility for elucidating the underlying biological processes responsible for disease development. The problem with this rationale is that the same pattern of data can be produced by many different underlying models of disease pathogenesis, and thus it can be risky drawing meaningful conclusions about underlying biology on the basis of a statistical interaction without functional data [18].

STRATEGIES FOR DETECTING EPISTASIS IN GENOME-WIDE ASSOCIATION STUDIES

Detecting gene interactions in GWA data has received little attention to date. Undoubtedly part of the reason is that the spectacular success in identifying single-locus effects has diverted attention elsewhere. However, part of the explanation is likely also to have been the daunting logistical difficulties associated with searching for interactions. Perhaps the simplest strategy conceptually for detecting interactions in genome-wide data is to perform an exhaustive search of all pair-wise combinations of loci across the genome [19, 20]. In situations where loci interact, performing an exhaustive pair-wise scan can be more powerful than a single-locus scan across the genome [19, 20]. This rather surprising result is due partly to the fact that the number of individuals required to detect association at a given level of power does not scale linearly with the number of comparisons performed, but scales roughly with the log of the number of comparisons performed [19].

However, there are a number of practical problems with an exhaustive pair-wise search across the genome. The first is the computational burden associated with performing such a large number of statistical tests. For example, an exhaustive pair-wise search of 500,000 markers, would entail $^{500,000}C_2 = 1.2 \times 10^{11}$ comparisons. Conducting this number of comparisons would be prohibitive on a standard desktop computer. However, with the advent of parallel computing, tasks like this are now possible within a reasonable time frame. For example, Marchini et al. [20] quote that an exhaustive pair-wise search of 300,000 loci in 1000 cases and controls takes ~33 hours on a ten node cluster. Although most genome-wide studies now involve larger numbers of individuals and more markers than those quoted by Marchini et al., continued improvements in computing are likely to bring computation times down even further. The existence of freely available software packages that implement exhaustive pair-wise searches of epistasis will also facilitate this. For example, the PLINK software package provides the epistasis option which fits an additive model for both the main effects and their interactions to case—control data [21].

A second practical issue is that storage of all these results generated by such a scan is prohibitive in itself. However, this issue can be overcome relatively easily by the investigator discarding comparisons that do not meet some pre-specified level of significance and/or are unlikely to represent interesting effects.

Finally, there is the thorny issue of multiple testing, and its effect upon power. For example, if one were to correct an exhaustive pair-wise scan of 500,000 markers using a simple Bonferroni adjustment, then a significance level of 4×10^{-13} would be required to keep the family-wise error rate at 0.05. At this level, one would require in excess of 7000 individuals to detect two loci jointly responsible for 1% of the phenotypic variance with 90% power. While this is clearly not impossible, it would be difficult given the size of most existing cohorts with genome-wide data and given that most complex trait loci discovered so far are probably responsible for considerably less than 1% of the total phenotypic variance. Part of the success in identifying individual single-locus effects has been due to the formation of large international consortia that combine the results from several GWA studies through meta-analysis of summary statistics or counts of alleles/genotypes. Employing the same strategy in the case of tests of interaction will be difficult given the huge number of possible comparisons and the large size of results files. One solution might involve transfer of the raw genotypic data from each of the individual scans to one centralized analysis center, although this may not be possible given strict privacy regulations governing the individual cohorts. Until these logistical problems are overcome, individual studies may find it difficult detecting interacting loci/epistatic interactions on their own.

TWO-STAGE STRATEGIES TO DETECT EPISTASIS

In order to reduce the computational and multiple testing burden associated with an exhaustive pair-wise scan across the genome, a number of two-stage procedures have been proposed [19, 20, 22–24]. The basic idea is that in the first stage a number of loci are identified, and then these loci are taken forward to a second stage where the possibility of interaction between them is investigated. One obvious method of selecting loci is to evaluate their performance using a single-locus test of association first. Loci that meet some (reduced) pre-specified threshold are then followed up using a two-locus test of association. For example, one could perform pair-wise tests between all loci that met the first-stage threshold, or perhaps test all markers in the genome pair-wise with those that met this first-stage threshold [19]. The potential advantage of this two-stage procedure is that the total number of comparisons and hence the computational and multiple testing burden might be reduced. A problem, however, is that some loci which are truly associated with disease might be inadvertently discarded through the first-stage screening procedure. This is particularly true for loci that interact epistatically but exhibit no main effect, although it is unclear how common these situations might be in reality [19]. There is therefore an argument for selecting loci on other grounds such as biological candidacy or performance in linkage scans, although this to some extent discounts one of the major advantages of genome-wide scans, their hypothesis-free nature [23].

OTHER SIMPLE TESTS FOR GENE-GENE INTERACTION

Several other simple tests of genetic interaction have been proposed in the literature [25–27]. One test examines the correlation between loci in affected individuals [27]. These tests imply a slightly different definition of the word "interaction" than the traditional Fisherian use of the term. If alleles at different loci influence risk of disease more than the product of their individual allelic risks, then we would expect a correlation between these alleles in affected individuals only. In other words, we expect there to be "linkage disequilibrium" between loci in cases but not controls. It is then possible to perform a chi-square test of independence between genotypes (a 4-degree-of-freedom test) or alleles (a 1-degree-of-freedom test). Quite surprisingly, a case-only design like this may in fact be more powerful than a test of interaction using logistic regression analysis of a traditional case-control design [27]. In addition, the method may pick up "non-linear" interactions between the alleles that are not captured by "linear" methods. However, the test assumes that genotypes/alleles at different loci are not correlated within control individuals. Clearly this will not be the situation for nearby loci in a genome-wide scan that are often correlated because of linkage disequilibrium.

To rectify this shortcoming, Zhao et al. proposed testing the difference in correlation between loci in affected individuals versus the correlation between loci in control individuals [26]. However, the authors showed that this test was often more powerful than the test of interaction using logistic regression, although this result might be due to the fact that the logistic regression test involves 4 degrees of freedom whereas Zhao et al.'s test only involves 1 degree of freedom. The test is particularly suited to GWA studies as it is very fast to calculate. A similar test has been implemented in the PLINK software package under the fast-epistasis option [21].

HIGHER-ORDER INTERACTIONS

The majority of this review has focused on interactions involving two loci. It is also possible to fit models of association that include higher-order interaction terms (e.g., third-order terms, fourth-order terms, etc.). However, there are a number of problems associated with detecting higher-order interactions, particularly in the context of genome-wide data. The first is that the full regression model which describes the relationship between the trait of interest and the genetic loci under scrutiny will require many interaction terms and therefore potentially many degrees of freedom. This fact implies that power will be reduced unless a particular interaction term is of interest. Second, the number of possible marker combinations increases quickly. For example, while an exhaustive pair-wise scan of 500,000 markers would "only" involve $^{500,000}C_2$ comparisons, an exhaustive search of third-order interactions would involve $^{500,000}C_3$ comparisons — several orders of magnitude more. As well as the practical difficulties associated with carrying out such an enormous number of comparisons, very low p-values would be required in order to demonstrate "genome-wide significance". Since there is currently no well-replicated example of two-locus epistasis in human complex disease genetics, it seems questionable attempting to detect higher-order interactions until confirmed examples of two-locus epistasis have been demonstrated first.

MORE SOPHISTICATED APPROACHES TO MODELING AND DETECTING INTERACTIONS

In this review we have concentrated on outlining simple methods for detecting gene-gene interactions. As interest in modeling epistasis grows, more sophisticated statistical techniques are being brought to bear on the problem, particularly within the context of GWA studies. These methods include, but are not limited to, penalized regression (e.g., Park and Hastie [28]), recursive partitioning [29] and data mining approaches including multifactor dimensionality reduction [30]. The application of these methods to genetic data is still in its infancy and has the potential to reap rich rewards within the context of genome-wide association studies. For a recent overview of some of these methods, see Cordell [18].

CONCLUSIONS

This review has endeavored to give the reader a brief introduction to some of the issues surrounding modeling gene-gene interactions in genetics research. We have examined the differences between biological and statistical epistasis, and shown the reader how the genetic variance in any system can be decomposed into single-locus and multi-locus interaction components. We have outlined some simple strategies for detecting genetic interactions within GWA analysis and given the reader a flavor of some of the practical difficulties associated with testing for epistasis. The discovery of complex trait loci via gene mapping is in its infancy. Loci responsible for the majority of the heritability of many complex traits and diseases still remain to be discovered. The existence of interacting loci which contribute significantly to this missing heritability is a possibility which deserves serious consideration and subsequent empirical investigation.

Reference

[1] R.B. Brem, J.D. Storey, J. Whittle, L. Kruglyak, Genetic interactions between polymorphisms that affect gene expression in yeast, Nature 436 (2005) 701−703.
[2] R.B. Brem, L. Kruglyak, The landscape of genetic complexity across 5,700 gene expression traits in yeast, Proc. Natl. Acad. Sci. USA 102 (2005) 1572−1577.
[3] T.F. Mackay, The genetic architecture of quantitative traits, Annu. Rev. Genet. 35 (2001) 303−339.
[4] J.D. Storey, J.M. Akey, L. Kruglyak, Multiple locus linkage analysis of genomewide expression in yeast, PLoS. Biol. 3 (2005) e267.

[5] O. Carlborg, C.S. Haley, Epistasis: too often neglected in complex trait studies? Nat. Rev. Genet. 5 (2004) 618–625.

[6] M.I. McCarthy, G.R. Abecasis, L.R. Cardon, D.B. Goldstein, J. Little, J.P. Ioannidis, et al., Genome-wide association studies for complex traits: consensus, uncertainty and challenges, Nat. Rev. Genet. 9 (2008) 356–369.

[7] B. Maher, Personal genomes: the case of the missing heritability, Nature 456 (2008) 18–21.

[8] T.A. Manolio, F.S. Collins, N.J. Cox, D.B. Goldstein, L.A. Hindorff, D.J. Hunter, et al., Finding the missing heritability of complex diseases, Nature 461 (2009) 747–753.

[9] W.G. Hill, M.E. Goddard, P.M. Visscher, Data and theory point to mainly additive genetic variance for complex traits, PLoS.Genet. 4 (2008). e1000008.

[10] H.J. Cordell, Epistasis: what it means, what it doesn' t mean, and statistical methods to detect it in humans, Hum. Mol. Genet. 11 (2002) 2463–2468.

[11] P.C. Phillips, Epistasis – the essential role of gene interactions in the structure and evolution of genetic systems, Nat. Rev. Genet. 9 (2008) 855–867.

[12] W. Bateman, Mendel's Principles of Heredity, Cambridge University Press, Cambridge, 1909.

[13] R.E. Everts, J. Rothuizen, B.A. van Oost, Identification of a premature stop codon in the melanocyte-stimulating hormone receptor gene (MC1R) in Labrador and Golden retrievers with yellow coat colour, Anim. Genet. 31 (2000) 194–199.

[14] R.A. Fisher, The correlation between relatives on the supposition of Mendelian inheritance, Philosophical Transactions of the Royal Society of Edinburgh 52 (1918) 399–433.

[15] P. Sham, Statistics in human genetics, Arnold, London, 1998.

[16] M. Lynch, B. Walsh, Genetics and Analysis of Quantitative Traits, Sinauer, Sunderland, Massachussets, 1998.

[17] W. Li, J. Reich, A complete enumeration and classification of two-locus disease models, Hum. Hered. 50 (2000) 334–349.

[18] H.J. Cordell, Genome-wide association studies: detecting gene-gene interactions that underlie human diseases, Nat. Rev. Genet. 10 (2009) 392–404.

[19] D.M. Evans, J. Marchini, A.P. Morris, L.R. Cardon, Two-stage two-locus models in genome-wide association, PLoS. Genet. 2 (2006) e157.

[20] J. Marchini, P. Donnelly, L.R. Cardon, Genome-wide strategies for detecting multiple loci that influence complex diseases, Nat. Genet. 37 (2005) 413–417.

[21] S. Purcell, B. Neale, K. Todd-Brown, L. Thomas, M.A. Ferreira, D. Bender, et al., PLINK: a tool set for whole-genome association and population-based linkage analyses, Am. J. Hum. Genet. 81 (2007) 559–575.

[22] J. Gayan, A. Gonzalez-Perez, F. Bermudo, M.E. Saez, J.L. Royo, A. Quintas, et al., A method for detecting epistasis in genome-wide studies using case-control multi-locus association analysis, BMC Genomics 9 (2008) 360.

[23] C. Herold, M. Steffens, F.F. Brockschmidt, M.P. Baur, T. Becker, INTERSNP: genome-wide interaction analysis guided by a priori information, Bioinformatics 25 (2009) 3275–3281.

[24] C. Kooperberg, M. Leblanc, Increasing the power of identifying gene × gene interactions in genome-wide association studies, Genet. Epidemiol. 32 (2008) 255–263.

[25] T. Wang, G. Ho, K. Ye, H. Strickler, R.C. Elston, A partial least-square approach for modeling gene-gene and gene-environment interactions when multiple markers are genotyped, Genet. Epidemiol. 33 (2009) 6–15.

[26] J. Zhao, L. Jin, M. Xiong, Test for interaction between two unlinked loci, Am. J. Hum. Genet. 79 (2006) 831–845.

[27] Q. Yang, M.J. Khoury, F. Sun, W.D. Flanders, Case-only design to measure gene-gene interaction, Epidemiology 10 (1999) 167–170.

[28] M.Y. Park, T. Hastie, Penalized logistic regression for detecting gene interactions, Biostatistics 9 (1) (2008) 30–50.

[29] R. Culverhouse, T. Klein, W. Shannon, Detecting epistatic interactions contributing to quantitative traits, Genet. Epidemiol 27 (2) (2004) 141–152.

[30] M.D. Ritchie, L.W. Hahn, N. Roodi, L.R. Bailey, W.D. Dupont, F.F. Parl, J.H. Moore, Multifactor-dimensionality reduction reveals high-order interactions among estrogen-metabolism genes in sporadic breast cancer, Am. J. Hum. Genet. 69 (1) (2001) 138–147.

CHAPTER

13

Copy Number Variant Association Studies

Vincent Plagnol, David Clayton

Juvenile Diabetes Research Foundation/Wellcome Trust Diabetes and
Inflammation Laboratory Department of Medical Genetics,
University of Cambridge, UK

OUTLINE

INTRODUCTION

Copy number variation is a segment of DNA for which differences in
number of copies have been found by comparing the genomes of

multiple individuals. Copy number variants (CNVs) include deletions or duplications of DNA regions but also more subtle changes such as inversions or variable number tandem repeats (VNTRs), the latter category including microsatellites. There is no strict size requirement to call a polymorphism CNV but CNVs are usually larger than a dozen bases and smaller variants are typically called small inserts or deletes (indels).

Early genetic studies, prior to the sequencing of the human genome in 2001, have demonstrated that the pattern of inheritance of several Mendelian disorders could be explained by CNVs. For example, the deletion of exons in the X chromosome DMD gene has been linked to Duchenne myopathy [1]. However, such studies focused on Mendelian traits and did not aim to identify genetic variants with a small effect on disease risk (odds ratio smaller than 1.5) which are typically found when studying the genetics of complex multi-factorial disorders [2].

Even as genome-wide association (GWA) studies become a common tool to analyze the genetic component of multiple disorders, the contribution of common CNVs to disease risk remains largely unexplored. The first wave of GWA studies exclusively targeted single-nucleotide polymorphisms (SNPs) because the early versions of genome-wide genotyping arrays (for example, the Affymetrix 500K mapping array of the Illumina Infinium HumanHap550K SNP array) did not include probes designed for CNV typing. Although, as will be seen, probes designed for SNP genotyping can provide some ability to type CNVs, the requirement for unique probe sequences surrounding targeted SNPs tends to exclude many portions of the genome where CNVs are present, so that very few CNVs can be typed using these early arrays.

Recent advances, however, are providing new opportunities to directly genotype CNVs in sufficiently large groups of subjects to detect the small effects that are typical of complex multi-factorial disorders. These technological advances include the mapping of common CNVs in the human genome [3], improvement of traditional genotyping arrays (which now include non-SNP probes designed to target CNVs), and the development of comparative genomic hybridization (CGH) arrays [3]. The specificity of CGH arrays consists of comparing the DNA sample of interest to a reference DNA sample in the same assay, thereby providing an internal control that has the potential to improve genotyping accuracy. Finally, next generation DNA resequencing technologies now use longer sequencing reads and often paired reads, and these data can be informative about CNV status. These technologies still do not have sufficient throughput to screen large case–control collections and we will therefore not discuss these technologies in this chapter. However, it is likely that in the following years such techniques will revolutionize CNV genotyping and GWA studies.

THE VALUE OF CNV ASSOCIATION STUDIES

Before further discussing the analysis of CNV association studies, their motivation should be discussed. Two main aims can be distinguished:

- First, under the working assumption that many CNVs cannot adequately be "tagged" by SNPs in linkage disequilibrium (LD) with them, disease associations with CNVs will be missed unless they are typed directly in large-scale genome-wide studies.
- Second, it is to be expected that many CNVs will be tagged by SNPs and, in some regions where an association signal has been found by GWA SNP studies, the causal variant may actually be a CNV. Thus, for the fine-mapping of associations in regions known to contain CNVs, it will be important to include assays directly targeting them.

It is the first of these aims which will most concern the present discussion. Fortunately, most recent genome-wide genotyping chips include a combination of CNV and SNP probes, so that both SNP and CNV association studies can be conducted simultaneously. Alternatively, dedicated CNV genotyping arrays can be designed. For example, the Wellcome Trust Case Control Consortium (WTCCC) has initiated such a study in a collection of 19,000 samples consisting of eight cases cohorts of 2,000 samples and two control cohorts of 1,500 individuals each.

CNVs occur relatively infrequently on the genome compared to SNPs. For example, the WTCCC CNV association study, in collaboration with the Genome Structural Variation Consortium (GSV), has designed a CNV genotyping array targeting up to 10,000 CNVs, a number much lower than the 500K SNPs included in early SNP GWA studies [2]. It is also not yet known how many of these can be adequately tagged by SNPs but a previous study [4] suggests that the tagging properties of the majority of CNVs are not radically different from SNPs. Moreover, complex multi-allelic CNVs that cannot be tagged by SNPs are often located in difficult DNA regions containing a large fraction of repetitive sequences. Owing to technical limitations, these CNVs are harder to type and are more likely to be missed by GWA studies. Thus it could be argued that the yield of new disease associations solely through CNV associations is likely to be rather modest.

However, it might also be argued that there is a much higher *a priori* probability of disease association for a CNV rather than an SNP. This debate will only be resolved by experience. A Bayesian statistical analysis of significance thresholds in the context of GWA [5] studies shows that the threshold for following up on a GWA finding should depend on the *a priori* probability that the variant is indeed associated, prior to observing the genotype data. If, indeed, experience shows that CNVs are more likely to be disease associated than SNPs, then a less stringent threshold should be used to decide whether a follow-up analysis should be conducted.

GWA SNP studies aim, primarily, to identify causal associations *indirectly* for LD mapping. This means that most SNPs on the array are targeted not as putative causal variants but as "tags" or markers of underlying causal variants. It is well known that this strategy is only effective in identifying common causal variants. The number of putatively causal SNPs which could be included on a genome-wide array is rather modest, being restricted to non-synonymous SNPs. Since the rationale for including these non-synonymous SNPs is to detect association *directly*, much rarer minor allele frequencies can be considered. The addition of CNVs to genome-wide arrays extends the class of variants which can be targeted *directly*. Rare CNV variants should be included because the vast majority of these cannot be identified by the indirect LD approach. Similarly, the design of CNV genotyping arrays should maximize the coverage of complex multi-allelic CNVs that are less likely to be tagged by SNPs and need to be targeted directly.

The ongoing development of LD mapping of human CNVs generated by large-scale resequencing in large samples (for example, the 1000 Genomes project) will identify CNVs with high *a priori* probability of functional effect and will also determine which of these can be tagged by SNPs. This information will be critical to design future CNV GWA studies. Frequent updates in the design of the next generation of CNV GWA arrays will be needed as our knowledge of human structural variation is rapidly improving.

DIFFERENCES BETWEEN SNP AND CNV ASSOCIATION STUDIES

From an analytical point of view, the analysis of CNVs and SNPs in GWA studies has many similarities. In both cases the genotype calling is usually done separately at each locus (SNP or CNV) but jointly for all samples using clustering algorithms (typically Gaussian mixtures). A key difference, however, is the type of signal generated by both arrays, as illustrated in Figure 13.1. SNP typing GWA studies employ two types of probe designed to capture the two SNP alleles (say A and G), and as a consequence the SNP signal summary is two-dimensional. On the other hand, CNV probes capture the number of copies of a given DNA region, and the signal is one-dimensional.

This difference has practical implications on data quality and DNA sample preparation. For SNPs the information for each sample at both

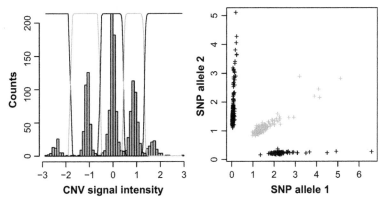

FIGURE 13.1   Comparing the distribution of signal intensity across samples for CNV and SNPs. In the CNV situation (left graph, Agilent 110K array to genotype 1,500 National Blood Services healthy donors that are part of the WTCCC CNV association study) the signal is one-dimensional. For SNP data (right graph, Illumina 550K data from the Type 1 Diabetes Genetics Consortium case-control study [6]) SNPs are genotyped using the signal from two types of probe (one per allele). Gray samples in the SNP intensity plot indicate heterozygous genotype calls

allelic probes is summarized by the ratio $\dfrac{I_A}{I_A + I_G}$ where I_A and I_G are the signal intensities for both SNP alleles. Because the probes targeting both alleles A and G are highly similar (differing by a single base) they will tend to behave in a similar manner when, for example, the DNA concentration or other experimental factors vary, thus providing an internal control. For example, it is clear from Figure 13.1 that within the AA genotype calls there is a large variability for the I_A signal but this is not an issue for the genotype calling as long as the I_G intensity remains sufficiently small. Similarly I_G and I_A intensities are strongly correlated within the heterozygous group.

In contrast, there is no equivalent internal control for CNV probes. Each probe provides a one-dimensional measurement per sample (as shown in Fig. 13.1), and for each sample the probe intensity rankings across the entire array are typically used as summary statistics for cross-sample normalization using quantile normalization [7]. Because different probes often react differently to varying experimental conditions, for example DNA concentration, these probe intensity rankings can be altered thus adding noise to the normalized data. Consequently, the requirements on DNA quality and the control of all potential sources of noise need to be more stringent for CNV GWA than SNP GWA studies. It is also more difficult to obtain CNV data that can be easily genotyped: given current technological limitations data quality is usually harder to call for CNVs than SNPs.

NORMALIZATION OF CNV INTENSITY DATA

In a similar manner to SNP GWA studies, normalization of genotyping array intensity data is necessary to enable clustering-based genotyping in large studies (as illustrated in Fig. 13.1). CNV arrays can be broadly categorized between one channel arrays, for example CNV probes on traditional SNP genotyping arrays (such as the Affymetrix 1M array of the Illumina 1M-duo beadchip) and two channel CGH arrays that also measure the hybridization of a reference DNA sample using a different dye.

One channel CNV arrays can be normalized across samples using, for example, a quantile normalization approach [7] to remove sample (chip) effects, i.e., variations in intensity affecting, to a greater or lesser degree, all probe intensities for a given sample. In quantile normalization, probe intensities are first ranked within the sample, and each intensity data point is then replaced by the appropriate quantile of the marginal distribution of probe intensities over all samples. Thus, the normalized probe intensities have an identical distribution in every sample. Other methods of normalization attempt to achieve the same aim in a variety of ways.

There is, however, more flexibility when normalizing two channel CNV CGH arrays. The log ratio of intensities of the sample of interest to those of the reference sample is typically the signal of interest in subsequent analysis. One approach to normalization is to quantile normalize the two channels separately before taking the log ratio of normalized intensities. Alternatively, quantile normalization can be applied separately for both channels before computing the log ratio of the intensities. There is no definitive argument in favor of either of the two approaches and the best normalization method may vary from one CNV to another. Ideally, several normalization approaches should be considered for each CNV although this may not always be practicable in genome-wide studies.

There is, however, one situation when CGH arrays with a clear rationale for separate normalizations of the sample channel are used. When the reference sample is homozygously deleted the reference signal is no more than background noise. Consequently, the log ratio of sample to reference raw intensity also becomes very noisy. In this situation the most efficient analysis ignores the reference channel, and takes forward the normalized sample channel intensity into subsequent analysis. An alternative to circumvent this problem is to use a pool of DNA samples for the reference sample, to minimize the probability of finding a DNA region completely absent of the reference sample. However, this may not fully address the issue when analyzing rare insertions because of the high frequency of the homozygous deleted genotype in the population.

NORMALIZATION OF SNP GENOTYPING DATA FOR CNV STUDIES

The normalization procedures described in the previous section are meant to correct for inter-sample variability. An additional refinement of normalization procedures originates from the fact that SNP probes from traditional SNP genotyping arrays can also provide information on DNA copy number. Given the wealth of data previously generated using SNP arrays, there is a strong rationale for devising methods to extract CNV data from SNP genome-wide arrays. However, because two types of probe are used for each SNP allele and the number of copies of each allele carried varies between subjects, the normalization problem becomes more difficult. This normalization procedure may be applied on a per sample basis, or across multiple samples. We now describe the principle of the normalization procedure applied by Illumina for genotyping chip data.

Figure 13.2 shows the principle of the affine transformation used by Illumina to normalize raw data from genotyping arrays. In contrast to Figure 13.1 showing data at a single SNP but for multiple individuals, Figure 13.2 shows the raw and normalized intensity data for a unique sample and 50,311 SNPs in the same sub-bead pool of the Illumina 550K genotyping platform. Illumina defines a sub-bead pool as a set of SNPs having similar properties, and these SNPs are usually clustered together. It is clear from the data in Figure 13.2 that the raw signal is asymmetric, most heterozygous SNPs having a red to green signal ratio greater than

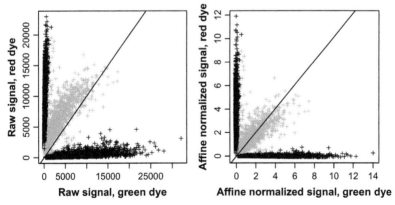

FIGURE 13.2 Affine transformation applied to 50,311 Affymetrix 500K SNPs from a unique sample (as opposed to Fig. 13.1 showing data across multiple samples but at the same locus). Heterozygous calls are shown in gray. The affine transformation (right) corrects for the dye-bias observed in the raw data (left) and brings the signal for homozygous SNPs to the X and Y axis

one. Following this affine normalization step, homozygous SNPs should be positioned along the transformed X and Y intensity axes and heterozygous SNPs on both sides of the $y = x$ line. In addition the overall intensities should be comparable for both alleles, thus allowing the use of the sum signal for both alleles $I_A + I_G$ to infer CNV status.

The Illumina affine transformation uses a method which is robust to outliers, although the details have not been published. However, studies have shown that it does not perfectly control for dye bias and additional normalization procedures have been developed [8]. Similar methods have been developed for Affymetrix arrays and are implemented in the BirdSuite software [4]. Normalization difficulties, together with the difficulties discussed below, mean that CNV data derived from SNP genotyping arrays should be treated with some caution.

EFFECT OF SNPS ON CNV PROBES

An additional complication is the potential effect of SNPs on CNV probe hybridization. CNV probes are usually relatively long and are designed to be robust to one base pair mismatch with the target sequence. Nevertheless, the presence of SNPs can sometimes be interpreted as a spurious CNV signal. Figure 13.3 shows an example of this situation: a dense tiling of the chromosome X using a custom designed Nimblegen 385K format CNV genotyping array yielded six overlapping CNV probes,

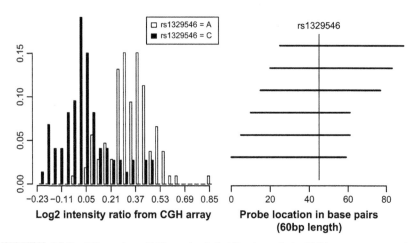

FIGURE 13.3 Effect of an SNP on the hybridization of six CNV genotyping probes located on the X chromosome for 220 male samples genotyped using the Nimblegen 385K format CNV genotyping array custom designed for dense tiling of the X chromosome. Probes including the reference allele C have better hybridization properties

all overlapping the location of the SNP rs1329546. The Nimblegen 385K array is a CGH genotyping array and the signal is summarized using the log ratio of the sample of interest by the reference sample. In this situation the distribution of the average log ratio across the six probes is significantly different among the 220 male samples in this study. The C allele was used in the probe design and samples with this allele have better hybridization properties for overlapping CNV probes. The resulting signal can easily be interpreted as a spurious CNV which is a concern for CNV discovery projects. Moreover, when testing a CNV for association, the presence of SNPs in the sequence can add noise to the data and affect data clustering.

In general, filtering out CNV probes containing SNPs is not necessary because the hybridization chemistry should be robust to such limited variability in the target sequence. Moreover, for all CNV genotyping arrays the probe design process accounts for previously reported SNPs. Therefore if a probe overlaps an SNP one expects that the effect on hybridization should be small. However, when considering specific CNVs of interest but for which the genotyping is difficult, it is useful to check whether some of the noise in the CNV signal can be removed by excluding a subset of CNV probes containing SNPs.

ASSOCIATION TESTS AND SOFTWARE FOR CNV DATA

It has been commonplace to use relatively simple methods to classify continuous signals described above into discrete CNV genotypes. This classification has been achieved by K-means clustering or even by simple binning. The data are then analyzed by comparing the distributions of CNV genotype between cases and controls using conventional contingency table methods, such as the Cochran-Armitage test [9] which, in this context, is sensitive to associations exhibiting a dose-response relationship between number of copies and disease risk. However, owing to the technical difficulty of genotyping CNVs, the distribution of signals for adjacent bins merge into each other and CNV calls are uncertain. Multi-allelic CNVs are particularly challenging since signal distributions typically become more diffuse for high numbers of DNA copies.

Failure to incorporate this uncertainty into association tests not only loses power, but can also inflate the false positive rate. A very simple method for dealing with this data is to dispense with the genotyping step and simply compare the distribution of signal intensities between cases and controls, for example using a t-test. However, this method is not robust to artifactual differences in intensities not controlled by the normalization procedure. A more robust approach is provided by formal statistical modeling using mixture models (usually Gaussian mixtures). This approach, rather than providing a single best copy number call, yields a set of probabilities for the possible copy number calls. Association tests can be incorporated into the mixture modeling, which is usually carried out using the EM algorithm. This algorithm repeats the following two-step procedure until convergence:

- **E-step:** Given current estimates of the conditional distribution of intensity signal given copy number and of the distribution of copy number given case—control status, calculate for each subject the "posterior probability" for each copy number. This then allows computation of the posterior expectation of the log-likelihood function for complete data, i.e., the log-likelihood if copy number were measured perfectly.
- **M-step:** Maximize this expected log-likelihood function with respect to the parameters of the model, yielding new estimates of the distributions to be used in the E-step of the next iteration.

In well-behaved cases, this method converges to the maximum likelihood solution although, in more complicated models, it can converge to false "local" maxima and it can be wise to run the algorithm multiple times from different starting points. Two implementations of this statistical procedure have been developed for case—control CNV association testing: CNVtools [10] and CNVassoc [11]. Note that the mixture modeling approach requires relatively large sample sizes, particularly when models are elaborated to describe complex features of the data, or when signal distributions are very diffuse, but such large collections are typically available in GWA studies.

Of course, when signals are unambiguous for copy number, very much the same answer will be obtained from contingency table analyses, t-tests of signal intensities, or mixture models. But the mixture modeling approach is preferred otherwise.

DIFFERENTIAL BIAS AND THE ADVANTAGE OF USING TRIO DATA

The fundamental problem with direct analysis of signal intensities, as briefly described above, is the phenomenon of *differential measure error*. Differences of signal intensities between case and control groups can arise for a number of reasons, such as differential DNA handling and sample collection and storage. Because these influences affect probe intensities unevenly, normalization cannot be relied upon to remove them. Such differential bias can be substantial when signals are processed into genotype calls [12], but are even more serious when signals are analyzed directly. Bias of contingency table analyses based on simple copy number calling methods can be reduced by calling copy number genotype separately for cases and controls. This allows for the distributions of signal given copy number to differ between cases and controls but, because the genotyping uncertainty is not allowed for in the subsequent contingency table analysis, the false positive rate will still be inflated. However, the formal mixture modeling approach can be extended to allow for such influences thus preserving the correct type 1 error rate, at least when the model has been correctly specified. Inevitably the power of tests suffers as a result, sometimes substantially. This is intuitive; when differences in signal distribution due to artifactual biases are allowed for, it becomes more difficult to find convincing evidence of differences due to shifts in the underlying distribution of copy number. Using this approach it is even possible to analyze copy number data comprising noisy signals derived from different platforms. The software CNVtools [10] optionally implements this extension of the mixture modeling approach.

Since protecting the association test against differential bias is costly in terms of statistical power, it would clearly be better if it were rendered unnecessary by the study design. This will often be difficult or impossible in the case–control study setting. However, cohort studies and family-based studies are much less likely to be affected by differential measurement error. Thus, when analyzing case–parent trios one can carry out an association test by using the paired t-test on the differences between the signal intensity in affected offspring to the mean of the signal intensities in his or her parents [13]. This class of CNV association test in the context of family data has been implemented in the PBAT suite of software.

SUMMARIZING SIGNAL ACROSS MULTIPLE PROBES

Because of the difficulty associated with generating CNV data, it is usually necessary to use multiple probes to capture information at the same CNV. However, clustering algorithms and mixture modeling used in the context of CNV data usually assume a one-dimensional signal. This is usually obtained as a simple mean, or trimmed mean, of the signals from different probes in the probe set for the CNV of interest. However, the information per probe is variable: some probes may perform poorly and should be excluded or at least down-weighted in the single summary signal used in the analysis.

A method to combine multiple probes into a one-dimensional measure has been proposed in Barnes et al. [10]. This approach first uses a principal component analysis to summarize the data across probes and generate an initial set of calls. These calls are then used in a second canonical correlation step [14; Chapter 4] to find the combination of probes maximally correlated to the posterior probability of genotype calls. This canonical correlation procedure can be interpreted as an approximate maximum likelihood estimation procedure, via the EM algorithm, for an extended mixture model in which the signal is multidimensional. Since this derivation has not appeared elsewhere it is set out in some detail below.

We first define notations. Let us consider n samples and for each sample the data consist of a p-dimensional signal vector, where p designates the number of probes. We assume that there are g different copy numbers. We denote by X the (n, p) matrix of probe signals and by Z the (n, g) matrix coding copy number status of the n subjects, where Z_{ij} if the copy number status of subject i is j and zero otherwise. The full statistical model specifies: (1) a model for the probability of probe signal X given copy number Z, (2) a model relating copy number, Z, to phenotype, and (3) the distribution of copy number in the population in which the study was conducted. In the simple case, X is taken as having just one column containing a single signal, assumed to be normally distributed given copy number Z. In the E-step of the EM algorithm, the expected value of Z given the data and the current best fit model is calculated; this coincides with the matrix of posterior probabilities of copy number status, the expected value of Z_{ij} being the probability that subject i has copy number status j.

Denote the ith rows of X and Z by x_i and z_i. To generalize the usual Gaussian mixture model we assume that the p-dimensional signal vector x_i for each sample is multivariate normally distributed with mean vectors which depend on copy number status, lying along a line in the p-dimensional space:

$$x_i \sim N(\mu_i, \Sigma)$$

$$\mu_i = \alpha + \beta \theta^t z_i$$

where α, β and θ are g-dimensional vectors. Note that we assume that the covariance matrix Σ of the p-dimensional signal does not depend on the genotype z_i. This model assumes that the mean signal location is located on a line parallel to β and with intercept α, and the mean signal location on this line is determined by the genotype call z_i and the vector θ. A graphical representation of this model is given in Figure 13.4. To fit this extended model it is necessary to maximize the

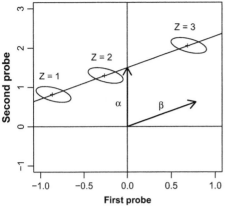

FIGURE 13.4 Schematic description of the statistical model linking two CNV probes with copy number status. The mean location of the two-dimensional signal is located along a line parallel to β and with intercept α. We assume three possible CNV genotypes ($Z = 1$, 2 or 3) and equal variance for these three copy numbers

likelihood with respect to α, β and θ together with the parameters of the model which relates copy number to phenotype. This will be carried out in the EM algorithm but we first consider the maximization of the complete data likelihood, i.e., the case where the matrix Z is known.

If the genotype matrix Z were known, the log-likelihood could be written as:

$$logL = -\frac{n}{2}(p\,log(2\pi) + log|\Sigma|) - \frac{1}{2}\sum_i (x_i - \mu_i)^t \Sigma^{-1}(x_i - \mu_i)$$

It is easy to show that given the other parameters the maximum likelihood estimate for α is $\hat{\alpha} = \bar{x} - \beta\theta^t\bar{z}$, so that substituting α in the likelihood amounts to centering the columns of the matrices X and Z. Thus, rewriting the log-likelihood using $\tilde{X}$ for the matrix with rows $(x_i - \bar{x})$ and $\tilde{Z}$ for the matrix with rows $(z_i - \bar{z})$, and omitting the first term which does not matter for this maximization, we obtain:

$$logL = \frac{1}{2}Tr\,(\tilde{X} - \tilde{Z}\theta\beta^t)\Sigma^{-1}(\tilde{X} - \tilde{Z}\theta\beta^t)^t$$

The maximization step requires constraints on β and θ for identifiability. The most convenient ones are:

$$\beta^t\Sigma^{-1}\beta = 1$$

$$\frac{1}{n}\theta^t\,\tilde{Z}^t\,\tilde{Z}\theta = 1$$

This maximization under constraints is done by adding Lagrange multipliers:

$$logL = \frac{1}{2}Tr\,(\tilde{X} - \tilde{Z}\theta\beta^t)\Sigma^{-1}(\tilde{X} - \tilde{Z}\theta\beta^t)^t - \frac{\lambda_\beta}{2}(\beta^t\Sigma^{-1}\beta - 1) - \frac{\lambda_\theta}{2}(\theta^t\,\overline{Z}^t\,\tilde{Z}\theta - n)$$

Differentiating with respect to β and θ yields the following estimating equations:

$$\Sigma^{-1}\tilde{X}^t\tilde{Z}\theta = (n + \lambda_\beta)\Sigma^{-1}\beta$$

$$\tilde{Z}\tilde{X}\Sigma^{-1}\beta = (1 + \lambda_\theta)\tilde{Z}^t\tilde{Z}\theta$$

Reorganizing these equations yields the following eigenvalue problem:

$$\tilde{X}^t\tilde{Z}(\tilde{Z}^t\tilde{Z})^\theta\tilde{Z}^t\tilde{X}\Sigma^{-1}\beta = \frac{n + \lambda_\beta}{1 + \lambda_\theta}\beta$$

where θ designates the generalized inverse operator (because the matrix $\tilde{Z}^t\tilde{Z}$ is not full rank). The multidimensional signal should then be summarized using the projection on the line $(\alpha + \beta\theta^tZ)$, but for each probe/axis of Figure 13.4 the distance to this line should be weighted by the probe variance using ϕX where $\phi = \beta^t\Sigma^{-1}$. We then have:

$$\mathbb{E}(\phi x_i) = \beta\Sigma^{-1}\alpha + (\beta^t\Sigma^{-1}\beta)\theta^t z_i = \beta\Sigma^{-1}\alpha + \theta^t z_i$$

and this expectation is directly proportional to $\theta^t z_i$. Moreover, if one estimates the covariance matrix Σ using $\hat{\Sigma} = \tilde{X}^t\tilde{X} - \tilde{X}^t\tilde{Z}(\tilde{Z}^t\tilde{Z})^\theta\tilde{Z}^t\tilde{X}$ the eigenvalue equation for this newly defined vector ϕ becomes:

$$(\tilde{X}^t\tilde{X})^{-1}\tilde{X}^t\tilde{Z}(\tilde{Z}^t\tilde{Z})^\theta\tilde{Z}^t\tilde{X}\phi = \lambda\phi$$

This is the usual canonical correlation eigenvalue problem for the matrices X and Z. Therefore, the multidimensional intensity vector X should be summarized as ϕX where ϕ is obtained using a canonical correlation analysis for the matrices X and Z.

In practice the matrix of genotype assignment Z is unknown but this procedure can be incorporated in the EM algorithm. The matrix Z is replaced with its expectation given the current parameter estimates, i.e., the probability matrix of posterior assignments. We also need the matrix $\mathbb{E}(\tilde{Z}^t\tilde{Z})$ but this matrix is unknown and in practice we approximate it using $\mathbb{E}(\tilde{Z}^t)\,\mathbb{E}(\tilde{Z})$. This assumption is accurate as long as most of the posterior probabilities are close to 0 or 1, which is likely to hold in situations where the data quality is sufficient for association testing.

In the estimation procedure suggested in Barnes et al. [10], the first principal component is a starting point for the EM algorithm. Ideally, more than one round of optimization by canonical correlations analysis should be used, the posterior probabilities being re-estimated at each step. However, in practice a unique step of the EM algorithm seems to provide good estimates for the probe weightings.

An additional difficulty arises for complex multi-allelic CNVs which usually result from multiple distinct mutation events, and different subsets of probes may be informative about different mutation events. In such a situation, the CNV signal needs to be de-convoluted, for example by separating CNV probes into subgroups. In this situation, the approach presented here will tend to down-weight all distinct subsets of probes but one, thus cleaning up the signal but at the same time preserving only a subset of the available information. To address this issue, a possible extension of this approach would maximize the data likelihood over two linear combinations of probes β_1 and β_2 instead of a unique vector β. Each linear combination of the probes would then capture information about a different CNV event.

References

[1] S.M. Forrest, et al., Preferential deletion of exons in Duchenne and Becker muscular dystrophies, Nature 329 (6140) (1987) 638–640.

[2] WTCCC, Genome-wide association study of 14,000 cases of seven common diseases and 3,000 shared controls, Nature 447 (7145) (2007) 661–678.

[3] R. Redon, et al., Global variation in copy number in the human genome, Nature 444 (7118) (2006) 444–454.

[4] J. Korn, et al., Integrated genotype calling and association analysis of SNPs, common copy number polymorphisms and rare CNVs, Nat. Genet. 40 (10) (2008) 1253–1260.

[5] S. Wacholder, et al., Assessing the probability that a positive report is false: an approach for molecular epidemiology studies, J. Nat. Cancer Institute 96 (6) (2004) 434–442.

[6] J. Barrett, et al., Genome-wide association study and meta-analysis find that over 40 loci affect risk of type 1 diabetes, Nat. Genet. 41 (6) (2009) 703–707.

[7] B.M. Bolstad, et al., A comparison of normalization methods for high density oligonucleotide array data based on variance and bias, Bioinformatics 19 (2) (2003) 185–193.

[8] J. Staaf, et al., Normalization of Illumina Infinium whole-genome SNP data improves copy number estimates and allelic intensity ratios, BMC Bioinformatics 9 (1) (2008) 409.

[9] P. Armitage, Tests for linear trends in proportions and frequencies, Biometrics 11 (3) (1955) 375–386.

[10] C. Barnes, et al., A robust statistical method for case-control association testing with copy number variation, Nat. Genet. 40 (10) (2008) 1245–1252.

[11] J. González, et al., Accounting for uncertainty when assessing association between copy number and disease: a latent class model, BMC Bioinformatics 10 (2009) 172.

[12] D. Clayton, et al., Population structure, differential bias and genomic control in a large-scale, case-control association study, Nat. Genet. 37 (11) (2005) 1243–1246.

[13] I. Ionita-Laza, et al., On the analysis of copy-number variations in genome-wide association studies: a translation of the family-based association test, Genet. Epidemiol. 32 (3) (2008) 273–284.

[14] W.J. Krzanowski, F.H.C. Marriott, Multivariate Analysis, Part 1: Distributions, Ordination and Inference (Kendall's Library of Statistics, No 1), Edward Arnold, (1994).

14

Family-based Association Methods

Jessica Lasky-Su[1]*, Amy Murphy*[1]*, Sungho Won*[2]*, Christoph Lange*[3]

[1] Department of Medicine, Channing Laboratories, Brigham and Women's Hospital and Harvard Medical School, Boston, MA, USA
[2] Department of Statistics, Chung-Ang University, Seoul, Korea
[3] Department of Biostatistics, Harvard School of Public Health, Boston, MA, USA; Institute for Genomic Mathematics, University of Bonn, Germany; German Center for Neurodegenerative Diseases (DZNE), BONN, Germany

OUTLINE

FBATS

Family-based association tests (FBATs) are an alternative study design to case—control or population-based studies that are implemented in human genetic research. The first popularized FBAT is the transmission disequilibrium test (TDT) [1], which compares the rates of the alleles that are transmitted and untransmitted to the affected offspring. The TDT is identical to the McNemar statistical test for matched studies. The success of the TDT motivated statisticians to develop methodological extensions of the test, which resulted in the development of a more generalized test. The TDT assumes an additive mode of inheritance and bi-allelic markers. The test was quickly extended to incorporate different modes of inheritance and multi-allelic marker data [2—5]. In contrast to the TDT, where parent—offspring trios are used, FBATs can apply any type of pedigree structure including missing parental data, multiple siblings, and extended pedigrees [3, 6—10, 11]. Many extensions have also been proposed to incorporate the use of quantitative traits. Extensions to quantitative traits are described [9, 12—16]. Analyses can also now be performed using copy number variants (CNVs) [17] in addition to more standard marker selections, such as single-nucleotide polymorphisms (SNPs). Another benefit of using family-based data is that it is robust to population stratification, and therefore the concerns of the control group not being representative of the case group is minimized.

The General FBAT Statistic

We now describe a general and adaptable version of the family-based association test that we refer to as the FBAT test statistic. To generate the FBAT statistic, three important characteristics of the original TDT are incorporated. First, the FBAT statistic is a conditional test, that is it conditions upon the parental genotype(s). In the case of missing parental data, the test conditions on the sufficient statistic for the genotype distribution in each family. Conditioning on the parental information enables one to bypass the need to estimate the genotype distribution of the data (e.g., the margins of the table in a case—control design) under the null hypothesis. By conditioning on the parental data we are therefore able to minimize the potential for population stratification. Second, the FBAT statistic also conditions on the offspring phenotype, making the test robust against the misspecification of the phenotypic assumptions. Third, the offspring genotype is the only random variable that is used in the FBAT statistic. The distribution of this variable under the null hypothesis can be calculated using Mendel's laws of transmission and therefore the primary requirement for a valid calculation of the FBAT statistic is Mendel's first law of transmission.

The FBAT approach compares the difference between the observed marker score and the expected marker score that is computed based on Mendelian transmissions conditional on the parental genotype(s) and the offspring phenotype. The FBAT statistic assesses the association between the phenotype and the genetic locus by the covariance between the phenotype and the Mendelian residuals. We define the covariance as:

$$U = \sum_{i,j} T_{ij}(X_{ij} - E(X_{ij}\,|S_i)) \tag{1}$$

In this equation i indexes the families and j indexes the non-founders in the family. The parameter T_{ij} denotes the coded phenotypic trait of interest (e.g., affections status) in the jth non-founder of the ith family. The offspring genotype is given by X_{ij} which is adjusted by its expected value $E(X_{ij}|S_i)$ under the null hypothesis, where S_i denotes the parental genotypes. Therefore, under the assumption of Mendelian transmission, the expected marker score $E(X_{ij}|S_i)$ is computed by using the parental genotypes to determine the expected offspring genotypes for the ith family. When parental genotypic data are missing, S_i denotes the sufficient statistics of the genetic distribution in the ith family. The "adjusted genotype", $(X_{ij} - E(X_{ij}|S_i))$, can therefore be interpreted as a Mendelian residual that measures the possible over-/under-transmission from the parents' genotypes to the offsprings' genotypes. In the case of two homozygous parents, the Mendelian residuals will always be zero and therefore these families do not contribute to the FBAT statistic. Any family with at least one Mendelian residual can be different from zero. When all of such families from a given study are added together, this number is referred to as "number of informative families". Depending on the phenotypic ascertainment condition, the coded phenotype, T_{ij}, can be centered or unadjusted. The proper selection of the coding function allows for the incorporation of qualitative, quantitative, time-to-onset and multivariate phenotypes into the FBAT statistic. Therefore, this basic equation (1) is applicable and can be extended to virtually any scenario whereby it selects the appropriate phenotypic and genotypic coding that reflects the underlying genetic model to be tested.

Large Sample Distribution of the FBAT Statistic under the Null Hypothesis

As described above, we compute the FBAT statistic distribution, U, by using the non-founder, i.e., offspring, genotype as the random variable while we condition on both the coded phenotype T_{ij}, and the sufficient statistic S_i. By definition, the expected value of the FBAT statistic is zero (i.e., $E(U) = 0$). Therefore, we must compute the variance of U conditional upon the offspring phenotype and S_i in order to normalize U under the null hypothesis. Given a univariate genotype and trait, we can write the FBAT statistic as follows:

$$Z = U/\sqrt{Var(U)} \quad \text{or} \quad X_{FBAT}^2 = U^2/Var(U)$$

where

$$Var(U) = \Sigma_i \Sigma_{j,j'} T_{ij}, T_{ij'} \, cov(X_{ij}, X_{ij'} | S_i, T_{ij}, T_{ij'}) \tag{2}$$

Similarly to the expected marker score, the covariance $cov(X_{ij}, X_{ij'} | S_i, T_{ij}, T_{ij'})$ also conditions upon the traits and the sufficient statistics when the null hypothesis is true. We now consider two different null hypotheses. Using the default null hypothesis of no association and no linkage, the covariance $cov(X_{ij}, X_{ij'} | S_i, T_{ij}, T_{ij'})$ is straightforward, as it does not depend on the phenotype $T_{ij'}$ and is computed using the Mendelian transmissions within a family. When the null hypothesis assumes no association in the presence of linkage, it is important to recognize that the transmissions to siblings within a family are correlated [11], making the derivation of the covariance difficult [6]. In this scenario, an empirical variance is used to estimate $Var(U)$ [11].

Z is asymptotically normally distributed, $N(0,1)$, and χ^2_{FBAT} has a χ^2 distribution with one degree of freedom. When multiple alleles and/or multiple traits are tested, the methodology can be extended easily, where U is a vector and $Var(U)$ is a variance/covariance matrix. The FBAT statistic then becomes a quadratic form $U^T Var(U)^- U$ and asymptotically follows a χ^2 distribution with degrees of freedom equal to the rank of $Var(U)$ [18]. When the number of families is small an exact test that does not rely on large sample theory is recommended [19].

Handling General Pedigrees and/or Missing Founders in the FBAT Approach

Based on its construction, the FBAT statistic is generalizable and may be applied to any complex pedigree, if the expected marker score,

$E(X_{ij} \mid S_i)$, and the variance/covariance structure, $\text{cov}(X_{ij}, X_{ij'} \mid S_i, T_{ij}, T_{ij'})$, can be computed. These quantities require the calculation of the marker densities $p(X_{ij} \mid S_i, T_{ij})$ and $p(X_{ij}, X_{ij'} \mid S_i, T_{ij}, T_{ij'})$, which are the individual and joint marker densities for the offspring, respectively, conditional on the parental genotypes/sufficient statistic and the offspring phenotype(s). In situations with nuclear families where both parents and one or multiple offspring are genotyped, the density $p(X_{ij} \mid S_i, T_{ij})$ is completely determinable assuming Mendel's law. Under the null hypothesis of no association and no linkage, the parental transmissions to all offspring are independent, thus the joint marker density is the product of the individual marker densities (i.e., $p(X_{ij}, X_{ij'} \mid S_i, T_{ij}, T_{ij'}) = p(X_{ij} \mid S_i, T_{ij}) * p(X_{ij'} \mid S_i, T_{ij'})$), and the computation of the expected marker score and its variance/covariance follows straightforwardly. When considering the hypothesis of no association in the presence of linkage, the transmissions from the parents to the offspring are no longer independent. The transmissions will depend on the recombination fraction, which is unknown. It is possible to remove the dependence on the recombination fraction by conditioning on the identity-by-descent patterns among offspring [6], but conditioning on identity-by-descent results in a number of uninformative families in the test statistic, which results in a substantial power loss. Consequently, it is recommended to estimate the variance/covariance structure directly by using empirical variance estimators.

The concepts for computing expected marker scores and their variance/covariance structure are also applicable to extended pedigrees in which the founder genotypes are known [6, 16]. For the analysis of extended pedigrees, the power of the FBAT statistic can be improved by computing the conditional marker distribution for the complete pedigree rather than dividing the pedigree into nuclear families and analyzing each nuclear family separately [6, 16].

For pedigrees with missing parental/founder genotypes, the computation of the expected marker scores and its variance is more involved. Rather than conditioning on the parental genotypes, the offspring genotypes distribution is calculated by conditioning on the sufficient statistics for the unobserved parental genotypes. The advantage of this approach is that no assumptions (e.g., minor allele frequency of the SNP) are needed about the unobserved parental genotypes. Making such assumptions would open the FBAT statistic to potential biases from the effects of population stratification and substructure. While the idea of conditioning on the sufficient statistic for the unobserved parental genotypes involves a number of technical details, it is fairly simple to compute the conditional distribution for the observed offspring genotypes using the algorithm developed by Rabinowitz and Laird [6].

The Between-family and Within-family Information

Besides the construction of a robust association test statistic, another important feature of the sufficient statistic is that it provides a decomposition of the genetic association into two components that are statistically independent under the null hypothesis. The first component is the so-called "between"-family component and contains information about the SNP-trait association at a population level. At the between-family level, the association is characterized by the proband's phenotype Y, and the sufficient statistics/parental genotypes. For example, when a quantitative trait is analyzed, regressing the offspring phenotype on the parental genotypes can be used to construct an estimate for the genetic effect size. This is often referred to as the conditional mean model approach. For a quantitative trait a standard linear regression can be used to model the offspring phenotype as a function of the expected genotype conditional on the sufficient statistic. Extensions of the conditional model have been developed to different trait types and ascertained samples. The second component, the so-called "within"-family component of the data, characterizes the SNP-trait association at the family level. The association at the within-family level is described by allele transmissions from the parents to their offspring, which defines the FBAT/TDT test statistic. In Van Steen-type testing strategies [20] to minimize the impact of the multiple testing problem for genome-wide association studies (GWAS), the between-component is typically used in the first step, the screening test, to identify the most promising associations. This information is then used in the second test to prioritize SNPs for FBAT testing.

Using Bayes rule, it can be shown that both components are statistically independent. The density of the distribution of the marker score X, the proband phenotype T, and the sufficient statistic/parental genotype P_1,P_2 can then be partitioned into two statistically independent components as follows: $p(X,T,P_1,P_2) = p(X \mid T,P_1,P_2) * p(Y,P_1,P_2)$. The density $p(T,P_1,P_2)$ corresponds to the first step of the testing strategy, the screening test, and the density $p(X \mid T,P_1,P_2)$ to the family-based association test that is applied in the second step, the testing step. The likelihood decomposition implies that the two steps of the testing strategy, the screening step and the testing step, are statistically independent under the null hypothesis. The information about the genetic association between each marker and the phenotype that is obtained in the screening step is then used in the testing step to maximize the statistical power of the testing strategy. It is important to note that the results of the second stage do not have to be adjusted for the effect size estimation in the first. The overall significance level will be maintained by controlling the type 1 error in the second stage. There are numerous ways in which the data from the screening step can inform the application of the FBAT statistic in the second step

[17, 20, 21]. Van Steen et al. [20] proposed to select a small number of SNPs based in the screening step, typically less than 100, and only test those SNPs in the testing step, using the FBAT statistic. The multiple comparison adjustment for less than 100 SNPs is far less stringent than the adjustment for multiple statistical tests in an analysis of all SNP genotypes in a GWAS (e.g., 500,000 and more). These results in substantial power gains of the Van Steen approach over other analysis approaches that test all genotyped SNPs can be substantial. Further advancements using the between- and within-family information are discussed in "Improving Family-based Association Analysis in a Robust Way", below.

GENETIC DATA

Specifying the Mode of Inheritance in the FBAT Statistic

When the genetic component of the FBAT statistic is SNP data, the coding of the genotype reflects the investigator-specified mode of inheritance. When the mode of inheritance is additive, X_{ij} reflects the number of target alleles, i.e., 0, 1 or 2. Under a dominant model, X_{ij} is 1 for subjects who have one or two copies of the target allele, and 0 otherwise. Under a recessive model, the recoded genotype will be 1 for subjects who carry two copies of the target allele, and 0 otherwise. For haplotypes or multi-allelic markers, X_{ij} will be defined as a vector with the element that reflects the coded genotype for each allele or haplotype.

Types of Genetic Data

FBATs have been developed for microsatellites, SNPs, and CNVs. Methods are currently under way to develop powerful family-based tests for rare allelic variants. We review these tests briefly. SNPs are the most commonly analyzed type of genetic data with FBATs. FBAT approaches have been generalized to incorporate microsatellites, and, as noted previously, haplotypes and multiple marker. However, other types of genetic data may be analyzed. Recent studies have demonstrated the importance of copy number variation [22−25] in modifying disease risk of certain diseases. Typically, CNVs are called from genotype intensities. However, for SNP chips, when standard calling algorithms are applied, calling CNVs may prove to be a difficult and error-prone process [26−28]. Because of this, the FBAT statistic developed for CNV testing within the PBAT framework is based on the probe intensity data rather than by the called CNVs [17]. Ionita-Laza et al. [17] demonstrated that the power from association tests based on genotype data and association tests based on intensity data are virtually identical. Furthermore, using intensity data may avoid bias due to incorrectly called CNV genotypes. In this analysis the intensities are treated as a continuous variable, and the FBAT statistic, in essence, is based on the covariance between probe intensity and the trait of interest. Since Mendelian transmissions cannot be identified in the intensity data, the empirical variance (as given by Lake et al. [11]) rather than the theoretical variance, is used to standardize the test statistic.

A rare variant may be defined as a single nucleotide change. It is differentiated from an SNP in that it occurs in less than 1% in the population of interest. The FBAT framework can also be used to analyze rare variants, as long as the number of families in the analysis is relatively large, such that the asymptotic properties of the test statistic hold (i.e., at least 20 informative families). If this requirement cannot be met, methodology has been developed to calculate an exact FBAT statistic [19, 29]. Using permutation tests is also a good option for rare genetic variants.

Handling Haplotypes and Multiple Markers in the FBAT Approach

When closely spaced markers are available for analysis, an approach that leverages LD between markers might be a better analysis strategy than an approach that evaluates each marker individually. In general, it is difficult to adequately adjust for between marker correlation when correcting for multiple comparisons in a single marker analysis. Thus, evaluating haplotypes/multimarkers may better incorporate the marker LD into the analysis. Additionally, by testing only a single locus at a time, the information available from other loci are not used. An optimal strategy should be to accommodate the simultaneous testing all of markers in a well-defined region of linkage disequilibrium.

For example, there are a few scenarios where haplotype or multimarker tests should be more powerful than single marker tests. If a true (and ungenotyped) disease susceptibility locus (DSL) is located in a region that is spanned by the genotyped markers, but the DSL is not in high disequilibrium with any one of the genotyped markers, it will likely be associated strongly enough to be identified by a single marker test. However, if the set of genotyped markers adequately captures the haplotype diversity in the region, a set of multiple markers may be strongly correlated with variation at the DSL. Another situation is where a haplotype or multimarker test might be preferred over a single marker test occurs when two or more of the genotyped markers are associated with the trait or phenotype under study. In contrast, if there is only a single putative locus in the region and it has been sufficiently represented or "tagged" by one of the typed markers, a haplotype or multimarker test can be less powerful than a single marker analysis.

As noted above, there are two available approaches: haplotype tests and multimarker tests. We define a haplotype as a set of alleles, which are located on the same chromosomal copy and that are inherited without recombination. If the haplotype phase (i.e., which alleles are located on the same copy of the chromosome) is known for each subject in the study, the collection of markers that identify that haplotype can be viewed as a single marker with multiple alleles. Therefore, the FBAT statistic can be computed similarly to the previously described method. Under the null hypothesis, the FBAT statistic will follow a chi-squared distribution with degrees of freedom equal to the number of markers in the analysis − 1. However, in most applications, the haplotypic phase is unknown and must be imputed. Even though it is generally easier to infer phase for family, it is not possible in determine with certainty the phase in all subjects, especially if parental genotypes are missing.

It is still possible to compute the FBAT statistic, even if the haplotype phase for subjects cannot be absolutely determined. The same principles that are used to determine the distribution of offspring genotypes when parental genotypes are missing can be applied to determining all possible haplotypes in the offspring. The haplotype distribution in offspring is computed by conditioning on both the parental genotypes/sufficient statistics and whether the phase of the haplotypes can be inferred [30]. The FBAT statistic may then be calculated by making the assumption that the marker set is a multi-allelic marker locus, whose alleles are defined by the phased haplotypes. Since this approach does not make any assumptions about population parameters (e.g., marker or haplotype frequencies, etc.) in order to infer haplotype phase, the haplotype analysis approach, like the single marker analysis, is also robust to population admixture and stratification. The FBAT statistic may be computed in a typical manner for haplotype analysis, where it may be applied to a specific target haplotype as a bi-allelic FBAT, or as a global haplotype test based on a multi-allelic FBAT. As previously noted, the presence of linkage can be accommodated by using the empirical variance estimator.

As the number of markers in a haplotype block increases, the advantages of a haplotype analysis tend to be outweighed by a few drawbacks to this analysis. As the size of the haplotype block increases, it is more difficult and numerically complex to infer phase. This is particularly true when the number of markers exceeds 5 to 10, and there are missing parental genotypes or extended pedigrees are analyzed. There also tends to be a larger proportion of rare (frequency <1%) haplotypes, which can dilute the power of the analysis. Furthermore, the assumption of non-recombination between the markers has to be carefully considered when analyzing a large span of markers.

In this situation (although this approach also applies to a smaller number of markers), a multi-marker FBAT can be an attractive alternative to a haplotype-based analysis. Rather than trying to determine the haplotype phase, multimarker FBATs directly account for between-marker LD by estimating their variance/covariance structure. To construct a multimarker FBAT, the univariate marker score X_{ij} in the standard univariate FBAT is generalized to be a vector X_{ij} whose elements are the genotypes representing each individual marker. The vector of expected marker scores, $E(X_{ij}|S_i)$, is calculated by computing each expected marker scores individually, by conditioning upon the corresponding parental genotypes/sufficient statistic. The between-marker LD is estimated by using the empirical estimator of $Var(X_{ij})$ in the calculation of $Var(U)$ in the FBAT statistic. The multimarker FBAT statistic then has a quadratic form. Under the null hypothesis, it has an asymptotic χ^2 distribution, where the degrees of freedom are given by the number of markers that are linearly independent [31]. Alternative approaches are also available [32, 33]. Extensions also exist for multimarker, multiphenotype analyses [34].

PHENOTYPES

FBATs have the great advantage of being able to use a variety of phenotypes including qualitative, quantitative, time-to-onset, and multivariate phenotypes. In this section we describe the statistical tests that have been developed to analyze this broad range of phenotypes.

Coding the Phenotype: Testing Binary Phenotypes in the FBAT Approach

Most often the initial phenotype used in a genetic analysis is affection status. When affection status is the phenotype and an additive genetic model is specified, the FBAT statistic that is equivalent to the classical TDT [16]. In this case, the only offspring information that is incorporated into the test statistic is information on affected subjects. This is done by setting $T_{ij} = 1$ for affected subjects and 0 otherwise. The FBAT test statistic can be extended to include unaffected individuals by defining T_{ij} differently.

T_{ij} is defined as $Y_{ij} - \mu$, where Y_{ij} is the original binary coding for affection status and μ is a user-defined offset parameter that can range between 0 and 1. In order to get the TDT from this definition of T_{ij}, μ is set equal to 0. In this setting, affected subjects then contribute $(1 - \mu)$ to the FBAT statistic and the unaffecteds contribute $-\mu$. By using an offset, the FBAT statistic becomes a contrast between transmissions to affected offspring, weighted by $(1- \mu)$, and unaffected offspring, weighted by μ. In population-based samples, the optimal offset choice is the prevalence of the disorder in the general population [18]. When samples are ascertained on a specific disease, this optimal offset remains approximately equal to the disease prevalence [35, 36]. There are many scenarios where the population prevalence of the disease of interest is unknown. This is not a problem, as the FBAT statistic achieves almost optimal power in a large range of values surrounding the actual population prevalence [36]. Additional screening methods have been developed for affected parent—offspring trios that further improve the power of these tests [37].

Quantitative Phenotypes

Similar to binary phenotypes, other complex phenotypes can also be incorporated into the FBAT statistic by specifying the appropriate coding function T_{ij}. As described above, the FBAT statistic is a conditional test, conditioning on both the parental genotypes and offspring phenotype. Therefore, regardless of the specification of the coding function, T_{ij}, the test will still be valid; however, selecting a poor choice for T_{ij} will reduce the overall power of the test. Therefore, care should be used in specifying T_{ij} and the choice of the coding function should be motivated by what is known without the phenotype. Quantitative traits can come in many forms, including clinical characteristics that are relevant to the disease outcome, endophenotypes of the disease, or relevant gene expression levels. We now review the FBAT statistic with quantitative phenotypes, time-to-onset phenotypes, and multivariate phenotypes.

Inherent in the phenotype, a quantitative trait contains more information by definition than the corresponding dichotomized trait [38]. Therefore it can be concluded intuitively that quantitative traits have more power than the corresponding dichotomous trait. However, quantitative traits usually depend on other non-genetic factors. If these factors are not accounted for in the analysis, the unadjusted result will be confounded with these factors and therefore the findings are likely to be diluted, thereby reducing the overall statistical power. In such cases, it is recommended to regress the phenotype of interest onto these other variables and use the residuals from this regression as the coded phenotype in the FBAT statistic. This strategy will reduce the variability in the phenotype that is attributable to non-genetic factors.

Time-to-Onset Traits

Another phenotype that is similar to affection status is a variable that measures the time-to-onset/age-at-onset of disease. Although similar to the binary coding for affection status, a time-to-onset variable will have greater statistical power than the corresponding binary variable. This is especially true for many diseases, where early onset forms of the disorder are associated with more severe disease and often have stronger genetic etiologies than later onset forms of the disease. For time-to-onset data, the standard log rank and Wilcoxon statistics were extended to use family data [39]. Further extensions of this methodology that up-weight and down-weight different periods of disease onset have also been developed [40].

Multivariate Phenotypes

Many complex diseases are diagnosed using several criteria, which may involve both qualitative and quantitative measurements. It is often important to consider all of these measurements to accurately characterize the disease. Therefore, another effective association test may include all of these phenotypes. Similarly, many diseases have multiple, relevant, correlated phenotypes, which are informative to the severity of disease. For example, measures of lung function, asthma diagnosis, and the number of asthma exacerbations are all useful measurements to assess the overall severity of an asthmatic individual. There are also cases where phenotypes cluster together, such as several related measures of lung function or repeated measurements of the same variable over time. Therefore, it may be useful to use all of these variables in one FBAT test. There are several methodologies that have been developed to analyze multiple phenotypes simultaneously, including FBAT-GEE (generalized estimating equations), FBAT-PC (principle components), and MFBAT (multivariate FBAT).

FBAT-GEE is a multivariate extension of the FBAT test that assesses the association to multiple phenotypes simultaneously [41]. While maintaining all of the advantages of the original FBAT statistic, FBAT-GEE does not require any distributional assumptions about the phenotypes, and allows phenotypes of different distributions to be used in one statistical test. A disadvantage of this methodology is that the number of degrees of freedom for the test increases for each added phenotype. Therefore, after many phenotypes are added to the analysis, the overall power of this test may not be optimal.

FBAT-PC is another multivariate test that is designed for quantitative traits that are measured repeatedly over time [42]. On a marker-by-marker level weights are generated for each quantitative trait (i.e., each time point) using generalized principal components. From these weights, FBAT-PC amplifies the genetic effects of each measurement by constructing an overall phenotype with maximal heritability. There are several advantages of this methodology. First, the weights are generated using the between-family information, which is independent of the within-family information. Therefore the FBAT-PC phenotypes are generated independently of the subsequent FBAT statistic. Second, because one overall phenotype is always generated from multiple phenotypes, the degrees of freedom for this test are always equal to one. Therefore, with increased time points, the power of the test increases without having to pay the penalty of more degrees of freedom. The power of FBAT-PC is consistently high relative to other family-based methods as the use of repeated adds valuable information to the analysis that is not incorporated into most of the cross-sectional analyses.

MULTIVARIATE PHENOTYPES AND GENOTYPES

Building on the FBAT-GEE and FBAT-PC methodology, a new methodology, omnibus family-based association test (MFBAT), was developed that uses multiple markers and multiple phenotypes simultaneously in one statistical test [34]. This is done by using the conditional mean model to construct a set of weights that adds various FBAT statistics together to generate one final FBAT statistic. Based on these simulations, we find that MFBAT substantially outperforms other methods including haplotypic approaches and multiple tests with single SNPs and single phenotypes. MFBAT is directly applicable to cases where there are multiple SNPs, such as candidate gene studies, and multiple phenotypes, such as expression data. Similar to FBAT-GEE, MFBAT can use phenotypes of varying distributions all on one statistical test, and similar to FBAT-PC, the degrees of freedom for the test are always one, thereby resulting in a highly powered test.

TESTING STRATEGIES FOR LARGE-SCALE ASSOCIATION STUDIES

As the genetics revolution continues to move forward, GWAS with hundreds or thousands to millions of SNPs are now readily available. As such, a new set of statistical challenges has emerged. Specifically, it is now commonplace to perform millions of tests in one overall analysis, which created a multiple testing problem. In addition, many GWAS studies are too small to detect modest genetic associations, which we now realize are the norm for most complex diseases. These statistical issues have led researchers to develop and refine methodologies that address both the multiple testing issues and the problem of underpowered studies. We now review these methodological developments.

Improving Family-based Association Analysis in a Robust Way

Fulker et al. [9] showed that there are two independent components that can explain the association between phenotypes and genotypes in a family design: the between-family and within-family components. In a family design, the presence of founders can prevent the effect of population stratification on a non-founder, making the within-family component robust to the population stratification. However, the between-family component remains sensitive to population stratification and requires additional adjustment to preserve the validity of the test. Several statistical techniques have been proposed that utilize both the within- and between-family component to develop a maximally powered statistical test.

Because GWAS suffer from insufficient power due to multiple testing, several strategies have been suggested that incorporate the between-family information in a screening stage prior to actually calculating the FBAT statistic. Van Steen et al. [20] suggested selecting only a subset of SNPs to be tested by using the between-family information in the conditional mean model framework to calculate the estimated power that all SNPs have to detect a genetic association. From this information, the SNPs with the greatest power are then selected and the FBAT statistical is calculated only for that subset of SNPs. Ionita-Laza et al. [17] generalized this screening algorithm by weighting each marker according to its informativity. In both approaches, informative SNPs are selected/weighted with between-family components and then conclusive inference is conducted with FBAT.

Alternatively, we can generate the robust statistics for screening statistics with genomic controls, STRUCTURE, EIGENSTRAT in a population-based approach. While the population-based approaches assume the constant level of population admixture, we can obtain the complete robustness with rank-based p-value in family design. If we let m be the number of markers in GWAS and S be the screening statistics, the rank-based p-value for the kth most significant screening statistics can be defined as:

$$\frac{k - \delta}{m}$$

where δ is the tuning parameter and should be carefully chosen for the validity. When δ is 0.5, the combined p-value is always valid under null hypothesis but $\delta = 0.7$ is recommended for GWAS because $\delta = 0.5$ often makes the overall statistics conservative. Overall significance level is obtained with the Fisher or Liptak method by combining the rank-based p-value from S and the distribution-based p-value from FBAT. It has been empirically/analytically shown that the Liptak method is more efficient when we consider a one-tailed test. If we let $pFBAT$ be the p-value of FBAT and Φ be the cumulative function of the standardized normal distribution, the Liptak method for the marker with the kth most significant S results in the p-value as:

$$\Phi(w_{FBAT}Z_{pFBAT} + w_S Z_{(k-\delta)/m})$$

where $w_{FBAT}^2 + w_S^2 = 1$. The efficiency of the Liptak method can be sensitive to the choice of weights (w_{FBAT} and w_S), and the optimal weights can be estimated without violating the validity (Won et al. [21]).

The rank-based p-value is asymptotically equal to the distribution-based approach under absence of population stratification if the number of markers is sufficient as in GWAS, and it exceeds the Van Steen and Ionita-Laza screening approach in GWAS with respect to efficiency. In addition, it is virtually as efficient as the population-based approach if covariances between family members are correctly considered. Under the presence of population stratification, it is intuitively better to estimate the population admixture if possible, and the rank-based p-value needs to be considered after applying the screening statistics that is most reasonably fit to the conditional mean model.

Incorporating Outside Controls into the Analysis

GWAS data are now available on large numbers of control individuals through various public and private databases. Recently, a new statistical test has proposed to integrate unselected controls in the genome-wide analysis of an ascertained family-based sample [43]. The general analysis follows three basic steps. First, a family-based association test is performed. Second, the between-family information from the family data is used in conjunction with the genotypes from properly matched controls in a case–control test. Third, a combined p-value is calculated using the p-values from the first two steps. By incorporating outside control individuals for free, this approach further increases the statistical power of the test at no financial cost. Since many genetic samples have been underpowered to begin with, this is an effective strategy to use in order to improve the overall power of the genetic data that are available. Depending on the number of outside control individuals incorporated into this analysis, this strategy has power estimates that are comparable to case–control studies [43].

GENE-ENVIRONMENT INTERACTION

Recently a gene-by-environment interaction method has been developed for family-based data. Vansteelandt et al. [44] proposed a gene-by-environment interaction method that is flexible in nature, allowing SNPs or haplotypes to be modeled, with dichotomous or continuous exposure variables. This approach uses causal inference methodology to derive estimating equations that are used to generate an estimate of the main genetic effect and the gene-by-environment interaction, from which a score test for the interaction term is derived. Specifically the interaction is tested conditional on (or after removing the effect of) the main genetic effect. Therefore this test is valid whether in the presence or absence of a main genetic effect.

SOFTWARE

Several family-based association analyses are now available. Most of this software is written and developed by individuals who developed the specific statistical methods. Several of these tools have gained great popularity among other genetic analysts. Commercial versions of several programs have also become available, which has provided researchers with more formal user support. The commercial programs also have user-friendly interfaces, which facilitates the proper use of these analytic methodologies. Table 14.1 shows an overview of the most popular packages and their functions.

DISCUSSION

In this chapter we reviewed the general family-based association test and discussed the many extensions that have been made to incorporate a variety of phenotype and genotype information. Although we have discussed the several advantages of family data, there are also several features of family-based designs that make them less attractive than their population-based counterparts. One feature is that, in general, the family-based design is less powerful and more expensive than similarly sized case–control study designs. In addition, the standard FBAT only uses the within-family information, and therefore if a basic FBAT analysis is performed, a sizeable portion of genetic data will remain utilized. Several analytic approaches now incorporate the between-family information, which therefore minimizes this problem.

TABLE 14.1 Software Programs Available for Family-based Association Tests

Program	Genetic analysis capability	Phenotypic analysis capability	Special features
ALP/PDT	single marker, haplotype	binary traits, quantitative traits, ranked traits, time-to-onset	X chromosome
FBAT	single marker, haplotype, multimarker	binary traits, quantitative traits, ranked traits, time-to-onset, multivariate traits	X chromosome, permutation test
GASSOC	single marker, multimarker	binary traits	permutation test
PBAT/ Golden Helix PBAT	single marker, haplotype, multimarker, CNV	binary traits, quantitative traits, ranked traits, time-to-onset, multivariate traits, gene-environment interaction	X chromosome, permutation test, covariate adjustment, many screening methods
PLINK	single marker	binary traits, quantitative traits	parent TDT, parent-of-origin
QTDT	single marker	quantitative traits	permutation tests

Nevertheless, it is usually less cost effective to use the family-based design because three individuals need to be genotyped for one unit of analysis, in contrast to case–control data where only two individuals are genotyped and population-based samples where one person is genotyped. In addition, FBATs have considerable sensitivity to genotyping errors [45, 46]. Because the test distribution is highly dependent on the parental genotypes being correct, genotyping errors can lead to false inferences.

Up until now we have discussed analyzing genotype data using SNPs or a small number of SNPs in one statistical analysis. One area of future research is to develop a systems biology approach to analyzing family data. With such an approach larger groups of SNPs/genes may be analyzed simultaneously. This has an advantage of evaluating SNPs in entire biological systems simultaneously, which is more a realistic view of how the biology operates. Also, modeling a large number of SNPs in addition to environmental factors will naturally incorporate both gene-environment (and gene-drug) interactions and gene-gene interactions into the analysis, both of which are believed to have important contributions to the etiology of most complex disorders.

References

[1] R.S. Spielman, R.E. McGinnis, W.J. Ewens, Transmission test for linkage disequilibrium: the insulin gene region and insulin-dependent diabetes mellitus (IDDM), Am. J. Hum. Genet. 52 (1993) 506–516.

[2] P.C. Sham, D. Curtis, An extended transmission/disequilibrium test (TDT) for multi-allele marker loci, Ann. Hum. Genet. 59 (Pt 3) (1995) 323–336.

[3] D. Curtis, P.C. Sham, A note on the application of the transmission disequilibrium test when a parent is missing, Am. J. Hum. Genet. 56 (3) (1995) 811–812.

[4] H. Bickeboller, F. Clerget-Darpoux, Statistical properties of the allelic and genotypic transmission/disequilibrium test for multiallelic markers, Genet. Epidemiol. 12 (6) (1995) 865–870.

[5] R.S. Spielman, W.J. Ewens, The TDT and other family-based tests for linkage disequilibrium and association, Am. J. Hum. Genet. 59 (5) (1996) 983–989.

[6] D. Rabinowitz, N. Laird, A unified approach to adjusting association tests for population admixture with arbitrary pedigree structure and arbitrary missing marker information, Hum. Heredity. 50 (2000) 211–223.

[7] R.S. Spielman, W.J. Ewens, A sibship test for linkage in the presence of association: the sib transmission/disequilibrium test, Am. J. Hum. Genet. 62 (2) (1998) 450–458.

[8] D.J. Schaid, H. Li, Genotype relative-risks and association tests for nuclear families with missing parental data, Genet. Epidemiol. 14 (6) (1997) 1113–1118.

[9] D.W. Fulker, S.S. Cherny, P.C. Sham, J.K. Hewitt, Combined linkage and association sib-pair analysis for quantitative traits, Am. J. Hum. Genet. 64 (1) (1999) 259–267.

[10] S. Horvath, N.M. Laird, A discordant-sibship test for disequilibrium and linkage: no need for parental data, Am. J. Hum. Genet. 63 (6) (1998) 1886–1897.

[11] S.L. Lake, D. Blacker, N.M. Laird, Family-based tests of association in the presence of linkage, Am. J. Hum. Genet. 67 (6) (2000) 1515–1525.

[12] G.R. Abecasis, L.R. Cardon, W.O. Cookson, A general test of association for quantitative traits in nuclear families, Am. J. Hum. Genet. 66 (1) (2000) 279–292.

[13] D. Rabinowitz, A transmission disequilibrium test for quantitative trait loci, Hum. Hered. 47 (6) (1997) 342–350.

[14] S. Horvath, X. Xu, N.M. Laird, The family based association test method: strategies for studying general genotype-phenotype associations, Eur. J. Hum. Genet. 9 (4) (2001) 301–306.

[15] N.M. Laird, S. Horvath, X. Xu, Implementing a unified approach to family-based tests of association, Genet. Epidemiol. 19 (Suppl. 1) (2000) S36–S42.

[16] N.M. Laird, C. Lange, Family-based designs in the age of large-scale gene-association studies, Nat. Rev. Genet. 7 (5) (2006) 385–394.

[17] I. Ionita-Laza, G.H. Perry, B.A. Raby, B. Klanderman, C. Lee, N.M. Laird, et al., On the analysis of copy-number variations in genome-wide association studies: a translation of the family-based association test, Genet. Epidemiol. 32 (3) (2008) 273–284.

[18] C. Lange, N.M. Laird, On a general class of conditional tests for family-based association studies in genetics: the asymptotic distribution, the conditional power, and optimality considerations, Genet. Epidem. 23 (2002) 165–180.

[19] K. Schneiter, N. Laird, C. Corcoran, Exact family-based association tests for biallelic data, Genet. Epidemiol. 29 (3) (2005) 185–194.

[20] K. Van Steen, M.B. McQueen, A. Herbert, B. Raby, H. Lyon, D.L. Demeo, et al., Genomic screening and replication using the same data set in family-based association testing, Nat. Genet. 37 (7) (2005) 683–691.

[21] S. Won, C. Lange, On the analysis of genome-wide association studies in family-based designs: a universal, robust analysis approach combing population-based and family-based components, PLoS. Genit. 5 (11) (2009) e1000741.

[22] T.L. Yang, X.D. Chen, Y. Guo, S.F. Lei, J.T. Wang, Q. Zhou, et al., Genome-wide copy-number-variation study identified a susceptibility gene, UGT2B17, for osteoporosis, Am. J. Hum. Genet. 83 (6) (2008) 663−674.

[23] T. Vrijenhoek, J.E. Buizer-Voskamp, I. van der Stelt, E. Strengman, C. Sabatti, A. Geurts van Kessel, et al., Recurrent CNVs disrupt three candidate genes in schizophrenia patients, Am. J. Hum. Genet. 83 (4) (2008) 504−510.

[24] C.R. Marshall, A. Noor, J.B. Vincent, A.C. Lionel, L. Feuk, J. Skaug, et al., Structural variation of chromosomes in autism spectrum disorder, Am. J. Hum. Genet. 82 (2) (2008) 477−488.

[25] S. Kathiresan, B.F. Voight, S. Purcell, K. Musunuru, D. Ardissino, P.M. Mannucci, et al., Genome-wide association of early-onset myocardial infarction with single nucleotide polymorphisms and copy number variants, Nat. Genet. 41 (3) (2009) 334−341.

[26] D. Komura, F. Shen, S. Ishikawa, K.R. Fitch, W. Chen, J. Zhang, et al., Genome-wide detection of human copy number variations using high-density DNA oligonucleotide arrays, Genome Res. 216 (12) (2006) 1575−1584.

[27] R. Redon, S. Ishikawa, K.R. Fitch, L. Feuk, G.H. Perry, T.D. Andrews, et al., Global variation in copy number in the human genome, Nature 444 (7118) (2006) 444−454.

[28] J.O. Korbel, A.E. Urban, F. Grubert, J. Du, T.E. Royce, P. Starr, et al., Systematic prediction and validation of breakpoints associated with copy-number variants in the human genome, Proc. Natl. Acad. Sci. USA 104 (24) (2007) 10110−10115.

[29] K. Schneiter, J.H. Degnan, C. Corcoran, X. Xu, N. Laird, EFBAT: exact family-based association tests, BMC Genet. 8 (2007) 86.

[30] S. Horvath, X. Xu, S.L. Lake, E.K. Silverman, S.T. Weiss, N.M. Laird, Family-based tests for associating haplotypes with general phenotype data: application to asthma genetics, Genet. Epidemiol. 26 (1) (2004) 61−69.

[31] C.S. Rakovski, X. Xu, R. Lazarus, D. Blacker, N.M. Laird, A new multimarker test for family-based association studies, Genet. Epidemiol. 31 (1) (2007) 9−17.

[32] X. Xu, C. Rakovski, N. Laird, An efficient family-based association test using multiple markers, Genet. Epidemiol. 30 (7) (2006) 620−626.

[33] C.S. Rakovski, S.T. Weiss, N.M. Laird, C. Lange, FBAT-SNP-PC: an approach for multiple markers and single trait in family-based association tests, Hum. Hered. 66 (2) (2008) 122−126.

[34] J. Lasky-Su, A. Murphy, M.B. McQueen, S.T. Weiss, C. Lange, An omnibus test for family-based association studies with multiple SNPs and multiple phenotypes, Eur. J. Hum. Genet. 18 (2010) 720−725.

[35] J.C. Whittaker, C.M. Lewis, The effect of family structure on linkage tests using allelic association, Am. J. Hum. Genet. 63 (3) (1998) 889−897.

[36] C. Lange, N.M. Laird, Power calculations for a general class of family-based association tests: dichotomous traits, Am. J. Hum. Genet. 71 (3) (2002) 575−584.

[37] A. Murphy, S.T. Weiss, C. Lange, Screening and replication using the same data set: testing strategies for family-based studies in which all probands are affected, PLoS Genet. 4 (9) (2008) e1000197.

[38] C. Lange, D.L. DeMeo, N.M. Laird, Power and design considerations for a general class of family-based association tests: quantitative traits, Am. J. Hum. Genet. 71 (6) (2002) 1330−1341.

[39] C. Lange, D. Blacker, N.M. Laird, Family-based association tests for survival and times-to-onset analysis, Stat. Med. 23 (2) (2004) 179−189.

[40] H. Jiang, D. Harrington, B.A. Raby, L. Bertram, D. Blacker, S.T. Weiss, et al., Family-based association test for time-to-onset data with time-dependent differences between the hazard functions, Genet. Epidemiol. 30 (2) (2006) 124−132.

[41] C. Lange, E.K. Silverman, X. Xu, S.T. Weiss, N.M. Laird, A multivariate family-based association test using generalized estimating equations: FBAT-GEE, Biostatistics 4 (2) (2003) 195−206.

[42] C. Lange, K. van Steen, T. Andrew, H. Lyon, D.L. DeMeo, B. Raby, et al., A family-based association test for repeatedly measured quantitative traits adjusting for unknown environmental and/or polygenic effects, Stat. Appl. Genet. & Mol. Biol. 3 (1) (2004) 1–27.

[43] J. Lasky-Su, S. Won, E. Mick, R. Anney, B. Franke, B. Neale, et al., On genome-wide association studies for family-based designs: an integrative analysis approach combing ascertained family-samples with unselected controls, Am. J. Hum. Genet. 86 (4) (2010) 573–580.

[44] S. Vansteelandt, D.L. Demeo, J. Lasky-Su, J.W. Smoller, A.J. Murphy, M. McQueen, et al., Testing and estimating gene-environment interactions in family-based association studies, Biometrics 64 (2) (2008) 458–467.

[45] D. Gordon, S.C. Heath, X. Liu, J. Ott, A transmission/disequilibrium test that allows for genotyping errors in the analysis of single-nucleotide polymorphism data, Am. J. Hum. Genet. 69 (2) (2001) 371–380.

[46] D. Gordon, J. Ott, Assessment and management of single nucleotide polymorphism genotype errors in genetic association analysis, Pac. Symp. Biocomput. (2001) 18–29.

Bioinformatics Approaches

Katherine S. Elliott

Wellcome Trust Centre for Human Genetics, University of Oxford, UK

Genome-wide association (GWA) studies generate numerous potential signals. Of these some will be discarded by a range of quality control procedures such as eliminating single-nucleotide polymorphisms (SNPs) with high rates of genotyping errors (see Chapter 7), and some will be found to represent the same signal due to linkage disequilibrium (LD). Once these have been taken into account, the researcher may then be left with a long list of signals to choose from to follow up with replication studies.

PRIORITIZING ASSOCIATION SIGNALS FOR FOLLOW-UP

There are a number of tools available to assist in annotating genes underlying association (and linkage) signals. Such tools provide information on the bibliography of genes, expression patterns and information about homologous and interacting genes. However, a word of

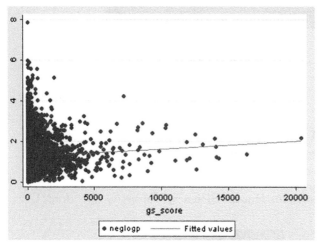

FIGURE 15.1 A plot of the negative-log (*p*-value) of the lowest *p*-value for association with type 2 diabetes per gene region plotted against GeneSniffer candidacy score for type 2 diabetes gene candidacy. The horizontal line shows there was a subtle but significant correlation (coefficient = 0.000326, $p < 10^{-3}$). Please refer to color plate section

warning: studies comparing disease candidacy scores with best disease association signals per gene for both type 2 diabetes and osteoarthritis showed only a very subtle correlation (Fig. 15.1). This indicates that biological candidacy is not a strong predictor of the location of strong association signals. However, tools that provide annotation of genes and regions surrounding SNPs provide an invaluable resource when reporting the results.

USING LINKAGE DISEQUILIBRIUM (LD) TO DEFINE REGIONS SURROUNDING SIGNALS

Of the hundreds of SNPs that have been identified by recent association studies for common human diseases, only a small fraction of these have proven to be the functional variants giving rise to disease. Once a signal has been identified, examining the regions surrounding the SNP in the context of LD is important since the most proximal gene or genes may not necessarily be the best candidate or likely contain the associated functional variant. The HapMap database (http://www.hapmap.org) provides LD information across the whole genome including the position of recombination hotspots. Given that linkage LD may extend for 100s of kb, the functional variant giving rise to the association signal may affect a gene that is some distance away and several genes removed from the signal SNP. To complicate matters further functional elements controlling gene expression may be 100ks of kb away. The SNP may sit in a region of clearly defined LD flanked by strong recombination hotspots spanned by a single gene such as the T2D signal at rs10946398 in *CDKAL1* [1]. Alternatively, the SNP may be in a gene-rich region of extensive LD where there are numerous potential candidates such as the region of LD surrounding the type 2 diabetes associated SNP rs5015480 [1] which contains two compelling candidate genes, *HHEX* adjacent to the SNP and *IDE* which is two genes away. A further scenario is the location of a signal SNP together with surrounding LD in an area not populated by any known genes. Signals have been found in these regions, dubbed "gene deserts", an example of which is the cluster of breast, colorectal and prostate cancer signals at 8q24 in a gene desert spanning over 1 Mb [2−7]. If this region is truly devoid of genes, this demonstrates the occurrence of causative SNPs within long distance control elements acting over distances of 100s of kb (the closest genes to the 8q24 cluster are *MYC* and *FAM84B*, over 300 kb away).

SOURCES OF BIOINFORMATION

When using bioinformatic approaches to assess biological candidacy a researcher will rely on extracting as much accurate information possible from online database sources. There are numerous databases (Table 15.1) from which to extract information to build a profile of biological candidacy, the largest and most important of which is the literature resource, PubMed [8].

TOOLS FOR ANNOTATING BIOINFORMATION

Evaluating the underlying biological potential of regions beneath these signals is daunting in the face of numerous databases containing varying amounts of information for each gene. The problem is no longer one of a paucity of biological data. The bottleneck lies in achieving meaningful integration between association or linkage data, local genomic annotations and associated biological information given a bewildering range of available data sources (each with distinctive patterns of reliability and completeness). Studying the available information in a methodological manner, gene by gene, manually, invariably takes a large amount of time and introduces human error. There are several tools designed to assist with this task, the most comprehensive of which is GeneSniffer.

TABLE 15.1 Publicly Available Databases and the Information they Provide

Information	Databases
literature	PubMed [8] (http://www.ncbi.nlm.nih.gov/sites/entrez?db=PubMed) Entrez Gene [9] (http://www.ncbi.nlm.nih.gov/sites/entrez?db=gene) Online Mendelian Inheritance in Man (OMIM) [10] (http://www.ncbi.nlm.nih.gov/sites/entrez?db=omim)
homologs	Homologs identified using BLAST [11] (http://blast.ncbi.nlm.nih.gov/Blast.cgi)
orthologs	Mouse Genome Informatics (MGI) [12] (http://www.informatics.jax.org/) Homologene [13] (http://www.ncbi.nlm.nih.gov/homologene/)
interactants	Human Protein Reference Database (HPRD) [14] (http://www.hprd.org/) Biomolecular Interaction Network Database (BIND) [15] (http://www.bind.ca/) BioGRID [16] (http://www.thebiogrid.org/) Reactome [17] (http://www.reactome.org/)
expression data	Array Express (EBI) [18] (http://www.ebi.ac.uk/microarray-as/ae/) Gene Expression Omnibus (GEO) [19] (http://www.ncbi.nlm.nih.gov/geo/) UniGene EST Profile [8] (http://www.ncbi.nlm.nih.gov/sites/entrez?db=unigene)

GeneSniffer

A researcher will usually arrive at the point of needing to analyze numerous candidate genes after conducting a scan for genetic association with a disease of interest. Therefore the most important input into GeneSniffer (www.genesniffer.org) is a list of disease-specific and relevant words and terms for the disease being studied (Fig. 15.2⑧). Ideally an expert of the disease or phenotype curates this list of words and terms which contains items such as the disease name and its symptoms, biological and biochemical markers characteristic of the disease pathogenesis, affected tissues, and previously identified genes associated or implicated in the disease pathogenesis. For example, for type 2 diabetes this list may contain words and terms such as "diabetes mellitus", "hypoglycaemia", "insulin intolerance", "glucose", "pancreas" and "PPARG". Care is taken to include terms which are not too generic since these may give spurious scores; for example "transcription factor" may be relevant to diabetes, but will also be found in many other non-diabetes-specific gene annotations as well and should therefore be avoided. These terms are weighted, between

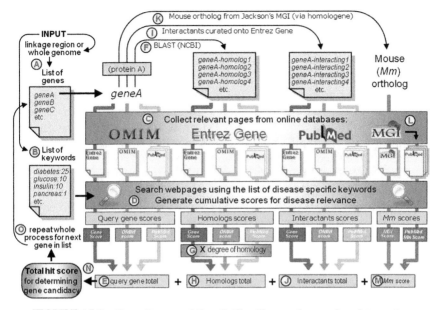

FIGURE 15.2 Flow diagram of GeneSniffer. Please refer to color plate section

1 and 25 according to disease relevance, again assigned according to the aims and interests of the initial study design. For example, "diabetes mellitus" would have a high score of 25 whereas less specific terms such as "glucose" and "pancreas" would have lower scores such as 5 and 1. The list can be of any length, but the more comprehensive it is, the greater the depth of the results obtained.

The second input into GeneSniffer is a list of genes. This can be a list of genes arising from a separate prioritization method such as microarray analyses; a chromosomal interval resulting from a linkage scan or the whole genome (Fig. 15.2Ⓐ). The input can also be a list of SNPs and GeneSniffer defines the region ±0.1 cM of the associated SNP which has an 85% likelihood of containing the linked functional variant (assuming the SNP follows HapMap reported patterns of LD). Fifty kilobase flanks are added to allow for close-by genes that may have regulatory variants within this region (eQTL studies suggest >90% of functional variants fall within the gene plus 50 kb flanking regions).

GeneSniffer starts with the first gene in the list and downloads appropriate webpages from Entrez Gene [9], OMIM [10] and PubMed [8] databases (Fig. 15.2Ⓒ) and interrogates the text using the list of disease-specific words (Fig. 15.2Ⓓ). Each term is scored once for its presence within each page (regardless of the number of times it may appear) and added to a cumulative score to make a total score for disease candidacy for each database page. PubMed searches are conducted using any one of the gene symbols or alternate symbols, plus specific words from the gene name. Non-specific/generic words (which are listed in a special exclusion list) such as "antigen", "binding" or "transcription" are omitted in an attempt to make the search as specific to each gene as possible. Per gene, the total scores from each of the papers is added together to give a PubMed score. The database scores are then weighted: Entrez Gene scores are multiplied by 20, and OMIM by 10 to account for the greater significance of manually checked and curated information in these databases. These scores are totaled with the PubMed score to give a score for the gene (Fig. 15.2Ⓔ).

Next homologs of the gene are identified by BLAST [11] (Fig. 15.2Ⓕ) and scored for content of their Entrez Gene, OMIM and PubMed entries, in the same way as described previously for the gene (Fig. 15.2Ⓒ,Ⓓ). These scores are weighted to generate an adjusted hit-score (Fig. 15.2Ⓖ). This weighting is calculated using the length of region matched and the length of the protein and homologous protein. These homolog hit scores are cumulated for all homologs (Fig. 15.2Ⓗ).

Interactants of the gene product are collected from NCBI's Gene database [9] which contains curations from HPRD [14], BIND [15], Bio-GRID [16] and Reactome [17] (Fig. 15.2Ⓘ). These are also screened for candidacy (Fig. 15.2Ⓒ,Ⓓ) and a total interactants score is generated (Fig. 15.2Ⓙ). The scores from the interactants are weighted depending on the number of interactants a particular gene product has, since genes with many interactants will have inflated scores.

The mouse ortholog is identified using NCBI's Homologene database [8] and the relevant page from Jackson's MGI database [12] is collected (Fig. 15.2Ⓚ) and screened using the list of keywords. Papers curated into the MGI database page are collected and screened (Fig. 15.2Ⓛ) and a total mouse ortholog candidacy score is generated (Fig. 15.2Ⓜ).

Finally the gene, homologs, interactants and mouse scores are totaled to give a grand total disease candidacy score for the gene (Fig. 15.2Ⓝ). The whole process is then repeated for the next gene in the list (Fig. 15.2Ⓞ) and so on until all of the genes in the list have been analyzed and scored.

The resultant output is presented in easily navigable linked webpages (in html format). A front page shows the list of genes with summary of the hitscores from the different parts of the analyses. This page gives details about the time of the run, region, number of genes and a menu to navigate to other regions of the genome analyzed (Fig. 15.2). Clicking on the gene symbol will then take the researcher to a page of gene-specific results (Fig. 15.3). At the head of this page is a summary of gene information, followed by the results of scanning the Entrez Gene and OMIM pages, including links to those pages so that researchers can view the

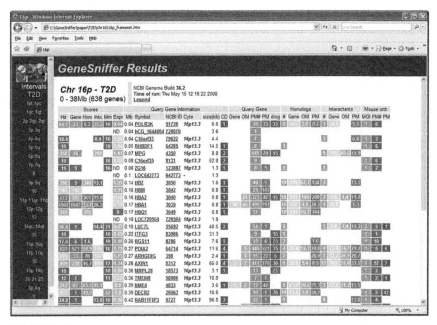

FIGURE 15.3 Screenshot of GeneSniffer's frontpage summarising results. Please refer to color plate section

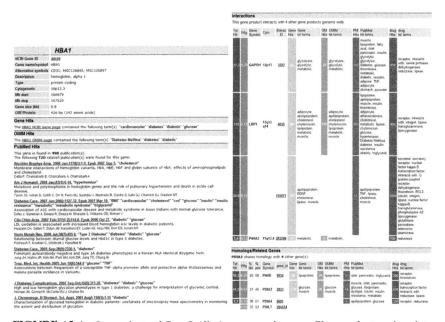

FIGURE 15.4 Screenshots of GeneSniffer's gene results page. Please refer to color plate section

context of any disease terms found. Any PubMed abstracts containing disease keywords are listed with the keywords found together with links to full abstracts at PubMed. Similarly, details of results from scanning the Jackson MGI page for the mouse ortholog are given together with scans of mouse PubMed abstracts. Any conserved protein domains found in the gene product are shown with linkouts and this is provided purely as information and not used to generate a candidacy score.

Expression information collected from NCBI's Unigene EST Profile [8], which estimates gene expression based on EST, counts from different cDNA library sources and the results are displayed as transcripts per million (TPM). The user specifies tissues of interest relevant to the disease being studied and GeneSniffer calculates the relative expression in these tissues compared to all other tissues. This figure is not used in the candidacy calculation, but is displayed in a column on the region summary page. A summary of the expression pattern of the gene is given in the gene information page listing TPM for all of the tissues available.

GeneSniffer provides the most comprehensive results of all tools known to date. However, due to its integrated databases it is unfortunately not available as a downloadable or stand-alone tool, but it can be run collaboratively. Programs which provide a fast interactive interface are detailed below and the best of these is PosMed.

- **PosMed** (http://omicspace.riken.jp/PosMed/): Developed by the Riken Institute in Japan [20] this program is targeted at identifying candidate genes for positional cloning studies. The tool uses any single keyword or keyphrase input such as "retinitis pigmentosa" or "diabetes" together with Boolean operators and searches single genes or interval(s) from human, mouse, rat, arabidopsis or rice. Results are presented in a table of ranked genes with p-values corresponding to the link between the gene and the keyword given. Links can be made through literature searching or interactions between genes.
- **GeneSeeker** (http://www.cmbi.kun.nl/GeneSeeker/): GeneSeeker allows searching in different databases simultaneously, given a known human genetic location and expression/phenotypic pattern. Gene names are returned corresponding to genes in the given location and expressed in the specified tissue. The input is a cytogenetic band(s) and a tissue which the program then uses to systematically search the OMIM [10], Medline, Swissprot (UniProt), Trembl (UniProt), GXD, Tbase, MLC, MGD, GDB and Mimmap databases. At the time of writing the site was not generating results.
- **Endeavour** (http://homes.esat.kuleuven.be/~bioiuser/endeavour/endeavour.php): Endeavour prioritizes candidates genes, based on a set of training genes, already known to be associated with the disease

of interest. First, information about the training genes is retrieved from numerous data sources including functional annotations, protein-protein interactions, regulatory information, expression data, sequence-based data and literature mining data, in order to build models. These models are then used to score the candidate genes and to rank them according to their scores. Lastly, the rankings per data source are fused into a global ranking using order statistics. Endeavour is available for human, mouse, rat, fruit fly and worm. The results are presented in a table of ranked hits, with a further breakdown of hits from each of the databases studied.

- **G2D** (http://coot.embl.de/g2d/): Candidate priorities are automatically established by data mining algorithms that extract putative genes in the chromosomal region where the disease is mapped, and evaluate their possible relation to the disease based on the phenotype of the disorder or their similarity to an already known related gene. If the phenotype has been linked to more than one locus, known or inferred interactions between proteins from two loci can also be examined.

References

[1] E. Zeggini, M.N. Weedon, C.M. Lindgren, T.M. Frayling, K.S. Elliott, H. Lango, et al., Replication of genome-wide association signals in UK samples reveals risk loci for type 2 diabetes, Science 316 (2007) 1336−1341.

[2] J. Gudmundsson, P. Sulem, A. Manolescu, L.T. Amundadottir, D. Gudbjartsson, A. Helgason, et al., Genome-wide association study identifies a second prostate cancer susceptibility variant at 8q24, Nat. Genet. 39 (2007) 631−637.

[3] D.F. Easton, K.A. Pooley, A.M. Dunning, P.D. Pharoah, D. Thompson, D.G. Ballinger, et al., Genome-wide association study identifies novel breast cancer susceptibility loci, Nature 447 (2007) 1087−1093.

[4] B.W. Zanke, C.M. Greenwood, J. Rangrej, R. Kustra, A. Tenesa, S.M. Farrington, et al., Genome-wide association scan identifies a colorectal cancer susceptibility locus on chromosome 8q24, Nat. Genet. 39 (2007) 989−994.

[5] I. Tomlinson, E. Webb, L. Carvajal-Carmona, P. Broderick, Z. Kemp, S. Spain, et al., A genome-wide association scan of tag SNPs identifies a susceptibility variant for colorectal cancer at 8q24.21, Nat. Genet. 39 (2007) 984−988.

[6] M. Yeager, N. Orr, R.B. Hayes, K.B. Jacobs, P. Kraft, S. Wacholder, et al., Genome-wide association study of prostate cancer identifies a second risk locus at 8q24, Nat. Genet. 39 (2007) 645−649.

[7] A. Tenesa, S.M. Farrington, J.G. Prendergast, M.E. Porteous, M. Walker, N. Haq, et al., Genome-wide association scan identifies a colorectal cancer susceptibility locus on 11q23 and replicates risk loci at 8q24 and 18q21, Nat. Genet. 40 (2008) 631−637.

[8] E.W. Sayers, T. Barrett, D.A. Benson, S.H. Bryant, K. Canese, V. Chetvernin, et al., Database resources of the National Center for Biotechnology Information, Nucleic Acids Res. 37 (2009) D5−D15.

[9] D. Maglott, J. Ostell, K.D. Pruitt, T. Tatusova, Entrez Gene: gene-centered information at NCBI, Nucleic Acids Res. 35 (2007) D26−D31.

[10] A. Hamosh, A.F. Scott, J.S. Amberger, C.A. Bocchini, V.A. McKusick, Online Mendelian Inheritance in Man (OMIM), a knowledgebase of human genes and genetic disorders, Nucleic Acids Res. 33 (2005) D514–D517.

[11] S.F. Altschul, W. Gish, W. Miller, E.W. Myers, D.J. Lipman, Basic local alignment search tool, J. Mol. Biol. 215 (1990) 403–410.

[12] J.T. Eppig, C.J. Bult, J.A. Kadin, J.E. Richardson, J.A. Blake, A. Anagnostopoulos, et al., The Mouse Genome Database (MGD): from genes to mice – a community resource for mouse biology, Nucleic Acids Res. 33 (2005) D471–D475.

[13] D.L. Wheeler, T. Barrett, D.A. Benson, S.H. Bryant, K. Canese, V. Chetvernin, et al., Database resources of the National Center for Biotechnology Information, Nucleic Acids Res. 36 (2008) D13–D21.

[14] S. Peri, J.D. Navarro, T.Z. Kristiansen, R. Amanchy, V. Surendranath, B. Muthusamy, et al., Human protein reference database as a discovery resource for proteomics, Nucleic Acids Res. 32 (2004) D497–D501.

[15] G.D. Bader, I. Donaldson, C. Wolting, B.F. Ouellette, T. Pawson, C.W. Hogue, BIND – the Biomolecular Interaction Network Database, Nucleic Acids Res. 29 (2001) 242–245.

[16] C. Stark, B.J. Breitkreutz, T. Reguly, L. Boucher, A. Breitkreutz, M. Tyers, BioGRID: a general repository for interaction datasets, Nucleic Acids Res. 34 (2006) D535–D539.

[17] L. Matthews, G. Gopinath, M. Gillespie, M. Caudy, D. Croft, B. de Bono, et al., Reactome knowledgebase of human biological pathways and processes, Nucleic Acids Res. 37 (2009) D619–D622.

[18] A. Brazma, U. Sarkans, A. Robinson, J. Vilo, M. Vingron, J. Hoheisel, et al., Microarray data representation, annotation and storage, Adv. Biochem. Eng. Biotechnol. 77 (2002) 113–139.

[19] T. Barrett, D.B. Troup, S.E. Wilhite, P. Ledoux, D. Rudnev, C. Evangelista, et al., NCBI GEO: archive for high-throughput functional genomic data, Nucleic Acids Res. 37 (2009) D885–D890.

[20] N. Kobayashi, T. Toyoda, Statistical search on the Semantic Web, Bioinformatics 24 (2008) 1002–1010.

Interpreting Association Signals

Nicole Soranzo [1,2], *Frank Dudbridge* [3]

[1] Wellcome Trust Sanger Institute, Genome Campus, Hinxton,
Cambridge, UK

[2] Department of Twin Research & Genetic Epidemiology,
King's College London, London, UK

[3] London School of Hygiene and Tropical Medicine, London, UK

OUTLINE

INTRODUCTION

In the last four to five years, genome-wide association studies (GWAS) have yielded hundreds of novel genetic loci underlying quantitative and disease phenotypes, with reproducible results. Previously, however, candidate-SNP and candidate-gene association studies had been marred by poor reproducibility and false-positive findings. As was later recognized, at the origin of such non-replications of candidate gene studies were often false positives in the original claims, due, for instance, to overliberal declaration of statistically significant associations. In some instances, failure to replicate a true locus may also have been observed, due for instance to an underpowered replication sample or heterogeneity arising from methodological differences between discovery and replication samples. Growing understanding of such phenomena has aided the design, analysis and interpretation of GWAS. In this chapter we discuss some of the known sources of bias that can affect replication and interpretation of genetic findings, and highlight recent recommendations for reporting genetic association studies in the context of genome-wide scans.

THE IMPORTANCE AND DEFINITION
OF REPLICATION

Importance

Independent replication represents the *sine qua non* for reliably claiming a novel genetic association. Prior to GWAS, genetic associations through candidate gene analysis had been marred by poor replicability. In one of the largest systematic efforts at the time, Hirschhorn et al. [1] reviewed published literature including over 600 different associations. They found that, of 166 associations tested in three or more independent studies, only six could be consistently replicated across all studies. A staggering half of the loci investigated yielded inconsistent replication results across different reports [1]. Lohmueller et al. [2] subsequently analyzed 301 published studies, covering 25 different putative disease loci, through meta-analysis. They showed that, although 11 of the 25 loci were genuinely associated with disease, the near totality of loci (24 out of 25) had lower odds ratio than the estimates based on an original discovery study. These early surveys provided empirical evidence supporting the need for caution in the interpretation of genetic associations deriving from early association reports.

A common problem in a large majority of early candidate gene studies was that discovery was based on small-sized samples. These were often orders of magnitude smaller than those shown by GWAS studies as necessary to detect loci of realistic effect size for complex diseases. Such small discovery sizes were underpowered to detect true effects, resulting in high false-positive rates.

A further widespread cause of false-positive associations was the limited understanding on how to properly estimate, and correct for, the true experiment-wide multiple hypothesis-testing burden. This resulted in the over-liberal declaration of statistical significance based on a p-value alone (typically defined by a p-value ≤ 0.05, whether or not a correction was applied for multiple markers in a gene). This would be appropriate when the alternative hypothesis − that the gene is associated − is reasonably plausible, but when the association of a gene is *a priori* less likely, stronger significance levels are appropriate. It now seems clear that overoptimistic confidence was placed on candidate genes so that p-values deemed significant at 0.05 have led to an unacceptably high false-positive rate.

Additional sources of bias and true heterogeneity were also found to contribute to the high rates of both false-positive and false-negative associations in these early discovery studies. Ranking and selection bias are known to affect estimates of odds ratios upwards (or downwards if OR < 1) when study power is low. Thus, replication samples based on such inflated effect sizes can lack sufficient power to replicate a true finding. Furthermore, researchers may have failed to recognize possible bias arising from selective or distorted reporting, which includes the use of improper statistical analyses, *ad hoc* selection of primary endpoints or other forms of "data massaging" aimed at producing statistical significance. Ioannidis et al. proposed a model to account for such forms of bias, and advanced the controversial statement that "the large majority of published research findings in the medical literature are false" [3]. While researchers later argued against the validity of some of the model's assumptions [4], such efforts were important in defining and assessing sources of bias affecting association studies, helping define the rigorous standards that were instrumental in the success of GWAS.

Some of the same sources of bias and heterogeneity discussed for candidate studies can also affect replication in GWAS. Owing to their high dimensionality, GWAS remain vulnerable to a range of errors and biases, especially those arising from differences in experimental and study design. Furthermore, GWAS are equally susceptible to low power associated with the modest effect sizes characteristic of most complex-trait susceptibility alleles. We will discuss in more detail some of these effects in the context of GWAS. However, it is important to stress here a key difference between the two. Candidate studies test sets of human genes or variants with a high biological candidacy prior. GWAS, on the contrary, represent hypothesis-free, mostly unbiased scans of the entire human genome. In this context, which ignores prior biological candidacy, it is easier to estimate unbiased metrics of experiment-wide (genome-wide) significance that allow rigorous (if overly conservative) correction for multiple testing. A second important point is that GWAS are aimed at discovery of novel loci, and the choice of markers to bring forward for replication is more often driven by practical considerations, particularly cost and optimum use of genotyping platforms, rather than theoretical considerations.

Aim and Definitions

In GWAS, the purpose of replication is to determine which of the findings arising from the primary association study reflect true reproducible associations in an independent study in a similar population with the same study design and analysis plan, and to allow a systematic appraisal of potential sources of error and bias that are responsible for the failure of replication in a study. It is important to make some distinctions in terminology.

A "true replication" indicates the evaluation of statistically significant association signals in additional independent study samples. Such replication efforts aim at reproducing the association signal identified in the initial study in non-overlapping study samples, and will typically involve the same genetic variant, or a variant with very similar or identical statistical properties (a "proxy") as defined by a high correlation coefficient (r^2).

"Technical validation" conversely indicates the re-analysis of all, or a subset of, genetic variants investigated in the association study, and is carried out on the original study samples, often using a different genotyping platform. The aim of technical validation is to detect errors in genotyping or genotype imputation that could lead to differential genotype calling and spurious associations. In high-throughput genome-wide association efforts, where large numbers of samples are analyzed in a single experiment, technical replication is also routinely used to confirm sample identity throughout the genotyping process.

Finally, evidence for "indirect replication" can be claimed when the replication study involves either a different allele at the same locus (for instance, different genetic variants or haplotypes in the same genomic region) or a phenotype trait that is closely correlated but not identical to the original study trait (for instance, an association with total cholesterol at a locus previously identified as associated with LDL cholesterol). While indirect replication can provide confirmatory evidence for the implication of a genetic locus in determining one or more traits, alone it does not provide sufficient evidence of replication of findings. However, often indirect replications can be important to gain additional insights into either the allelic architecture of a locus of interest, or of shared or unique genetic determinants of correlated biological traits or processes. Finally, it is perhaps obvious, but worth stating, that a replication involving the same allele, but with opposite effects (e.g., the risk allele in one study is protective in the second study), should not be considered as a direct replication. Such a scenario is possible if there are large standard errors on estimated effects, or under some population genetics scenarios [5], but one should generally be skeptical about such an outcome.

POWER OF REPLICATION STUDIES

Knowledge of the true (unbiased) effect size of an SNP can provide clues as to the causes of an unsuccessful replication of a true locus, or for designing replication studies with sufficiently powered sample sizes. Two types of bias that contribute to inflating naïve estimates of odds ratios or other effect sizes from genome scans are "ranking" and "significance" bias [6, 7], discussed in detail below. Similarly, sampling schemes designed to increase the power of an association study by enriching the genetic effects present in the sample (for instance, designs using extreme phenotypes or cases with a high degree of familial aggregation) tend to have upwardly biased effect sizes. In this section we discuss in detail some of these effects, and summarize methods proposed to properly account for these sources of bias in association studies.

Principles of Selection Bias

It is widely recognized that naïve estimates of odds ratios or other effect sizes from genome scans are upwardly biased, a phenomenon that affects both whole-genome linkage and association scans [6–8]. This occurs because the same data are used both to select the genetic regions and to estimate their effects. It is important to distinguish two different sources of bias that can lead to an overestimation of the true effect sizes of an SNP. One form of such ascertainment bias is known as "significance bias". This arises when estimation of effect size is performed only for effects that are statistically significant: the expected value of an estimator, conditional on it being significant, is then typically higher than its unconditional expectation represented by the population value. This is the same principle that underlies publication bias in scientific literature. This effect depends strongly on the power of the initial test for association. If the power is high, most random draws from the distribution of genotype counts will result in a significant test for association; thus, the ascertainment effect is small. On the other hand, if the power is low, conditioning on a successful association scan will result in a large ascertainment effect. Ranking bias [9] is a form of selection bias of estimated effect sizes that arises because of the selection of SNPs with the most extreme test statistics (highly ranked results). Thus, if following a genome scan SNPs are ranked by their p-values, then the expected value of an estimator (conditional upon it being the most significant) is again greater than its unconditional expectation. Ranking bias applies not only to the most significant SNP, but also to the second most significant, third most and so on, and is present even when there is no selection by significance. These biases are often referred to as "winner's curse", a term originating from economic theory of auctions: although its original sense applies to ranking bias (there is only one winner of an auction), it is often also applied to significance bias. In addition to significance and ranking biases, ascertainment or sampling schemes designed to increase the power of an association study by enriching the genetic effects present in the sample will also tend to produce upwardly biased effect sizes, compared to those in the population at large. Such designs include, for example, selection of subjects with extreme phenotypes, or a family history of disease. These biases are not discussed further here.

Several authors have proposed methods for partially correcting for significance bias to infer bias-corrected estimates based on the initial sample data. To address significance bias, Zöllner and Pritchard proposed to maximize the likelihood of the genotype data conditional on it passing a significance threshold [10]. The likelihood function is defined on the individual genotype data in terms of genotype frequency and penetrance parameters. The method produces a point estimate and confidence region for the parameter estimates, which are used to assess the impact of the detected variant, as measured, for example, by the attributable risk. Through simulations the authors show that it is possible to reduce the bias in the parameter estimates even when the original association study had low power, and that the uncertainty of the estimate decreases with increasing sample size, independent of the power of the original test for association. A disadvantage of this procedure is that, although it reduces the bias in risk estimation, it is only developed for case–control data and cannot be performed with standard statistical software.

The approach proposed by Ghosh and colleagues [11] is similar, but uses the summary estimate of genetic effect and its standard error as reported by standard statistical software, rather than the individual genotype data. Advantages of their approach are that it is applicable to a wide range of designs (not just case–control), it can be applied to previously published studies, and it is easy to implement. Zhong and Prentice [12] develop this approach further by proposing a weighted combination of corrected and uncorrected estimators, and versions that are applicable to multistage GWAS. The methods from Ghosh et al. [11] and Zhong [12] assume the summary odds ratio to be normally distributed with its variance equal to the sample estimate. In both cases these approaches do not fully correct the selection bias, with both methods showing a tendency to overcorrect.

Sun and Bull [13] and Yu and colleagues [14] proposed the use of a bootstrap procedure and cross-validation to correct for selection bias in linkage and association scans. Their simulations demonstrate that the results from the discovery stage can improve sample size calculations for the replication stage adaptively. Despite the adaptive use of discovery stage data, the proposed method maintains the nominal global type I error for final analyses on the basis of either pure replication with the replication stage data only or a joint analysis using information from both stages. Simulation studies show that sample-size calculations accounting for the impact of regression to the mean with the bootstrap procedure are more appropriate than the conventional method is. Again, however, this approach only reduces bias without achieving unbiased estimation, and the need to analyze large numbers of bootstrap samples is a practical limitation.

Unbiased estimates of SNP effects can be obtained from replication studies carried out in independent sample sets [6]. Because of the inconvenience of implementing novel methods for reducing bias, and their incomplete performance, this pragmatic solution is the one employed by many published studies. However, we note that this approach is inefficient as it ignores information containing the initial scan. Since discovery datasets are typically very large, use of such information could greatly increase the precision of the estimate of effect sizes. To exploit information contained in discovery samples, Bowden and Dudbridge recently proposed an unbiased estimator combining information from both the initial scan and the replication study [15]. This approach starts with a standard weighted combination of the estimates from each stage and then applies a correction for both rank and significance bias in the initial scan. This procedure is easy to implement, it allows for multiple associations arising from a scan and is robust to misspecification of a significance threshold. The estimator is both unbiased and guaranteed to be more efficient than the estimate from replication data alone.

While replication data can yield unbiased estimates, such information is not useful for estimating sample sizes for replication studies themselves. Again, pragmatic researchers typically assume that the true effect size is considerably lower than the estimated parameter value (unless the primary GWAS has the power to detect very weak genetic effect), and will choose validation samples that are at least twice the size of the original study to ensure replication of the most significant loci. The methods described above can make this process easier by reducing the bias in the primary GWAS.

A final point worth mentioning is that, while replication studies should be designed around corrected estimates, in most association studies the causal locus will have higher odds ratios than the marker variant when these two variants are in incomplete linkage disequilibrium (LD) with each other. In these settings the effect of the odds ratio correction may be reversed if the replication sample uses a denser set of markers than the original data.

Two-stage Genome Scans

A variation on the previous designs involves two-stage genome scans. In these studies, an initial whole genome scan is performed on a subset of the subjects available for genotyping. The most promising SNPs are then taken forward for genotyping in the remaining subjects, with the number of such SNPs chosen to be large enough to include most of the true positives, but small enough to exclude most of the true negatives. Much research has gone into developing the most efficient approaches when the total genotyping or cost constraints are fixed in advance [16, 17]. Of course, more than two stages could be employed, and the same principles apply to scans with three or more stages.

There are important differences between two-stage designs and formal replication of GWAS hits. The two-stage approach is designed as an economically efficient version of a genome scan, and is only intended to generate hypotheses at its conclusion. The first stage of a two-stage approach does not generate strong hypotheses, so it is inappropriate to analyze the second stage in isolation, and fallacious to claim that the multiple testing is reduced. Fortunately, methods are available to test the data from both stages jointly, and it has been demonstrated that these are more powerful than analysis of the second stage alone [18–21]. In contrast, replication studies are intended to confirm the hypotheses generated by an earlier scan, and as such should be analyzed separately from the scan data. This is particularly important for eliminating any bias in the initial scan. Thus, while stronger significance levels might be obtained by combined analysis of scan and replication data, these would not be free from biases in the scan data, and true independent replication carries greater weight in establishing the association of a genetic variant.

Causes of Heterogeneity in Effect Estimates

In addition to power considerations discussed before, other features of study designs can generate significant heterogeneity in effects, which may hamper the success or interpretation of a replication study. One is if variable patterns of LD exist between the genotyped SNP and untyped causal alleles in different study samples. As discussed in Chapter 3, local LD patterns may vary significantly across different ethnicities at a large fraction of the genome as a result of events in the history of the populations under study, particularly drift, inbreeding, natural selection or admixture. Genome-wide human variation mapping projects such as the HapMap (www.hapmap.org) and 1000 Genomes (www.1000genomes.org) projects, provide detailed information on allelic associations in different populations, which aid the interpretation of association studies. Extreme differences in the frequency of the causal alleles and local LD patterns can arise not only among different ethnicities, but also in populations of similar ethnicity which have experienced significant drift or inbreeding in their past, such as founder populations.

Significant heterogeneity can also arise from alternative case-ascertainment strategies, with respect, for instance, to family history, age of onset or distribution of risk factors. Timpson and colleagues [22] recently re-examined 1924 case and 2938 control subjects from the Wellcome Trust Case Control Consortium genome-wide association study to characterize how differences in the BMI distribution of type 2 diabetic case subjects affected genome-wide patterns of type 2 diabetes association. After stratifying case subjects (into "obese" and "non-obese") according to median BMI (30.2 kg/m^2), they showed reproducible heterogeneity in odds ratios for two loci (*FTO* and *TCF7L2*), and nominal evidence for a third locus (*SLC30A8*), demonstrating the impact of differences in case ascertainment on the power to detect and replicate genetic associations in GWAS.

Family-based samples might have stronger genetic effects depending on their ascertainment criteria. Antoniou and Easton [23] explored the efficacy of utilizing cases with a family history of the disease, together with unrelated controls, for a polygenic model of breast cancer risk. They showed that, relative to a standard case–control association study with cases unselected for family history, the sample size required to detect a common disease susceptibility allele was typically reduced by more than twofold if cases with an affected first-degree relative were selected, and by more than fourfold if cases with two affected first-degree relatives were utilized. This highlights that, for given diseases, association studies based on cases with a strong family history may be substantially more efficient than population-based studies. The relative efficiency obtained by using familial cases appeared greater for rarer alleles. Evangelou, et al. [24] empirically evaluated differences in effect size estimates in 93 investigations where both unrelated case–control and family-based designs had been employed. They found that estimated odds ratios differed beyond chance between the two designs in only 4% of the cases, with most heterogeneity being consistent with differences in power. This suggests that the relative usefulness of unrelated case–control and family-based designs needs to be evaluated in a context-specific manner for different diseases.

Significant heterogeneity can arise from non-additive interactions with other genetic variants or environmental exposures [25–29]. If these interactions are not modeled explicitly, but are absorbed into the marginal effect of a tested variant, then differences in the distribution of other risk factors between populations will alter the marginal effect size, creating heterogeneity in the estimated effect.

Finally, it should be noted that technical differences between modeling assumptions and the true disease model may lead to effect heterogeneity. For example, the logistic model is commonly applied to binary disease endpoints, but the effect of a genetic variant may act on the relative risk rather than the odds ratio scale. Then, differences in the background risk of disease would lead to differences in the odds ratio between studies. The choice of the wrong genetic model (e.g., additive in place of dominant or recessive) can represent a further source of heterogeneity and bias.

GUIDELINES FOR REPORTING ASSOCIATION RESULTS

Inadequate study reporting even in well-conducted studies hampers assessment of a study's strengths and weaknesses, and hence the integration of evidence. Recently a multidisciplinary group including epidemiologists, geneticists, statisticians and journal editors developed a set of guidelines called the STrengthening the REporting of Genetic Association studies (STREGA) statement, which represents an extension of previously developed STROBE guidelines (STrengthening of REporting of OBservational studies in Epidemiology). STROBE/STREGA recommendations are based on empirical evidence on the reporting of observational studies and are not intended to support or oppose the choice of study design, but rather to set criteria for standardized reporting of methods that enhance transparency, completeness and quality and allow readers to assess potential sources of bias in published studies. STREGA guidelines include 22 separate items, disseminated through coordinated publications [30] and at http://www.medicine.uottawa.ca/public-health-genomics/web/eng/strega.html. In this section we highlight some recommendations in key areas of concern, while we direct the readers to the original publications for the complete list of recommendations.

- *Title and abstract* must respectively indicate the study design and provide informative and balanced summary of results.
- *Methods* must detail the study design clearly (e.g., first report, replication or meta-analysis), describe possible sources of bias derived from ascertainment of study samples (for instance, population

stratification or criteria for inclusion/exclusion of study samples at different *analysis* steps) and define genetic and non-genetic exposures. Laboratory methods must describe quality control procedures applied to data production (e.g., assessment of genotyping error rates), reproducibility across methodologies and concordance among repeat measurements. The study must clearly define genetic exposures (genetic variants) using a widely used nomenclature system, and identify variables likely to be associated with population stratification. Statistical analyses must specify protocols for phenotype data handling and modeling, describe statistical methods used to assess and control for confounding and explain handling of possible sources of bias due to data handling. In particular, authors must state methods used to deal with missing data, genotyping errors, population stratification, relatedness among subjects, deviations from Hardy-Weinberg equilibrium and to address multiple comparisons or to control risk of false positive findings. Software tools must be cited together with software version used and options (or settings) chosen.

- *Results* must contain detailed indicators of primary and secondary analyses, reports on the adjustments applied (for instance, for multiple comparisons, risk factors and/or confounders, etc.). Analysis results must indicate if the same variant or haplotype was analyzed, or alternatively declare their statistical association with the lead variant. Details on ascertainment of study individuals, including ethnicity, must be provided. Statistical analyses must be based on comparable statistical model (additive, dominant, etc.); trait modeling, including trait transformation and adjustments for covariates and risk factors, must also be provided.
- *Discussion* must summarize key results with reference to study objectives, discuss limitations of the study (taking into account sources of potential bias or imprecision and direction and magnitude of any potential bias), give a cautious overall interpretation of results considering objectives, limitations, multiplicity of analyses, results from similar studies, and other relevant evidence and finally discuss the generalizability (external validity) of the study results.

FOLLOW-UP OF ASSOCIATIONS

While GWAS have been instrumental in identifying novel biological functions associated with many diseases, the identification of causative variants responsible for the observed association has proven more challenging. Due to the extended nature of LD in Europeans, association signals identified by genome-wide scans often map to large genomic regions of hundreds of kilobases in size and thousands of variants with similar statistical properties. Furthermore, due to the sparse nature of many tag-based commercial genotyping arrays, in the majority of cases it is reasonable to assume that the associated marker will only be a statistical proxy for an ungenotyped causative variant. This observation suggests the need for further approaches for the systematic discovery, evaluation and prioritization of genetic variants within association signals. Several complementary approaches can be utilized for dissecting the underlying association signals, some of which are listed below. It has to be remembered that, ultimately, the identification of genetic variants will require the functional validation of the effect of each variant through *ad hoc* experiments. The use of multiple approaches, however, has the potential to greatly increase the efficiency of this process by reducing the number of hypotheses to be tested in costly and lengthy functional experiments.

Fine-mapping

Fine-mapping approaches indicate a comprehensive exploration of variation around strong signals of association through additional genotyping and/or resequencing. In fine-mapping, additional genetic variants in regions of association are identified from sequence variant repositories (dbSNP, HapMap, 1000 Genomes project, www.1000genomes.org) or through targeted resequencing of association intervals (defined by recombination hotspots boundaries), in a subset of individuals to compile exhaustive catalogues of genetic variants within associated intervals. Cases, controls or a mixture of each can all be used for variant discovery. The statistical association of each newly identified variant is then re-assessed in the original case–control study. A great opportunity is now provided by the 1000 Genomes project, which aims to systematically catalogue genetic variants down to 1% frequency or below in multiple worldwide populations. The project will soon provide us with the opportunity to impute genetic variants across association intervals to systematically evaluate common and rare genetic variants. A possible application of this project will be the development of next-generation genome-wide SNP arrays with up to 10 M variants that will allow the immediate survey of genetic variants in genome-wide scans. In the meanwhile, the development of disease-specific, high-density custom arrays will allow the assessment of disease associated loci at great depth in thousands of individuals.

Trans-ethnic Fine-mapping

A promising avenue for fine-mapping of association signals is to exploit differences in local LD patterns in individuals of different ethnicities. This approach exploits differences in LD patterns across individuals to reduce the size of the association interval. The use of African populations in this context is particularly advantageous as LD typically extends for shorter intervals in these populations, thus reducing the degree of statistical association between SNPs [31, 32]. The success of these approaches will require careful attention to possible sources of heterogeneity in disease associations across these different population samples; for instance, different exposure to environmental risk factors.

Analysis of GxE and GxG Interactions

A better understanding of disease variants will come with systematic efforts to define the impact of gene-gene and gene-environment interactions for disease-associate variants. Full interaction scans are still prohibitively time-consuming, but can usefully be limited to SNPs showing modest marginal signals of association [33].

Phenotype Correlations

The systematic comparison of association patterns for diseases with similar etiology or correlated quantitative traits can provide a further refinement of association signals. To this extent, the development of methods for quantitative assessment of associations for correlated traits will be of great interest, by providing tools to systematically exploit association.

Bioinformatic and Experimental Functional Annotation of Association Intervals

Two difficulties are the prioritization of genes within association signals. Pathway-based analyses based on known interaction can be of limited use due to incomplete nature of these datasets. Of more interest is the use of methods that apply unknown or unclassified correlations to search for over-representation of related genes among significant hits [34, 35]. Similar approaches can be employed for prioritizing signals for replication from the bulk of mid-ranking association p-values.

The juxtaposition of association signals with experimental functional annotation datasets in relevant experimental systems promises to improve formulating hypotheses on causal mechanisms. The first attempts to functionally characterize the human genome have involved the analysis of genome-wide gene expression profiles and the analysis of correlations with SNPs to identify eQTLs [36]. Further gains can be expected from the analysis of tissue-specific expression [37, 38] and the integration of genetic data with multiple genome-wide functional datasets [39]. For instance, custom array and RNAseq datasets will be important for detecting SNPs affecting genetic expression through allele-specific regulation [40]. ChIP-ChIP and ChIPseq datasets will be useful for the identification of SNPs located in regulatory elements such as transcription factor binding sites, and for better understanding tissue-specific effects. Analysis of chromatin structure through FAIRE and other experimental techniques will allow the identification of additional regulatory elements. The further use of bioinformatics approaches should allow the identification of sets of variants that are more likely to interact with each other, and to guide the evaluation of statistical interactions for SNP effects.

Evolutionary Analyses

The overlaying of evolutionary patterns, notably evolutionary conservation and natural selection, to association regions provides a further interesting opportunity to aid the identification of candidate variants. The underlying assumption is that natural selection arises in relation to variants that modify phenotypes in a detectable way. Therefore, the presence of natural selection can be a strong pointer to a variant being a functional and potentially affecting disease. Similarly, variants within genomic regions that are conserved across species are also likely to have functional consequences, even if the role of such regions is not obvious. The promise of these approaches has increased with the development of methods that improve the resolution of selection signals to individual genetic variants [41].

CONCLUSIONS

We have discussed some of the principles and biases that affect the interpretation of genetic association signals. We have shown how in large part, the difficulties of replication occur because most genuine associations have modest effects; hence, there is generally incomplete power to detect associations in any given study. Replication studies planned on upwardly biased effect size estimates may be underpowered to detect a genuine but weak association. Furthermore, non-replication may be caused by heterogeneity due to biases or true differences in genetic effects in different populations. Such common systematic biases that differently affect the observed effects across various studies may include population stratification, misclassification of phenotype, genotyping error, and selection biases (publication and selective reporting biases). On the contrary, true differences in genetic effects may include differential LD of the identified genetic marker with the true causative variant in different populations, or latent population-specific gene-gene or gene-environment interactions. While the reproducibility of many GWAS is a testament to our ability to account for at least some of these effects, a refinement of these approaches is likely to further increase the resolution of current genetic datasets. Furthermore, the in-depth exploration of the nature of heterogeneity in genetic association patterns is likely to greatly improve our understanding of factors underlying disease and quantitative trait genetics.

References

[1] J.N. Hirschhorn, K. Lohmueller, et al., A comprehensive review of genetic association studies, Genet. Med. 4 (2) (2002) 45–61.
[2] K. Lohmueller, C. Pearce, et al., Meta-analysis of genetic association studies supports a contribution of common variants to susceptibility to common disease, Nat. Genet. 33 (2003) 177–182.
[3] J.P. Ioannidis, Why most published research findings are false, PLoS Med 2 (8) (2005) e124.
[4] S. Goodman, S. Greenland, Why most published research findings are false: problems in the analysis, PLoS Med. 4 (4) (2007) e168.
[5] G.M. Clarke, L.R. Cardon, Aspects of observing and claiming allele flips in association studies, Genet. Epidemiol. 34 (3) (2009) 266–274.
[6] H.H. Goring, J.D. Terwilliger, et al., Large upward bias in estimation of locus-specific effects from genomewide scans, Am. J. Hum. Genet. 69 (6) (2001) 1357–1369.
[7] C. Garner, Upward bias in odds ratio estimates from genome-wide association studies, Genet. Epidemiol. 31 (2007) 288–295.
[8] J.N. Hirschhorn, M.J. Daly, Genome-wide association studies for common diseases and complex traits, Nat. Rev. Genet. 6 (2) (2005) 95–108.
[9] N. Jeffries, Ranking bias in association studies, Hum. Hered. 67 (2009) 267–275.
[10] S. Zollner, J.K. Pritchard, Overcoming the winner's curse: estimating penetrance parameters from case-control data, Am. J. Hum. Genet. 80 (4) (2007) 605–615.
[11] A. Ghosh, F. Zou, et al., Estimating odds ratios in genome scans: an approximate conditional likelihood approach, Am. J. Hum. Genet. 82 (2008) 1064–1074.
[12] H. Zhong, R.L. Prentice, Bias-reduced estimators and confidence intervals for odds ratios in genome-wide association studies, Biostatistics 9 (4) (2008) 621–634.
[13] L. Sun, S.B. Bull, Reduction of selection bias in genomewide studies by resampling, Genet. Epidemiol. 28 (4) (2005) 352–367.
[14] K. Yu, N. Chatterjee, et al., Flexible design for following up positive findings, Am. J. Hum. Genet. 81 (3) (2007) 540–551.
[15] J. Bowden, F. Dudbridge, Unbiased estimation of odds ratios: combining genomewide association scans with replication studies, Genet. Epidemiol. 33 (5) (2009) 406–418.
[16] J.M. Satagopan, E.S. Venkatraman, et al., Two-stage designs for gene-disease association studies with sample size constraints, Biometrics 60 (3) (2004) 589–597.
[17] D. Thomas, R. Xie, et al., Two-stage sampling designs for gene association studies, Genet. Epidemiol. 27 (4) (2004) 401–414.
[18] R.L. Prentice, L. Qi, Aspects of the design and analysis of high-dimensional SNP studies for disease risk estimation, Biostatistics 7 (3) (2006) 339–354.
[19] H. Wang, D.C. Thomas, et al., Optimal two-stage genotyping designs for genome-wide association scans, Genet. Epidemiol. 30 (2006) 356–368.

[20] H.H. Muller, R. Pahl, et al., Including sampling and phenotyping costs into the optimization of two stage designs for genome wide association studies, Genet. Epidemiol. 31 (8) (2007) 844–852.

[21] A.D. Skol, L.J. Scott, et al., Optimal designs for two-stage genome-wide association studies, Genet. Epidemiol. 31 (7) (2007) 776–788.

[22] N.J. Timpson, C.M. Lindgren, et al., Adiposity-related heterogeneity in patterns of type 2 diabetes susceptibility observed in genome-wide association data, Diabetes 58 (2) (2009) 505–510.

[23] A. Antoniou, D. Easton, Polygenic inheritance of breast cancer: implications for design of association studies, Genet. Epidemiol. 25 (2003) 190–202.

[24] E. Evangelou, T. Trikalinos, et al., Family-based versus unrelated case-control designs for genetic associations, PLoS Genet. 2 (2006) e123.

[25] G. Clarke, K. Carter, et al., Fine mapping versus replication in whole-genome association studies, Am. J. Hum. Genet. 81 (2007) 995–1005.

[26] P. Gorroochurn, S.E. Hodge, et al., Non-replication of association studies: "pseudo-failures" to replicate? Genet. Med. 9 (6) (2007) 325–331.

[27] J.P. Ioannidis, Non-replication and inconsistency in the genome-wide association setting, Hum. Hered. 64 (4) (2007) 203–213.

[28] J.P. Ioannidis, N.A. Patsopoulos, et al., Heterogeneity in meta-analyses of genome-wide association investigations, PLoS One 2 (9) (2007) e841.

[29] R. Moonesinghe, M. Khoury, et al., Required sample size and nonreplicability thresholds for heterogeneous genetic associations, Proc. Natl. Acad. Sci. USA 105 (2008) 617–622.

[30] J. Little, J.P.T. Higgins, et al., STrengthening the REporting of Genetic Association studies (STREGA) – an extension of the STROBE statement, PLoS Med. 6 (2) (2009) e1000022.

[31] C.A. McKenzie, G.R. Abecasis, et al., Trans-ethnic fine mapping of a quantitative trait locus for circulating angiotensin I-converting enzyme (ACE), Hum. Mol. Genet. 10 (10) (2001) 1077–1084.

[32] S.A. Tishkoff, F.A. Reed, et al., The genetic structure and history of Africans and African Americans, Science 324 (5930) (2009) 1035–1044.

[33] J. Marchini, P. Donnelly, et al., Genome-wide strategies for detecting multiple loci that influence complex diseases, Nat. Genet. 37 (4) (2005) 413–417.

[34] P. Holmans, E.K. Green, et al., Gene ontology analysis of GWA study data sets provides insights into the biology of bipolar disorder, Am. J. Hum. Genet. 85 (1) (2009) 13–24.

[35] S. Raychaudhuri, R. Plenge, et al., Identifying relationships among genomic disease regions: predicting genes at pathogenic SNP associations and rare deletions, PLoS Genet. 5 (2009) e1000534.

[36] B.E. Stranger, A.C. Nica, et al., Population genomics of human gene expression, Nat. Genet. 39 (10) (2007) 1217–1224.

[37] A.S. Dimas, S. Deutsch, et al., Common regulatory variation impacts gene expression in a cell type-dependent manner, Science 325 (5945) (2009) 1246–1250.

[38] E. Grundberg, T. Kwan, et al., Population genomics in a disease targeted primary cell model, Genome Res. 19 (11) (2009) 1942–1952.

[39] A.C. Nica, E.T. Dermitzakis, Using gene expression to investigate the genetic basis of complex disorders, Hum. Mol. Genet. 17 (R2) (2008) R129–R134.

[40] B. Ge, D.K. Pokholok, et al., Global patterns of cis variation in human cells revealed by high-density allelic expression analysis, Nat. Genet. 41 (11) (2009) 1216–1222.

[41] S.R. Grossman, I. Shylakhter, et al., A composite of multiple signals distinguishes causal variants in regions of positive selection, Science 327 (5967) (2010) 883–886.

Delineating Signals from Association Studies

Benjamin F. Voight

Medical Population Genetics, The Broad Institute of Harvard and MIT,
Cambridge, MA, USA; Center for Human Genetics Research,
Massachusetts General Hospital, Cambridge, MA, USA

O U T L I N E

INTRODUCTION

In recent years, genome-wide studies have revealed an impressive number of genetic loci which associate to human traits [1, 2]. Moving these associations forward into biological insight is truly the next step in genomic studies, augmenting the core goals of genetic mapping efforts. Armed with biological insight, pathways relevant to the trait or disease of interest can be highlighted, for example, in the case of ciliary dysfunction in Bardet-Biedl syndrome [3], or autophagy in Crohn's disease [4].

Association signals are, by their very nature, rarely (if at all) causal for the trait studied. Establishing conclusive evidence of causality is arguably the only way biological insight will ever be achieved [5]. For genome-wide association (GWA) signals, causality is often hidden both in terms of pinpointing the exact single-nucleotide polymorphism (SNP) variant which is causal for the trait, because of linkage disequilibrium and incomplete ascertainment of genetic variation in the region of interest. Causality is also hidden in terms of the specific gene or genes which ultimately are responsible for the trait. An intermediate step along the path of genetic trait dissection to biological insight is to try to delineate and refine association signals to pinpoint the exact SNP and genes implicated in the trait. For example, it would be useful to narrow the regional occurrence of many genes in an association interval *by proximity* to one or a few genes which can be demonstratively and convincingly shown to relate to the trait of interest.

To provide clarity in this regard, fine-mapping studies of the loci of interest are proposed. Successful application of these studies are able to refine signals into candidates for casual variants and genes; both pieces of information can then be translated into functional studies whose aim is to understand the mechanism by which causal variants perturb the causal genes, thereby achieving the central goal of the mapping effort. The design of these studies using existing data is straightforward, and the associated caveats of these designs are relatively well understood [3]. Realizing that these studies are a bottleneck in translating whole genome associations into biological insight, intense effort in the community has been targeted to provide innovations in high-throughput sequencing, which are already beginning to make new datasets and technologies available which will transform the design and speed to which these studies can be performed [7, 8]. As a result, one can expect a wave of fine-mapping or other studies which seek to obtain this information in the years to come.

This chapter will begin with a description of the goals and requirements of fine-mapping effort and principles in attempting to select loci of interest. The chapter will then move on to consider the design in the context of next-generation sequencing technologies and will describe a case study demonstrating the analytical procedure employed by these studies. Finally, it will conclude with future considerations for fine-mapping and will describe a complementary approach — exonic resequencing by DNA pooling.

Definitions and Goals

Before going into significant detail, it is important to begin by enumerating the definition of what constitutes *delineating* an association signal — when can victory in fact be declared? Already, it needs to be presupposed that, at the locus of interest, there exists an association signal exceeding a genome-wide significance threshold, one which is the target of a refinement process. Additional considerations are required for hybrid designs which *combine* fine-mapping with replication or validation, where an association has yet to be demonstrated conclusively [9]. For these loci, there are several objectives of this broad process:

1. Pinpoint the set of genetic variants likely to be causal and generate the association signal with high posterior probability.
2. Identify the specific gene or genes in the region which play a causal role in the phenotype.
3. Translate the collection of casual variants into a biological function and/or interpretation.
4. Expand the search for causal variation by testing additional genetic variants at the locus.
5. Apply haplotype analysis to refine and interpret the influence of collections of genetic variants on the phenotype.

To achieve these objectives, this chapter will focus primarily on the *locus fine-mapping* approach. Later, a complementary approach will be described which specifically focuses on points 2 and 4 above, and involves *pooled, exonic resequencing* of genes in an associated interval.

Which Loci Should I Target?

Analysis of whole genome datasets are discovering new loci associated with disease at an expeditious pace: i.e., in just three short years, associations to type 2 diabetes (T2D) went from three loci to 41; Crohn's disease (CD) went from a handful to 30, and those notwithstanding new loci were expected to be reported by the end of calendar year 2010. As these lists continue to grow, given a finite budget to initiate follow-up studies, selecting the most tractable candidates for fine-mapping is critical to maximize the biological understanding extracted. Not every one of these loci is equally tractable to dissect genetically, and there are many practical considerations to ponder. For one, defining the size of the region will give a sense of the amount of genetic variation one is intending to catalog under some reasonable population genetic assumptions, and consequently how much genotyping on discovered variation one might be expected to perform. One way to define regions is to start at the initial association signal, and move along the chromosome in the 5′ and 3′ directions to the nearest recombination hotspots flanking the association (based on those estimated from the HapMap), adding an extra length of physical distance beyond the hotspot boundaries in the event that the hotspot fails to completely break down the haplotype structure for the given interval. For example, a particularly attractive fine-mapping target based on the size of the region relates to an association for T2D on chromosome 9p21, which happens to fall within a very narrow region (approximately 8 kb) flanked by two intense recombination hotspots (see Saxena et al. [10], Fig. 2A). In contrast, a much less attractive locus in this regard maps proximal to the genes *HHEX*, *IDE*, and *KIF11* [11] and encompasses nearly 400 kb under the same criteria (see Saxena et al. [10], Fig. 2E).

A second important consideration is the extent to which annotated features associated with the locus are present, namely, expressed sequences, predicted transcripts, or annotated genes, and if so, how many reside in the region of interest. Occasionally, an association signal will map to a coding change exactly, and for some traits such features have been identified: *SH2B3* for blood pressure and cardiovascular disease [12], *THADA* for T2D [13], *PTPN22* for CD [14], rheumatoid arthritis [15], and type 1 diabetes [16], and *ITGAM* for systemic lupus erythematosus (SLE) [17]. In general, however, examples like these where primary association exposes a gene for causality tend to be exceptional. As an example in the most extreme alternative case for Crohn's disease, six of the 30 loci map to regions without any known protein-coding genes, notwithstanding half of them containing more than one gene in strong linkage disequilibrium (LD) with the associated variant [3]. The most attractive candidates in this regard tend to have one or only a few genes nearby, e.g., ADAMTS9 in type 2 diabetes [13], which limits the number of plausible candidates one might have to entertain in downstream functional or genetic experiments.

Another major factor to consider is the expected statistical power one has available to perform studies on the target locus. Ultimately, power here depends on the effect size of the underlying casual variant (which one can approximate given that the initial signal is a close proxy for this variant), frequency of the target SNP, and the number of samples one has to invest in the experiment. If the goal is to improve the resolution of an existing signal, or to identify new associated variants independent of the previously established one, the larger the initial effect size, the fewer the samples one will require to conclusively answer either of these two questions. It should be noted that many online resources (for example, the Genetic Power Calculator) allow quick and easy computation of the expected power to detect association for given sample sizes under different genetic models, which includes consideration of LD [18].

A corollary to this consideration is the extent to which other traits also harbor associations in the region. As of August 4th, 2009, there were 2055 entries registered in the National Human Genome Research Institute NHGRI GWA studies catalog, 923 of which exceeded a genome-wide threshold ($p < 5 \times 10^{-8}$). For example, 9p21 is attractive in this regard due to distinctive associations to type 2 diabetes and cardiovascular disease occurring in a close physical proximity but in linkage equilibrium with each other [10, 19−22]. Given the evidence that diabetics are at increased risk for cardiovascular complications [23], a detailed understanding of the casual risk factors and implicated genes in this region would be particularly insightful for both phenotypic communities.

Besides statistical genetics approaches, bioinformatic tools provide an additional way to interrogate the collection of loci of interest, in order to prioritize regions with likely candidate gene or genes of interest, or to identify potential causal genes. These tools generally interrogate large and often but diverse aspects of data. A non-exhaustive list of tools utilized in this context includes:

1. The Human Mutational Database (HGMD). An updated collection of mutations in all genes that relate to a phenotype in humans.
2. Gene Relationships Among Implicated Loci (GRAIL). A text-based mining tool which takes a list of disease regions and automatically assesses the degree of relatedness of implicated genes based on word commonality from 250,000+ PubMed abstracts [24].
3. Protein-protein networks. These datasets are based on screens which identify proteins with direct interactions with each other.
4. REACTOME. An expertly curated database for a variety of biological pathways and networks.
5. Sylamer. A tool designed to detect enrichment of specific subsequence features across many large collections of sequences, designed particularly for detecting micro-RNAs [25].
6. Expression Screening. A method which integrates information from thousands of microarray datasets to identify genes that are consistently co-expressed with a target pathway across biological contexts [26, 27].

One general implication here is that each locus comes with individual challenges and potential traction to be genetically dissected *a priori*. An important aspect of any study design is to consider (and weight appropriately) as much information from a diverse portfolio of resources available for disposal. Often, there will be no "perfect" locus to fine-map, the best selection of considerations should be accumulated collectively, any of which can increase the likelihood of a successful mapping experiment: one which improves resolution, conclusively implicating a set of SNPs or genes likely to relate to the trait of interest. Once a set of loci has been identified as an ideal target, design of the fine-mapping experiment can begin, to which we will now turn.

LOCUS FINE-MAPPING: DESIGN AND A CASE STUDY

Design of a Fine-mapping Study

While not all studies are defined in precisely the same way, the fine-mapping approach has basic formulation principles which are easily described. The first step is to *define the scope* of the experiment. Initially, this is done simply by defining the interval around the association signal which will be mapped (see above for the discussion on a regional definition based on flanking recombination hotspots). A consideration of the depth of minor allele frequencies which are sought to be captured (1% or greater, 5% or greater, etc.) is also required, due the expense and power considerations to detect and validate associations given the sample size one has available for genotyping. Once the initial scope has been defined, the second step is to *collect a catalog of existing known variation*. This involves extracting a set of validated and known SNPs that reside in public databases such as dbSNP or the HapMap project. It is important to recognize that, at least for these specific resources, ascertainment bias exists in the frequency spectra of variation present in these sets toward common variation; i.e., lower frequency variation is generally under-represented relative to population genetic expectations. Because of this bias in the frequency spectrum, it is often desirable to augment this catalog with a *collection of newly discovered variants*, identified via resequencing of the target region (or exons of genes in the region). For this purpose, it is typical to select a small number of cases and controls from the cohort, or alternatively from a population-based sample without phenotypic ascertainment (e.g., the HapMap population which matches the ethnicity of the study). While resequencing is generally a sufficient starting point in collecting a comprehensive catalog of genetic variation, this step is quite expensive and time consuming. In future, this process will be augmented or replaced by content obtained from the 1000 Genomes project (1KG), which is described below.

Once a catalog has been formulated, *tagging* of this catalog is performed. There are many practical ways to consider tagging, ranging from the most expensive (perform genotyping on all variations that have been identified) to the least expensive option (minimal tagging of all common variations in the region). Often, a blend of inefficient and efficient tagging is often employed, in the following ways. For SNPs that are in modest LD ($r^2 > 0.3$) with the primary association signal, all are potential candidates for causality and ought to be included (rather than proxies allocated for them). Similarly, genetic variations lying in functional regions, in principle, are candidates for causality and ought to be typed directly. These include, but are not limited to, coding changes, splice-site variants, mutations in poly-A tails, variants in 3' or 5'-UTR sequences, mutations in known transcription factor binding sites, perhaps even mutations in regions highly conserved across species. Each of these classes of variation is highly likely to be a functional candidate and should be prioritized for genotyping, even disregarding of the observed frequency of the variant. Once SNPs in LD with the primary associated locus and functional variation have been selected, the remaining variation in the region can be tagged using standard practices which efficiently capture as many of the SNPs as possible while minimizing genotyping efforts [28]. One other consideration might be to provide redundant tags at regular intervals across the region, to protect against genotyping failure. Finally, the last step is simply to *genotype* the selected SNP variants. If motivated, and the appropriate reference datasets are available, imputation (see discussion below) can be used to directly fill in ungenotyped SNP data.

1KG and the Future of Fine-mapping

A determining factor in the success of the fine-mapping approach is a comprehensive catalog of genetic variation at the interrogated locus. Without such a catalog, the primary objective of the study (pinpointing a confidence set of genetic variation implicated in the trait) remains, at best, incomplete and potentially inconclusive. Until very recently, variant discovery in a small set of samples resequenced along a tractable span of DNA sequence was the only way in which such a catalog can be constructed. Using existing-generation sequencing technologies, this step is labor intensive and nearly cost prohibitive which places an incontrovertible limit on the number of samples resequenced. These clear limitations motivate the ascertainment of a deeper catalog of genetic variation across the genome in a large sample of individuals (across multiple ethnic groups), similar in scope to that of the International HapMap project [29]. Generating this rich catalog of sequence variation demands new technologies capable of efficiently sequencing the entire genomes of hundreds of samples at relatively low cost while maintaining high accuracy in base pair calls [7].

The mutual interest in developing these technologies, in combination with the direct tangible benefit for future genetic studies, consequently inspired formation of the 1000 Genomes project [8]. This project is an international research consortium aiming specifically to sequence approximately 1,200 people from major continental groups worldwide, receiving support from the National Human Genome Research Institute (NHGRI), part of the National Institutes of Health (NIH), the Beijing Genomics Institute Shenzhen in China, and the Wellcome Trust Sanger Institute in Hinxton, England. One stated aim of the project is to achieve a nearly complete catalog of common human genetic variants (minor allele frequency 1% or higher) by generating high-quality sequence data for >85% of the genome of the targeted samples. This catalog will fill in not only the spectra of common SNP variation, of which only a subset has been ascertained and reside in databases like dbSNP, but also copy number and short insertion/deletion polymorphisms. Importantly, these data are destined for the public domain, similar to that of the HapMap project. Already, the datasets from the high-coverage pilot for genome sequences was made publicly available in late December 2008, and data and variant calls from the second, low-coverage pilot of 280 samples proceed in earnest. Consequently, it can be anticipated that datasets, able to be leveraged for fine-mapping studies, will be available in relatively short order.

An Example Case Study: Fine-mapping at the IRF5 Locus

A particularly illustrative example of the challenges in fine-mapping a complex disease locus involves an early study of the interferon regulatory factor 5 gene, *IRF5*, variation of which at this locus has been shown to associate to SLE. Multiple groups have attempted to dissect multiple independent associations, in both Caucasian and Asian populations [30–34]. Initially, a common variant association was reported for a specific SNP (rs2004640), an intronic variant leading to an exon 1B alternative splice form of the gene [35]. After this initial discovery, *IRF5* was targeted for dissection and resequenced in a case–control cohort. Many new sequence variants were discovered, and of those three new variants in strong LD with one another were revealed (group 1), which strongly associated with the trait even more than the previously studied SNP rs2004640, and were independent of it [30]. After logistic regression analysis, conditioning on one of these newly discovered variants, a new group of correlated variants was shown to associate to SLE, which did in fact include the previously described SNP (group 2). Subsequent regression analysis conditional on SNPs from both group 1 and 2 uncovered yet another set of correlated SNPs associated with SLE independent of the previous sets (group 3). Finally, the collection of independent effects from each of these three groups seemed to explain all of the residual association at this locus, at least in that study.

Upon inspection of SNPs in each of these groups with functional considerations in mind, reasonable biological insight was hypothesized for risk contributed by group 2 (altered splice form of the gene) and group 3 (mRNA instability cause by disruption of a polyA+ signal sequence), but not obviously so for SNPs in group 1. This led to an additional conditional logistic regression not based on statistical association data but with functional considerations in mind. This revealed a new set of independent SNPs and tagging of a putatively disruptive insertion/deletion (in/del) polymorphism in exon 6 of the gene. Haplotype analysis was able to show that a haplotype containing all three risk variants demonstrated the highest risk of disease, with modest or protective effects dependent for haplotypes containing fewer of these functional alleles [30]. To augment an already complex story, a new group's fine-mapping effort in parallel identified additional SNP and length polymorphism variants in *IRF5* that also associated to SLE [31]. The primary associations in that study could be reduced to two variants: a marker from group 1 of the previous study (involving the disruptive exon 6 in/del) and another in/del only a short distance upstream of the first exon; the latter hypothesized to explain the associations reported for group 2 and group 3 of the previous study.

Mapping this locus in different ethnic groups consistently implicates this gene in the etiology of SLE, though inconsistently so with the previously listed causal risk factors identified in Caucasians. In a study of African-Americans, previously reported associations in studies of European populations were confirmed, with a majority of the haplotypic effects accounted for by a single SNP [34]. In Asian populations, risk factors shown to be polymorphic in Europeans were found to be monomorphic, but an entirely new collection of risk factors associating to SLE in these groups was instead identified as associated to SLE [32, 33].

Taken collectively, studies at this locus provide some insight into the complexities of the fine-mapping process with practical guidelines to take forward to future studies. First, future studies ought to begin with a careful statistical dissection of the locus of interest. In the case of *IRF5*, this was a critical step in disentangling effects which were independent with a potentially known functional cause, from those without an obvious cause. A key requirement that allowed this analysis to work was that the study itself was large enough (2,158 case and 3,596 control chromosomes were analyzed) and thus well powered to detect the effects examined (odds ratio for one copy was 1.46, two copies was 2.96 for the high risk haplotype) [30]. Second, the interpretation of statistical results in the context of functional information was crucial to understanding a plausible mechanism for risk, of which functional hypotheses were tested and supported by experimental data. Finally, with respect to multi-ethnic studies, even if a common gene is implicated as causal, there is no strict requirement for the same SNPs or same risk variants to be causal, though there are cases of SNP associations which do meet this criteria (see discussion below). Future work (and associated lessons learned) at this locus will continue to be relevant for future studies which seek to tackle fine-mapping of common disease loci.

DELINEATING ASSOCIATION SIGNALS IN THE FUTURE

Considered Complexities for Fine-mapping

The above example of fine-mapping at the *IRF5* locus was an early example of success these studies can have at revealing biological mechanisms underlying disease at specific loci of interest. Despite this success, other challenges loom on the horizon. One open question is how the fine-mapping approach can advance understanding in regions where no obvious coding sequence is apparent, if they can at all (see the examples in CD, above). Alternatively, these studies may demonstrate conclusively situations where a single SNP potentially implicates multiple genes as causal, like the case of the sortilin 1 (*SORT1*) locus [36]. Both of these examples break the previous genetic mapping paradigm of "one gene, one mutation". Future studies should continue to test and reject this parsimonious hypothesis first, before positing more complicated explanations for the data.

Another challenge to these studies relates to the statistical power in performing the fine-mapping experiment in the first place. For GWA loci that are only reproducibly associated in very large samples (tens of thousands), it is unlikely that insightful fine-mapping experiments can be performed on much smaller sets of data, especially for loci where the effect size estimated is modest. As stated above, one aspect of the fine-mapping experiment is to find the SNP(s) which most strongly associate to (and therefore provide the best tags for) the underlying casual variant (s). Robustly differentiating between random statistical fluctuations due to chance and systematic small changes odds ratios or in LD patterns provided by different SNP tags is likely to be extremely challenging unless one is sufficiently powered to make that assessment. This motivates the need for *collaborative* fine-mapping studies, analogous to consortia driving the genome-wide locus discovery efforts through meta-analysis, where groups of studies keen to target specific loci combine the results from analysis of genotyping data for an agreed-upon panel of SNP variation at the locus. Already, phenotypic consortia relating to metabolic, cardiovascular, and related traits are beginning these discussions, with the aim to develop new genotyping technologies designed to engage in the fine-mapping experiment for a large number of loci in a large number of samples in a cost efficient way, leveraging information obtained from the 1000 Genomes project.

One alternative proposal which can be pursued in parallel to collecting large sample sizes is to perform analysis in populations which have LD patterns more suitable for fine-mapping. For example, owing to population history, LD on average decays much quicker in African populations relative to Europeans, and as such fine-mapping studies in these populations is an attractive option to improve resolution at a locus as a practical alternative. Already, genome-wide trait mapping studies are under way across multiple ethnic groups, and these studies will not only discover a whole host of new loci underlying human phenotypes, but also aim to answer a fundamental question: how similar is the genetic architecture of disease across continental groups? On the one hand, a shared etiology for specific loci in specific genes has only been shown for a handful of cases, and the extent to which the same genes (or the same SNPs) are implicated in disease in different populations is a question that has not yet been fully answered. On the other hand, there are examples where some clarity has been obtained. In one example, for a locus related to T2D, *TCF7L2*, even though multiple SNPs at that locus associated in a sample of Caucasians [37], only one SNP so far has shown consistent association and magnitude of effect in a West African population [38]. In another example of mapping obesity and related traits which implicated the *FTO* gene, between two SNPs associated in near-perfect LD with each other in Caucasians, only one seemed to show any support in an African sample, suggesting an ancestral origin for the mutation and better resolution of the signal [39]. Discouragingly, none of these variants seem to associate to obesity in a well-powered sample of Asian descent [40], which perhaps exemplifies the requirement of shared etiology at the locus in order to reap the benefits of population-based LD differences.

Finally, while the costs of genotyping and sequencing continue to plummet, an open question regards the extent to which imputation — the process whereby unknown SNP genotypes are inferred from haplotype patterns in a reference sample — will be able to minimize genotyping efforts [41]. One attractive option is to pursue the fine-mapping design as described above. However, rather than genotype the sample for the entire set of discovered SNPs of interest, genotype a subset of the discovered variants and, in conjunction with a sequenced (and partially phased) data set as a reference, infer the remainder of ungenotyped SNPs in the sample. Future work will explore strategies to optimize the prediction of low frequency genetic variants (the bulk of genetic variation likely to be discovered by sequencing experiments) and design a fine-mapping experiment which tags these variants, given the cost constraints of genotyping. Innovations are already being advanced which improve imputation given multiple reference samples as well as their size [42].

Exonic Pooled Resequencing: A Complementary Approach

The fine-mapping procedure is an experimental method in which the primary association at the target locus of interest can be investigated and understood, as well as any other variation common in the population (1% frequent and greater). One quantity that remains unknown at present is the extent to which rare variants are implicated in the genetic architecture of phenotypic traits. Rare variants have been hypothesized to contribute to the impact a locus has on a trait (in terms of the inter-individual population variance explained), and perhaps even to explain (from partially to completely) the initial common variant association that led one to the locus in the first place. Regardless of prior hypotheses regarding the contribution of these alleles to the architecture of traits, collecting this information and constructing powerful tests to directly test this hypothesis is a laudable goal. In principle, the identification of rare functional variants can not only help identify causal genes, but can give insight into the mechanism of causality (e.g., loss of function at this gene is associated with specific direction of change in the trait) [43].

In order to assess the rare variant hypothesis, collections of rare variation at target genes are required, and this is primarily accomplished by resequencing studies. The primary goal of these studies is the detection of additional genetic variation that associates with disease, which is not captured by existing variation catalogs. The maximally powered design with unlimited resources would simply be to sequence the entire genomes of every case and control sample available. With current technologies, this approach is cost prohibitive. Two strategies are employed to reduce costs, namely: (1) limit the amount of territory targeted, in order concentrate the sequencing coverage on specific regions (exons), and (2) performing the experiment in *pools* of cases and controls. This provides an assay of genetic variation in thousands of samples effectively. In this more cost efficient experiment, the goal changes slightly to variation *detection*, followed by association testing on validated SNP discoveries in a replication sample large enough to be powered to detect the purported genetic effect. Eventually, the technology will be sufficiently accurate using methods (already under development) that will allow some initial association analysis to be performed on the pools directly, given proper design features of the experiment.

The design of the pooling experiment principally involves two features. The first feature deals with the sample ascertainment question. Based on previous design and quality control considerations, it is useful first to consider if sequencing should be done in pools of cases, controls, or both. Pooling in cases is likely to ascertain risk alleles which are disease causing; control samples will likely ascertain genetic variation expected from close to a random population sample. Providing case and control pools into the experiment might be considered if one wished to estimate case and control frequencies for given variants, as an initial scan for high priority risk or protective variants. A second consideration is to, if possible, maximize the genetic load of factors modulating the trait using sample ascertainment. One way this could be done would be to account for non-genetic risk factors during the ascertainment process, and condition admittance into the study based on this. For example, suppose an environmental exposure was a risk factor for a genetically heritable trait. In that case, ascertainment of sample with minimal environmental exposure would consequently enrich the frequency of genetic factors altering the trait of interest. Another way to enrich for genetic load would be to search for phenotypic extremes in the liability or trait distribution. This latter tactic is a mainstay and has in fact worked successfully in the context of sequencing and variant discovery [43]. The second feature addresses the way in which samples are pooled. In fact, there exist specific design proposals which, when overlapping designs are utilized, can unambiguously determine genotype of the sample given the variants detected, improving the quality of the sequencing experiment [44, 45].

Once genetic variation has been detected, two types of analysis can be performed to assess association with disease. The most straightforward analysis is simply a *variant by variant* approach, where each SNP is tested in turn for association to the trait of interest, analogous to what has been performed in the whole genome association study context. This approach is well understood with previously established statistical thresholds and associated analytical methods, assuming the variation tested is common (greater than 1%), or minimally that the number of samples tested is sufficiently large that asymptotic properties of the association analysis hold. These properties tend to fail individually for variants observed once, twice, or only a few times in the sample, and population genetics intuition suggests that most genetic variation discovered will be of the rare variety. In those cases, *analysis of burden* has been proposed as a methodological approach to assess the relationship between a collection of rare alleles and trait [46–48]. In either case, once a SNP or collection of rare SNPs have been identified, genotyping in a large number of samples will be required to conclusively demonstrate reproducible association. However, once this has been done, a strong candidate gene will emerge from the region, which can then be interpreted in the context of additional genetic variation in the region (see the case study for *IRF5*, above). Already, pooled sequencing experiments following this design strategy have discovered unambiguous associations, firmly establishing casual genes for common genetic traits [49], and likely the number of examples of this will grow in the not too distant future.

CONCLUDING REMARKS

The end result of this experiment represents a means to an end. The next set of difficult experiments take the results of the *improved* genetics resolution at the locus, advancing only the most interesting SNPs into a function assay in which the SNP can be tested for causality. The target goal of these functional experiments is to elucidate genetic mechanism based on identification of causal variants and/or genes. In this regard, there are many ways in which the beginnings of a mechanism can be understood: *in vitro* enhancer or silencer assays; RNAi knockdown experiments; measuring gene expression in cells containing mutations or haplotypes of interest, in cases and control populations; site-specific mutagenesis experiments; creation of knockout mice of the causal gene or genes, and so on. All of these experiments are time-consuming and labor-intensive; consequently, achieving the goal of fine-mapping and reducing the collection of genetic variation required to be taken into these assays can only improve inference regarding disease biology, as well as the pace at which genetic data can be translated into biological insight with the goal of improving human health.

WORLD WIDE WEB URL LINKS

HGMD: http://www.hgmd.cf.ac.uk/ac/index.php
GRAIL: http://www.broadinstitute.org/mpg/grail/
REACTOME: http://www.reactome.org/
Sylamer: http://www.ebi.ac.uk/enright/sylamer/
Genetic Power Calculator: http://pngu.mgh.harvard.edu/~purcell/gpc/

Acknowledgments

The author would like to thank Jessica Shae and Vineeta Agarwala for helpful discussion about fine-mapping design and functional experimental considerations, Hyun Min Kang for careful work and practical guidelines in selecting SNP tags in fine-mapping studies leveraging data generated by the 1000 Genomes project, and Ben Neale for helpful comments after a careful reading of the chapter.

References

[1] L.A. Hindorff, P. Sethupathy, H.A. Junkins, E.M. Ramos, J.P. Mehta, F.S. Collins, et al., Potential etiologic and functional implications of genome-wide association loci for human diseases and traits, Proc. Natl. Acad. Sci. USA 106 (2009) 9362–9367.

[2] M.I. McCarthy, G.R. Abecasis, L.R. Cardon, D.B. Goldstein, J. Little, J.P. Ioannidis, et al., Genome-wide association studies for complex traits: consensus, uncertainty and challenges, Nat. Rev. Genet. 9 (2008) 356–369.

[3] S.J. Ansley, J.L. Badano, O.E. Blacque, J. Hill, B.E. Hoskins, C.C. Leitch, et al., Basal body dysfunction is a likely cause of pleiotropic bardet-biedl syndrome, Nature 425 (2003) 628–633.

[4] J.D. Rioux, R.J. Xavier, K.D. Taylor, M.S. Silverberg, P. Goyette, A. Huett, et al., Genome-wide association study identifies new susceptibility loci for Crohn disease and implicates autophagy in disease pathogenesis, Nat. Genet. 39 (2007) 596–604.

[5] G.P. Page, V. George, R.C. Go, P.Z. Page, D.B. Allison, "Are we there yet?": deciding when one has demonstrated specific genetic causation in complex diseases and quantitative traits, Am. J. Hum. Genet. 73 (2003) 711–719.

[6] J.P. Ioannidis, G. Thomas, M.J. Daly, Validating, augmenting and refining genome-wide association signals, Nat. Rev. Genet. 10 (2009) 318–329.

[7] D.R. Bentley, Whole-genome resequencing, Curr. Opin. Genet. Dev. 16 (2006) 545–552.

[8] B.M. Kuehn, 1000 genomes project promises closer look at variation in human genome, JAMA 300 (2008) 2715-2715.

[9] G.M. Clarke, K.W. Carter, L.J. Palmer, A.P. Morris, L.R. Cardon, Fine mapping versus replication in whole-genome association studies, Am. J. Hum. Genet. 81 (2007) 995–1005.

[10] R. Saxena, B.F. Voight, V. Lyssenko, et al., Genome-wide association analysis identifies loci for type 2 diabetes and triglyceride levels, Science 316 (2007) 1331–1336.

[11] R. Sladek, G. Rocheleau, J. Rung, C. Dina, L. Shen, D. Serre, et al., A genome-wide association study identifies novel risk loci for type 2 diabetes, Nature 445 (2007) 881–885.

[12] C. Newton-Cheh, T. Johnson, V. Gateva, M.D. Tobin, M. Bochud, L. Coin, et al., Genome-wide association study identifies eight loci associated with blood pressure, Nat. Genet. 41 (2009) 666–676.

[13] E. Zeggini, L.J. Scott, R. Saxena, B.F. Voight, J.L. Marchini, T. Hu, et al., Meta-analysis of genome-wide association data and large-scale replication identifies additional susceptibility loci for type 2 diabetes, Nat. Genet. 40 (2008) 638–645.

[14] J.C. Barrett, S. Hansoul, D.L. Nicolae, J.H. Cho, R.H. Duerr, J.D. Rioux, et al., Genome-wide association defines more than 30 distinct susceptibility loci for Crohn's disease, Nat. Genet. 40 (2008) 955–962.

[15] A.B. Begovich, V.E. Carlton, L.A. Honigberg, S.J. Schrodi, A.P. Chokkalingam, H.C. Alexander, et al., A missense single-nucleotide polymorphism in a gene encoding a protein tyrosine phosphatase (ptpn22) is associated with rheumatoid arthritis, Am. J. Hum. Genet. 75 (2004) 330–337.

[16] N. Bottini, L. Musumeci, A. Alonso, S. Rahmouni, K. Nika, M. Rostamkhani, et al., A functional variant of lymphoid tyrosine phosphatase is associated with type I diabetes, Nat. Genet. 36 (2004) 337–338.

[17] S.K. Nath, S. Han, X. Kim-Howard, J.A. Kelly, P. Viswanathan, G.S. Gilkeson, et al., A nonsynonymous functional variant in integrin-alpha(m) (encoded by ITGAM) is associated with systemic lupus erythematosus, Nat. Genet. 40 (2008) 152–154.

[18] S. Purcell, S.S. Cherny, P.C. Sham, Genetic power calculator: design of linkage and association genetic mapping studies of complex traits, Bioinformatics 19 (2003) 149–150.

[19] R. McPherson, A. Pertsemlidis, N. Kavaslar, A. Stewart, R. Roberts, D.R. Cox, et al., A common allele on chromosome 9 associated with coronary heart disease, Science 316 (2007) 1488–1491.

[20] A. Helgadottir, G. Thorleifsson, A. Manolescu, S. Gretarsdottir, T. Blondal, A. Jonasdottir, et al., A common variant on chromosome 9p21 affects the risk of myocardial infarction, Science 316 (2007) 1491–1493.

[21] L.J. Scott, K.L. Mohlke, L.L. Bonnycastle, C.J. Willer, Y. Li, W.L. Duren, et al., A genome-wide association study of type 2 diabetes in Finns detects multiple susceptibility variants, Science 316 (2007) 1341–1345.

[22] E. Zeggini, M.N. Weedon, C.M. Lindgren, T.M. Frayling, K.S. Elliott, H. Lango, et al., Replication of genome-wide association signals in UK samples reveals risk loci for type 2 diabetes, Science 316 (2007) 1336–1341.

[23] C.S. Fox, M.J. Pencina, P.W. Wilson, N.P. Paynter, R.S. Vasan, R.B. D'Agostino, Lifetime risk of cardiovascular disease among individuals with and without diabetes stratified by obesity status in the Framingham heart study, Diabetes Care 31 (2008) 1582–1584.

[24] S. Raychaudhuri, R.M. Plenge, E.J. Rossin, A.C. Ng, International Schizophrenia Consortium, S.M. Purcell, et al., Identifying relationships among genomic disease regions: predicting genes at pathogenic SNP associations and rare deletions, PLoS Genet. 5 (2009) e1000534.

[25] S. van Dongen, C. Abreu-Goodger, A.J. Enright, Detecting microrna binding and sirna off-target effects from expression data, Nat. Methods 5 (2008) 1023–1025.

[26] J.M. Baughman, R. Nilsson, V.M. Gohil, D.H. Arlow, Z. Gauhar, V.K. Mootha, A computational screen for regulators of oxidative phosphorylation implicates SLIRP in mitochondrial RNA homeostasis, PLoS Genet. 5 (2009) e1000590.

[27] R. Nilsson, I.J. Schultz, E.L. Pierce, K.A. Soltis, A. Naranuntarat, D.M. Ward, et al., Discovery of genes essential for heme biosynthesis through large-scale gene expression analysis, Cell Metab. 10 (2009) 119–130.

[28] P.I. de Bakker, R. Yelensky, I. Pe'er, S.B. Gabriel, M.J. Daly, D. Altshuler, Efficiency and power in genetic association studies, Nat. Genet. 37 (2005) 1217–1223.

[29] International HapMap Consortium, The International HapMap Project, Nature 426 (2003) 789–796.

[30] R.R. Graham, C. Kyogoku, S. Sigurdsson, I.A. Vlasova, L.R. Davies, E.C. Baechler, et al., Three functional variants of IFN regulatory factor 5 (IRF5) define risk and protective haplotypes for human lupus, Proc. Natl. Acad. Sci. USA 104 (2007) 6758–6763.

[31] S. Sigurdsson, H.H. Goring, G. Kristjansdottir, L. Milani, G. Nordmark, J.K. Sandling, et al., Comprehensive evaluation of the genetic variants of interferon regulatory factor 5 (IRF5) reveals a novel 5 bp length polymorphism as strong risk factor for systemic lupus erythematosus, Hum. Mol. Genet. 17 (2008) 872–881.

[32] H.D. Shin, I. Kim, C.B. Choi, S.O. Lee, H.W. Lee, S.C. Bae, Different genetic effects of interferon regulatory factor 5 (IRF5) polymorphisms on systemic lupus erythematosus in a Korean population, J. Rheumatol. 35 (2008) 2148–2151.

[33] A. Kawasaki, C. Kyogoku, J. Ohashi, R. Miyashita, K. Hikami, M. Kusaoi, et al., Association of IRF5 polymorphisms with systemic lupus erythematosus in a Japanese population: support for a crucial role of intron 1 polymorphisms, Arthritis Rheum. 58 (2008) 826–834.

[34] J.A. Kelly, J.M. Kelley, K.M. Kaufman, J. Kilpatrick, G.R. Bruner, J.T. Merrill, et al., Interferon regulatory factor-5 is genetically associated with systemic lupus erythematosus in African Americans, Genes Immun. 9 (2008) 187–194.

[35] R.R. Graham, S.V. Kozyrev, E.C. Baechler, M.V. Reddy, R.M. Plenge, J.W. Bauer, et al., A common haplotype of interferon regulatory factor 5 (IRF5) regulates splicing and expression and is associated with increased risk of systemic lupus erythematosus, Nat. Genet. 38 (2006) 550–555.

[36] S. Kathiresan, O. Melander, C. Guiducci, A. Surti, N.P. Burtt, M.J. Rieder, et al., Six new loci associated with blood low-density lipoprotein cholesterol, high-density lipoprotein cholesterol or triglycerides in humans, Nat. Genet. 40 (2008) 189–197.

[37] S.F. Grant, G. Thorleifsson, I. Reynisdottir, R. Benediktsson, A. Manolescu, J. Sainz, et al., Variant of transcription factor 7-like 2 (TCF7L2) gene confers risk of type 2 diabetes, Nat. Genet. 38 (2006) 320–323.

[38] A. Helgason, S. Palsson, G. Thorleifsson, S.F. Grant, V. Emilsson, S. Gunnarsdottir, et al., Refining the impact of TCF7L2 gene variants on type 2 diabetes and adaptive evolution, Nat. Genet. 39 (2007) 218–225.

[39] S.F. Grant, M. Li, J.P. Bradfield, C.E. Kim, K. Annaiah, E. Santa, et al., Association analysis of the FTO gene with obesity in children of Caucasian and African ancestry reveals a common tagging SNP, PLoS One 3 (2008) e1746.

[40] H. Li, Y. Wu, R.J. Loos, F.B. Hu, Y. Liu, J. Wang, et al., Variants in the fat mass-and obesity-associated (FTO) gene are not associated with obesity in a Chinese Han population, Diabetes 57 (2008) 264–268.

[41] J. Marchini, B. Howie, S. Myers, G. McVean, P. Donnelly, A new multipoint method for genome-wide association studies by imputation of genotypes, Nat. Genet. 39 (2007) 906–913.

[42] B.N. Howie, P. Donnelly, J. Marchini, A flexible and accurate genotype imputation method for the next generation of genome-wide association studies, PLoS Genet. 5 (2009) e1000529.

[43] J. Cohen, A. Pertsemlidis, I.K. Kotowski, R. Graham, C.K. Garcia, H.H. Hobbs, Low LDL cholesterol in individuals of African descent resulting from frequent nonsense mutations in PCSK9, Nat. Genet. 37 (2005) 161–165.

[44] S. Prabhu, I. Pe'er, Overlapping pools for high-throughput targeted resequencing, Genome Res. 19 (2009) 1254–1261.

[45] Y. Erlich, K. Chang, A. Gordon, R. Ronen, O. Navon, M. Rooks, et al., DNA Sudoku – harnessing high-throughput sequencing for multiplexed specimen analysis, Genome Res. 19 (2009) 1243–1253.

[46] S. Morgenthaler, W.G. Thilly, A strategy to discover genes that carry multi-allelic or mono-allelic risk for common diseases: a cohort allelic sums test (CAST), Mutat. Res. 615 (2007) 28–56.

[47] B. Li, S.M. Leal, Methods for detecting associations with rare variants for common diseases: application to analysis of sequence data, Am. J. Hum. Genet. 83 (2008) 311–321.

[48] B.E. Madsen, S.R. Browning, A groupwise association test for rare mutations using a weighted sum statistic, PLoS Genet. 5 (2009).

[49] S. Nejentsev, N. Walker, D. Riches, M. Egholm, J.A. Todd, Rare variants of IFIH1, a gene implicated in antiviral responses, protect against type 1 diabetes, Science 324 (2009) 387–389.

A Genome-wide Association Case Study on Obesity

C.M. Lindgren [1,2]

[1] Wellcome Trust Centre for Human Genetics, University of Oxford, UK
[2] Oxford Centre for Diabetes, Endocrinology and Metabolism,
University of Oxford, UK

Overweight and obesity are two words receiving widespread attention with global projections of more than 2.16 billion overweight and 1.12 billion obese individuals by 2030 [1]. In the United States, the prevalence of obesity (body mass index, BMI $\geq$ 30) increased 24% between 2000 and 2005 and during this period extreme obesity (BMI $\geq$ 40) and super-obesity (BMI $\geq$ 50) increased by more than 50 and 75%, respectively [2]. It is thus clear that obesity is presenting an important, worldwide clinical and public health burden which is associated with social and personal stigmatization as well as an increase in both morbidity and mortality [3], increased risk of diabetes, cardiovascular disease, osteoarthritis and cancer at population levels [4, 5]. Current intensive efforts to reduce obesity by diet, exercise, education, surgery and drug therapies are obviously failing to provide an effective, long-term solution to this epidemic.

The impact of environmental factors (i.e., food intake and energy expenditure) on obesity is significant but still poorly understood and it is clear that the condition has an equally noteworthy underlying genetic component. Twin studies have indicated heritability estimates of ~70% for BMI, both in children and adults [6–8] and other measures of obesity (skinfold thickness, waist circumference, total fat mass and regional fat distribution) show similar values [6–15]. There are differences among ethnic and racial groups in obesity and recent efforts in admixture mapping have demonstrated that obesity correlates closely with the percentage of ancestry deriving from ethnic groups with elevated prevalence of obesity [16, 17]. Given this, dissection of the genetic etiology of obesity is pursued with the hope of unraveling pathways and mechanisms that control obesity, that in turn would lead to better prevention, management and therapy.

Until 2006, the main approaches used to track down common variants influencing obesity involved either hypothesis-free genome-wide linkage mapping in families with multiple obese subjects or association studies within "candidate" genes using case–control samples or parent–offspring trios. The former suffered from being seriously underpowered for any sensible susceptibility models, as linkage is best placed to detect variants with high penetrance. As far as we can tell, common variants with high penetrance do not contribute substantially to the risk of common forms of obesity and few, if any, robust signals have emerged from such efforts [18]. The latter candidate-gene association approach has historically been compromised by difficulties associated with choosing credible gene candidates. Selection was typically based on a particular hypothesis about the biological mechanisms that are putatively involved in obesity pathogenesis but, because the function of much of the genome is poorly characterized, it remains almost impossible to make fully informed decisions. In addition, all too often these candidate-gene studies were conducted in sample sets that were far too small to offer confident detection of variants with the kinds of effect sizes that are now known to be realistic. With hindsight, it is easy to appreciate why these approaches yielded so few examples of genuine obesity-susceptibility variants.

As a consequence, the efforts to identify the genetic variants that are proven to contribute to differences in predisposition to obesity have largely been characterized by slow progress and limited success in common forms of obesity, which is in sharp contrast to the more successful gene identification in monogenic and syndromic forms of obesity [18]. The 2005 edition (the most recent at the time of writing) of the "Human Obesity Gene Map" gives an excellent overview of this when it lists 11 single gene mutations; 50 loci related to Mendelian syndromes relevant to human obesity; 244 knockout or transgenic animal models, and 127 candidate genes, of which slightly less than 20% are supported by five or more studies related to obesity [18]. Further, 253 quantitative trait loci (QTL) for obesity-related phenotypes from 61 genome-wide linkage scans have been reported and of these 20% have been supported by more than one study [18].

Over the past three years it has become possible, from both technical and economic perspectives, to undertake hypothesis-free genome-wide association (GWA) testing in samples of sufficient size to generate convincing association results. The advent of the GWA approach was the result of at least two components: the first was the human genome sequence which subsequently enabled the catalog of patterns of human genome-sequence variation through the efforts of the International HapMap Consortium (http://www.hapmap.org) [19]. One of the important messages generated by the HapMap was that (in non-African-descent populations at least) extensive correlations between neighboring single nucleotide polymorphisms (SNPs), i.e., linkage disequilibrium (LD), could constrain the number of independent genetic tests required to survey the genome, such that ~80% of all common variation could be sampled with as few as 500,000 carefully chosen SNPs [20, 21]. Second, new genotyping methods were developed that addressed the technical challenges required to perform massively parallel SNP typing at high accuracy and low cost [22].

Hereto results from a total of 11 "high-density" GWA scans (i.e., $\geq$300,000 SNPs, offering genome-wide coverage >65%) for obesity (as defined by BMI) have been published (Table 18.1). These studies together have yielded at least 17 loci that are robustly associated with obesity (i.e., show p-values $< 5 \times 10^{-8}$ in imputed datasets or $<5 \times 10^{-7}$ in directly genotyped data only) (Table 18.2) as well as a handful of promising associations that beg further investigation (Table 18.2).

The first gene that was identified to be unequivocally associated with common, non-syndromic obesity [23], FTO (fat mass and obesity associated), was initially identified as a result of a GWA of type 2 diabetes (T2D) [24] where it was the second strongest associated loci but where the entire association was completely abolished when adjusting for T2D. The association of the FTO region with obesity explains ~1% of the BMI heritability and the increase of 2-3 kg of weight gain seen in adults homozygous for the risk allele [23]. Interestingly, FTO is reported to operate on fat mass and has been suggested to encode a 2-Oxoglutarate-Dependent Nucleic Acid Demethylase involved in regulation of food intake [25]. In parallel it has been reported to be involved in decreased lipolytic effect in adipocytes [26]. Thus the role of the FTO loci still begs further investigation. Further, it is not clear whether the association is acting through FTO or the adjacent FTM gene.

With reports of the first dichotomous trait associations [27] and the discovery of FTO [23, 28, 29] and the first other quantitative trait loci (such as HMGA2 for height [30]) came the realization that the effect sizes would be smaller than anticipated, and that success would require analyses of far larger sample sizes than previously considered. This objective has catalyzed large-scale international collaboration and meta-analysis of existing data in various cohorts.

TABLE 18.1 Overview of GWA scans or meta-analysis thereof for obesity phenotypes

Study name (if any)	Number of samples in discovery cohort	Origin of discovery cohort	Phenotype	Reference	Comment
The Wellcome Trust Case Control Consortium	1,924	Northern European	BMI-quantitative analysis	[23]	10K+500K arrays
LOLIPOP	>4,000	European	BMI-Waist circumference (WC) quantitative analysis	[29]	300K array
	2,684	Indian Asians	Insulin resistance and related quantitative phenotypes	[32]	Meta-analysis of obesity related traits, 500K markers
	16,876	Northern European	BMI-quantitative analysis	[31]	Meta-analysis of obesity related traits
The CHARGE consortium	31,373 individuals	European	WC-quantitative analysis	[38]	Meta-analysis of obesity related traits
The GIANT consortium	38,580 individuals	European	WC and waist:hip-ratio (WHR)-quantitative analysis	[39]	500K markers
	775 cases and 3197 unascertained controls	Eurpoean	Extreme obesity/BMI	[37]	350K markers
	1,380 and 1,416 age-matched normal-weight control	Europeans	Early-onset and morbid adult obesity	[36]	300K markers
DeCODE	37,347	Europeans + African-Americans	BMI-quantitative analysis	[34]	Meta-analysis of obesity related traits
The GIANT consortium	>32,000	Europeans	BMI-quantitative analysis	[35]	500K markers
KARE	8,842	Asian	BMI, WHR-quantitative analysis	[46]	

TABLE 18.2

Closest gene	Chromosomal location	Phenotype	Associated lead SNP	Proposed molecular or cellular function	Additional phenotypes	Expression[a]	Ref.[b]
Gene desert; GNPDA2 is one of three genes nearby	4p13	BMI	rs10938397	—	Associated with T2D[c]	A	4
MTCH2 (locus with 14 genes)	11p11.2	BMI	rs10838738	Cellular apoptosis	—	A, H, L	4
SEC16B, RASAL2	1q25	BMI	rs10913469	—	—	L	5
ETV5 (locus with three genes, strongest association in ETV5)	3q27	BMI	rs7647305	—	—	None	5
Locus containing NCR3, AIF1 and BAT2	6p21	BMI	rs2844479, rs2260000, rs1077393	—	Associated with weight, not BMI	NCR3: A, H AIF1: H BAT2: H	5
BDNF (locus with four genes, strongest association near BDNF)	11p14	BMI	rs4074134, rs4923461, rs925946, rs10501087, rs6265	BDNF expression is regulated by nutritional state and MC4R signaling	Associated with T2Dc. Individuals with WAGR syndrome with BDNF deletion have BMI > 95th percentile. Bdnf knockdown in mouse hypothalamus causes hyperphagia and obesity	H	5

(Continued)

Table 18.2—cont'd

Closest gene	Chromosomal location	Phenotype	Associated lead SNP	Proposed molecular or cellular function	Additional phenotypes	Expression[a]	Ref.[b]
FAIM2 (locus also contains BCDIN3D)	12q13	BMI	rs7138803	Adipocyte apoptosis	—	A, H	5
PRL	6p22.2–p21.3	BMI	rs4712652	—	—	Pituitary only	6
PTER	10p12	BMI	rs10508503	—	—	H, L	6
MAF	16q22–q23	BMI	rs1424233	Transcription factor involved in adipogenesis and insulin-glucagon regulation	—	No data	6
NPC1	18q11.2	BMI	rs1805081	Intracellular lipid transport	Npc1-null mice show late-onset weight loss and poor food intake. NPC1 interferes with function of raft-associated insulin receptor signaling	A, H	6
FTO	16q22.2	BMI	rs9939609, rs6499640, rs8050136, rs3751812, rs7190492, rs8044769	Neuronal function associated with control of appetite	Associated with T2Dc	A, H	1, 2, 3, 4, 5, 6
NEGR1	1p31	BMI	rs2815752, rs3101336, rs2568958	Neuronal outgrowth	—	A	4, 5

Gene	Locus	Trait	SNPs	Function	Associated with T2D[c]		References
TMEM18 (closest gene)	2p25	BMI	rs6548238, rs2867125, rs4854344, rs7561317	Neural development		None	4, 5
SH2B1 (locus with 19–25 genes)	16p11.2	BMI	rs7498665, rs8049439, rs4788102, rs7498665	Neuronal role in energy homeostasis	Sh2b1-null mice are obese and diabetic	A, H	4, 5
KCTD15	19q13.11	BMI	rs11084753, rs29941	—	—	A, H	4, 5
MC4R	18q22	BMI	rs17782313, rs12970134	Hypothalamic signaling	Haploinsufficiency in humans is associated with morbid obesity. MC4R-deficient mice show hyperphagia and obesity	A, H, L	4, 5, 6, 7, 8
LYPLAL1; ZC3H11B	1q41	WHR	rs2605100	—	—		
TFAP2B	6p12	WC, BMI	rs987237		—		
MSRA	8p23.1	WC	rs7826222		—		
NRXN3	14q31	WC, BMI	rs10146997				

[a] A, adipocyte; H, hypothalamus; L, liver (according to https://biogps.gnf.org/).
[b] 1. Frayling et al. [23]; 2. Scuteri et al. [29]; 3. Dina et al. [28]; 4. Willer et al. ($P < 5\times10^{-8}$) [35]; 5. Thorleifsson et al. ($P < 1.6\times10^{-7}$) [34]; 6. Meyre et al. ($P < 5\times10^{-7}$) [36]; 7. Loos et al. ($P < 5\times10^{-8}$) [31]; 8. Chambers et al. [32].
[c] Associated with T2D as determined by Willer et al. [35] and Thorleifsson et al. [34]. WAGR, Wilms' tumor, Aniridia, genitourinary anomalies and mental retardation.

Through the first such large-scale obesity collaboration, there has been a strong association between SNPs located 188 kilobases (kb) downstream from the melanocortin 4 receptor gene (MC4R) (Table 18.2) and BMI in ~17,000 samples of European descent (Table 18.1) [31]. It was reported in parallel to be associated with obesity and waist circumference in ~15,000 individuals of Indian Asian or European ancestry (Table 18.1) [32]. The risk variant has subsequently been associated with larger intake of high energy and fat [33]. Also, the increased BMI reported in children is consistent with early onset obesity caused by MC4R mutations [31].

Even larger GWA meta-analysis efforts, through the Genetic Investigation of Anthropometric Trait (GIANT) Consortia and through deCODE using large cohorts comprising 32,000 and 25,000 individuals, respectively (Table 18.1), followed rapidly and brought eight new obesity loci to the community (Table 18.2) [34, 35]. Six of these loci were reported by Willer et al. [35] and several of the likely causal genes in the associated regions are highly expressed or known to act in the central nervous system (CNS), suggesting, as in rare monogenic forms of obesity, the role of this pathway in predisposition to overall obesity [35]. Both studies confirm the MC4R and FTO associations and four of the loci reported in Willer et al. [35] were overlapping with the loci reported in Thorleifsson et al. [34] who in addition reported five novel loci (Table 18.2).

In parallel, a smaller GWAS by Meyre et al. [36] focused on more extreme forms of obesity (early-onset and morbid adult obesity) and provided a list of complementary loci showing overlap with the other two studies in the FTO and MC4R associations but also identified three additional associations (Table 18.2) [36]. Cotsapas et al. also published a GWAS of 775 cases and 3,197 unascertained controls at ~550,000 autosomal markers [37]. They found nominal evidence of association to the FTO gene and six of 12 other loci previously reported to influence BMI as well as one of three associations reported by Meyre et al. This suggests that variants influencing BMI also contribute to severe obesity, which then would represent a condition at the extreme of the phenotypic spectrum rather than a distinct condition.

As described above, efforts to identify common and rare variants influencing BMI and risk of obesity have emphasized a key role of neuronal regulation of overall or general obesity [23, 28, 29, 31, 34–37] but until recently provided fewer clues to processes that are specifically responsible for individual variation in central obesity and fat distribution [32, 38, 39]. Measures of central and general adiposity are highly correlated: BMI has $r^2 \sim 0.9$ with waist circumference (WC) and ~0.6 with waist-hip ratio (WHR). Also, WC and WHR are correlated with measures of intra-abdominal fat measured by magnetic resonance imaging (MRI) in obese women ($r^2 \sim 0.6$ and 0.5, respectively) [40]. Several lines of evidence indicate that individual variability in patterns of fat distribution involves local, depot-specific and body-shape processes, which are likely independent of the mechanisms that control overall energy balance and general obesity. First, anthropometric measures of central adiposity are highly heritable [41] and, after correcting for BMI, heritability estimates remain high (~60% for WC and ~45% for WHR) [12–14]. Second, there are substantial gender-specific differences in fat distribution, and these appear to reflect specific and separate genetic influences [42]. Third, inherited lipodystrophies, which are monogenic syndromes, demonstrate that DNA variants can have specific effects on the development and/or maintenance of specific regional fat depots and body shape [43]. Lastly, three GWAS studies of central obesity (WC and WHR) have been published [32, 38, 39] (Table 18.1) which have identified four novel, common obesity loci (Table 18.2). Of these, at least one appears to be independent from a general obesity effect and specific to central obesity and fat distribution; *LYPLAL1* [39].

In summary, the GWAS strategy has obvious limitations in identification of rare and low frequency variants [44, 45] as anticipated; nevertheless this does not negate the fact that the GWAS efforts have moved forward positions of the field and the loci that have been robustly associated to common forms of obesity over recent years. Caution as to the interpretation of the results and associations is, however, warranted as the associated genetic variants are not always within known genes and occasionally span large chromosomal areas containing a large number of genes. Thus, in most cases, we cannot say for certain that we have identified "the smoking gun", this is a challenging task that lies ahead of us. Large-scale fine-mapping and resequencing efforts are needed to catalog and evaluate the genetic variation from the full allele spectra in associated regions. The use of the different patterns of linkage disequilibrium in samples from different ethnicities in these efforts is an obvious attempt to home in on regions with higher likelihood of containing the etiological variant. Dissecting function and proving causation of genetic variants is not necessarily a straightforward task, even for relatively simple cases of monogenic diseases. A challenging task lies before the research community to, robustly, translate these findings into characterization of function and consequences on physiology.

References

[1] T. Kelly, W. Yang, C.S. Chen, K. Reynolds, J. He, Global burden of obesity in 2005 and projections to 2030, Int. J. Obes. (Lond.) 32 (9) (2008) 1431–1437.

[2] R. Sturm, Increases in morbid obesity in the USA: 2000–2005, Public Health 121 (7) (2007) 492–496.

[3] A.M. Wolf, What is the economic case for treating obesity? Obes. Res. 6 (Suppl. 1) (1998) 2S–7S.

[4] K.M. Flegal, B.I. Graubard, D.F. Williamson, M.H. Gail, Cause-specific excess deaths associated with underweight, overweight, and obesity, JAMA 298 (17) (2007) 2028–2037.

[5] T. Pischon, H. Boeing, K. Hoffmann, M. Bergmann, M.B. Schulze, K. Overvad, et al., General and abdominal adiposity and risk of death in Europe, N. Engl. J. Med. 359 (20) (2008) 2105–2120.

[6] A.J. Stunkard, T.T. Foch, Z. Hrubec, A twin study of human obesity, JAMA 256 (1) (1986) 51–54.

[7] M. Turula, J. Kaprio, A. Rissanen, M. Koskenvuo, Body weight in the Finnish Twin Cohort. Diabetes Res, Clin. Pract. 10 (Suppl. 1) (1990) S33–S36.

[8] J. Wardle, S. Carnell, C.M. Haworth, R. Plomin, Evidence for a strong genetic influence on childhood adiposity despite the force of the obesogenic environment, Am. J. Clin. Nutr. 87 (2) (2008) 398–404.

[9] P.T. Katzmarzyk, L. Perusse, D.C. Rao, C. Bouchard, Familial risk of overweight and obesity in the Canadian population using the WHO/NIH criteria, Obes. Res. 8 (2) (2000) 194–197.

[10] C. Malis, E.L. Rasmussen, P. Poulsen, I. Petersen, K. Christensen, H. Beck-Nielsen, et al., Total and regional fat distribution is strongly influenced by genetic factors in young and elderly twins, Obes. Res. 13 (12) (2005) 2139–2145.

[11] P.P. Moll, T.L. Burns, R.M. Lauer, The genetic and environmental sources of body mass index variability: the Muscatine Ponderosity Family Study, Am. J. Hum. Genet. 49 (6) (1991) 1243–1255.

[12] K.M. Rose, B. Newman, E.J. Mayer-Davis, J.V. Selby, Genetic and behavioral determinants of waist-hip ratio and waist circumference in women twins, Obes. Res. 6 (6) (1998) 383–392.

[13] J.V. Selby, B. Newman, C.P. Quesenberry Jr., R.R. Fabsitz, D. Carmelli, F.J. Meaney, et al., Genetic and behavioral influences on body fat distribution, Int. J. Obes. 14 (7) (1990) 593–602.

[14] J.V. Selby, B. Newman, C.P. Quesenberry Jr., R.R. Fabsitz, M.C. King, F.I. Meaney, Evidence of genetic influence on central body fat in middle-aged twins, Hum. Biol. 61 (2) (1989) 179–194.

[15] A.J. Stunkard, T.I. Sorensen, C. Hanis, T.W. Teasdale, R. Chakraborty, W.J. Schull, et al., An adoption study of human obesity, N. Engl. J. Med. 314 (4) (1986) 193–198.

[16] D.T. Redden, J. Divers, L.K. Vaughan, H.K. Tiwari, T.M. Beasley, J.R. Fernandez, et al., Regional admixture mapping and structured association testing: conceptual unification and an extensible general linear model, PLoS Genet. 2 (8) (2006) e137.

[17] R.C. Williams, J.C. Long, R.L. Hanson, M.L. Sievers, W.C. Knowler, Individual estimates of European genetic admixture associated with lower body-mass index, plasma glucose, and prevalence of type 2 diabetes in Pima Indians, Am. J. Hum. Genet. 66 (2) (2000) 527–538.

[18] T. Rankinen, A. Zuberi, Y.C. Chagnon, S.J. Weisnagel, G. Argyropoulos, B. Walts, et al., The human obesity gene map: the 2005 update, Obesity (Silver Spring) 14 (4) (2006) 529–644.

[19] International HapMap Consortium, A haplotype map of the human genome, Nature 437 (7063) (2005) 1299–1320.

[20] J.C. Barrett, L.R. Cardon, Evaluating coverage of genome-wide association studies, Nat. Genet. 38 (6) (2006) 659–662.

[21] I. Pe'er, P.I. de Bakker, J. Maller, R. Yelensky, D. Altshuler, M.J. Daly, Evaluating and improving power in whole-genome association studies using fixed marker sets, Nat. Genet. 38 (6) (2006) 663–667.

[22] J.B. Fan, M.S. Chee, K.L. Gunderson, K.L., Highly parallel genomic assays, Nat. Rev. Genet. 7 (8) (2006) 632–644.

[23] T.M. Frayling, N.J. Timpson, M.N. Weedon, E. Zeggini, R.M. Freathy, C.M. Lindgren, et al., A common variant in the FTO gene is associated with body mass index and predisposes to childhood and adult obesity, Science 316 (5826) (2007) 889–894.

[24] E. Zeggini, M.N. Weedon, C.M. Lindgren, T.M. Frayling, K.S. Elliott, H. Lango, et al., Replication of genome-wide association signals in UK samples reveals risk loci for type 2 diabetes, Science 316 (5829) (2007) 1336–1341.

[25] T. Gerken, C.A. Girard, Y.C. Tung, C.J. Webby, V. Saudek, K.S. Hewitson, et al., The obesity-associated FTO gene encodes a 2-oxoglutarate-dependent nucleic acid demethylase, Science 318 (5855) (2007) 1469–1472.

[26] K. Wahlen, E. Sjolin, J. Hoffstedt, The common rs9939609 gene variant of the fat mass-and obesity-associated gene FTO is related to fat cell lipolysis, J. Lipid Res. 49 (3) (2008) 607–611.

[27] Genome-wide association study of 14,000 cases of seven common diseases and 3,000 shared controls, Nature 447 (7145) (2007) 661–678.

[28] C. Dina, D. Meyre, S. Gallina, E. Durand, A. Korner, P. Jacobson, et al., Variation in FTO contributes to childhood obesity and severe adult obesity, Nat. Genet. 39 (6) (2007) 724–726.

[29] A. Scuteri, S. Sanna, W.M. Chen, M. Uda, G. Albai, J. Strait, et al., Genome-wide association scan shows genetic variants in the FTO gene are associated with obesity-related traits, PLoS Genet. 3 (7) (2007) e115.

[30] M.N. Weedon, G. Lettre, R.M. Freathy, C.M. Lindgren, B.F. Voight, J.R. Perry, et al., A common variant of HMGA2 is associated with adult and childhood height in the general population, Nat. Genet. 39 (10) (2007) 1245–1250.

[31] R.J. Loos, C.M. Lindgren, S. Li, E. Wheeler, J.H. Zhao, I. Prokopenko, et al., Common variants near MC4R are associated with fat mass, weight and risk of obesity, Nat. Genet. 40 (6) (2008) 768–775.

[32] J.C. Chambers, P. Elliott, D. Zabaneh, W. Zhang, Y. Li, P. Froguel, et al., Common genetic variation near MC4R is associated with waist circumference and insulin resistance, Nat. Genet. 40 (6) (2008) 716–718.

[33] L. Qi, P. Kraft, D.J. Hunter, F.B. Hu, The common obesity variant near MC4R gene is associated with higher intakes of total energy and dietary fat, weight change and diabetes risk in women, Hum. Mol. Genet. 17 (22) (2008) 3502–3508.

[34] G. Thorleifsson, G.B. Walters, D.F. Gudbjartsson, V. Steinthorsdottir, P. Sulem, A. Helgadottir, et al., Genome-wide association yields new sequence variants at seven loci that associate with measures of obesity, Nat. Genet. 41 (1) (2009) 18–24.

[35] C.J. Willer, E.K. Speliotes, R.J. Loos, S. Li, C.M. Lindgren, I.M. Heid, et al., Six new loci associated with body mass index highlight a neuronal influence on body weight regulation, Nat. Genet. 41 (1) (2009) 25–34.

[36] D. Meyre, J. Delplanque, J.C. Chevre, C. Lecoeur, S. Lobbens, S. Gallina, et al., Genome-wide association study for early-onset and morbid adult obesity identifies three new risk loci in European populations, Nat. Genet. 41 (2) (2009) 157–159.

[37] C. Cotsapas, E.K. Speliotes, I.J. Hatoum, D.M. Greenawalt, R. Dobrin, P.Y. Lum, et al., Common body mass index-associated variants confer risk of extreme obesity, Hum. Mol. Genet. 18 (18) (2009) 3502–3507.

[38] N.L. Heard-Costa, M.C. Zillikens, K.L. Monda, A. Johansson, T.B. Harris, M. Fu, et al., NRXN3 is a novel locus for waist circumference: a genome-wide association study from the CHARGE Consortium, PLoS Genet. 5 (6) (2009) e1000539.

[39] C.M. Lindgren, I.M. Heid, J.C. Randall, C. Lamina, V. Steinthorsdottir, L. Qi, et al., Genome-wide association scan meta-analysis identifies three Loci influencing adiposity and fat distribution, PLoS Genet. 5 (6) (2009) e1000508.

[40] E.G. Kamel, G. McNeill, M.C. Van Wijk, Usefulness of anthropometry and DXA in predicting intra-abdominal fat in obese men and women, Obes. Res. 8 (1) (2000) 36–42.

[41] K. Schousboe, G. Willemsen, K.O. Kyvik, J. Mortensen, D.I. Boomsma, B.K. Cornes, et al., Sex differences in heritability of BMI: a comparative study of results from twin studies in eight countries, Twin Res. 6 (5) (2003) 409–421.

[42] M.C. Zillikens, M. Yazdanpanah, L.M. Pardo, F. Rivadeneira, Y.S. Aulchenko, B.A. Oostra, et al., Sex-specific genetic effects influence variation in body composition, Diabetologia 51 (12) (2008) 2233–2241.

[43] A. Garg, Acquired and inherited lipodystrophies, N. Engl. J. Med. 350 (12) (2004) 1220–1234.

[44] D.B. Goldstein, Common genetic variation and human traits, N. Engl. J. Med. 360 (17) (2009) 1696–1698.

[45] J.N. Hirschhorn, Genomewide association studies – illuminating biologic pathways, N. Engl. J. Med. 360 (17) (2009) 1699–1701.

[46] Y.S. Cho, M.J. Go, Y.J. Kim, J.Y. Heo, J.H. Oh, H.J. Ban, et al., A large-scale genome-wide association study of Asian populations uncovers genetic factors influencing eight quantitative traits, Nat. Genet. 41 (5) (2009) 527–534.

Case Study on Rheumatoid Arthritis

Stephen Eyre, Wendy Thomson

Epidemiology Unit School of Translational Medicine, University of Manchester, UK

Rheumatoid arthritis (RA) is a complex disease, believed to be caused by a combination of genetic and environmental factors. Prior to the Genome Wide Association Study (GWAS) era, in keeping with many other autoimmune diseases and complex genetic traits, the methods used to elucidate the genetic component of RA had focused on linkage studies and candidate genes studies. Although many hundreds of such studies have been reported in the literature they are fraught with methodological issues such as small sample sizes, inappropriately matched controls and overinterpretation of results.

Indeed over 30 years of work has led to the identification of just three robustly validated susceptibility loci for RA, namely *HLA*, *PTPN22* and *STAT4*. In addition, there were a few well-powered high profile publications providing convincing preliminary evidence for association to a number of loci but which remain to be validated; in particular, Peptidylarginine Deiminase Type IV (PADI4), Solute Carrier Family 22 Member 4 (SLC22A4), Fc Receptor-Like Protein 3 (FCRL3) and Major Histocompatibility Complex Class II Transactivator (MHC2TA).

CANDIDATE GENE ON STUDIES

Replicated Loci

HLA

As with many autoimmune diseases, the first candidate genes to be tested for association with RA were the HLA loci; indeed the first association was reported, with the serologically defined HLA-DRw4, as far back as 1978 [1]. Further characterization of the gene structure for HLA class II molecules and subsequent sequencing of the *HLA-DRB1* locus, which encodes the β-chain of the HLA-DR molecule, enabled more refined genotyping to be undertaken. Subsequent studies revealed RA associations with multiple different *HLA-DRB1* alleles and this in turn led to the shared epitope (SE) hypothesis — the identification of a conserved sequence within the third hypervariable region (amino acids 70–74) of *HLA-DRB1* which is shared between RA associated alleles [2]. The HLA region is believed to contribute around 30–50% of the total genetic contribution to RA and while the *HLA-DRB1* SE is thought to account for much of this, evidence suggests that other loci within the major histocompatibility complex may also play a role.

Protein Tyrosine Phosphatase Non-receptor Type 22 (PTPN22)

Protein tyrosine phosphatase non-receptor type 22 (*PTPN22*) was first identified as a susceptibility locus for type 1 diabetes (T1D) [3]. The minor allele of a non-synonymous single nucleotide polymorphism (SNP) (rs2476601, 1858C-T, R620W) has subsequently been shown to be an autoimmune locus, in that it is associated with multiple autoimmune diseases, including RA [4, 5]. Indeed, the minor T allele has been associated with RA in multiple populations, including UK [6], Canadian [7],

Finnish [8], German [9], Spanish [10] and Swedish [5] populations. However, *PTPN22* has not been associated with RA in Asian populations, probably because the 1858T allele is extremely rare in this population [11]. To date this variant remains the second highest genetic risk factor for RA, conferring an odds ratio (OR) of ~1.8.

Signal Transducer and Activator of Transcription 4 (STAT4)

More recently, a study examining candidate genes under a peak of linkage on chromosome 2q identified the signal transducer and activator transcription (*STAT*)*1−STAT4* region as a susceptibility locus for RA [12]. Further characterization identified association with a four SNP haplotype, located within intron three of the *STAT4* gene, the strongest effect being with rs7574865. This association has since been confirmed in a number of studies in patients of European descent [13−15] and interestingly has also been replicated in a Korean population [16]. As for *PTPN22*, *STAT4* has also been associated with multiple autoimmune diseases.

STAT4 appears to confer a relatively modest risk of RA. A recent meta-analysis of 14 studies of RA cases of European and Asian descent, giving a total population of 15,609 cases and 15,793 controls, calculated the overall OR for the minor T allele of *STAT4* rs7574865 as 1.27 (95% CI: 1.2−1.34) with no significant difference by ethnicity or anti-CCP antibody status [17].

High-profile Candidate Gene Studies Published Prior to the Advent of GWAS

Peptidylarginine Deiminase Type IV (PADI4)

Under an RA peak of linkage on chr1p36, the genes encoding the peptidylarginine deiminase citrullinating enzymes (*PADI* genes) are excellent candidates for involvement in RA as individuals with RA frequently possessing auto-antibodies to citrullinated peptides. In a candidate gene study a Japanese group, using what at the time was a relatively well-powered study design (830 cases versus 736 controls), identified *PADI4* as a potential susceptibility locus for RA [18]. Functional work in this study went on to suggest that a 4 SNP haplotype could stabilize mRNA levels.

Replication studies of this initial finding implicated a different genetic make-up at this locus between European Caucasians and Asian populations. In Japanese and Korean populations replication has been consistent, unlike European populations were the evidence is not as clear [5, 19−21]. Indeed, a meta-analysis of the available data indicated that the gene plays a role in RA susceptibility in Far Eastern populations, but none in populations of European descent [22].

Solute Carrier Family 22 ember 4 (SLC22A4)

Using a candidate gene study design to investigate the cluster of cytokine genes on 5p31, the same Japanese group attempted to identify genotype associations in a similarly powered study to their *PADI4* discovery. They identified association to the organic cation transporter *SLC22A4*, and again supported this finding with functional work, suggesting that allelic changes resulted in differential binding of the transcription factor *RUNX1* [23]. To date, no other genetic studies, even ones in similar Far Eastern populations, have managed to validate this initial finding [24–26].

Fc Receptor-like Protein 3 (FCRL3)

The Japanese group had further success with another candidate gene approach, this time focusing on the *FCRL3* gene [27]. FCRL3 is an immune molecule, expressed in B-cells, which had been implicated in other autoimmune disorders and a mouse model of RA. The initial paper included replication data and functional work suggesting that the most associated SNP was in the promoter region and operated via the differential binding of NF-kB. Similarly to *PADI4*, validation of this initial finding dichotomized into Far Eastern populations supporting the finding, while populations of European descent did not show association of this locus with RA, at least not at the SNP suggested in the Japanese study [28–32].

Major Histocompatibility Complex Class II Transactivator (MHC2TA)

The final high-profile candidate gene study in RA was published by a Swedish group who followed up findings of a rat model of MHC expression in which a QTL was localized to the *MHC2TA* gene [33]. The gene was then tested directly as a candidate gene for RA and other autoimmune diseases. An SNP in the promoter region was found to be associated with RA, and further functional work indicated that this SNP was linked to the down-regulation of the gene itself.

The genetic study was reasonably sized for the time (1,300 cases versus 700 controls), providing a modest statistical association by post-GWAS standards but extensive validation efforts by other groups have failed to replicate the initial finding [34–37].

GENOME-WIDE ASSOCIATION STUDIES

The GWAS era has revolutionized the study of complex genetic traits and has led to the identification of many robustly replicated susceptibility loci for many complex conditions. Indeed, the October 2009 release of the Catalog of Published Genome Wide Association Studies contained details from 417 published GWAS with 1,873 SNP associations to diseases or quantitative traits ($p < 1 \times 10^{-5}$). Two well-powered GWAS in RA have really driven the field to date arising from the Wellcome Trust Case Control Consortium (WTCCC) study and the North American Rheumatoid Arthritis Consortium (NARAC)/Swedish Epidemiological Investigation of Rheumatoid Arthritis (EIRA) study.

GWAS in RA

Wellcome Trust Case Control Consortium (WTCCC) Study

RA was one of seven diseases involved in the first, pioneering WTCCC GWAS study [38]. Approximately 2,000 RA cases were included and compared against the 3,000 common controls in the association study. Using the Affymetrix 500k SNP chip, whole genome association studies revealed many areas of statistically significant differences between case and control allele frequencies. Results were split into tiers of significance. For RA, two loci were in tier 1 ($p < 5 \times 10^{-7}$) with the best evidence for significant association, nine in the second tier ($p = 10^{-5} - 5 \times 10^{-7}$) and 49 in the third tier ($p = 10^{-4} - 10^{-5}$). As expected the HLA region and *PTPN22* were the two loci in the top tier of significance, and required no further validation. The second and third tiers, although including many plausible loci such as IL2RA and IL2RB, had to be replicated in larger case and control cohorts in order to validate the initial findings.

NARAC/EIRA Study

The North American Rheumatoid Arthritis Consortium (NARAC) and the Swedish Epidemiological Investigation of Rheumatoid Arthritis (EIRA) first published a GWAS with 317,503 SNPs in 1,522 seropositive cases and 1,850 controls [39]. This study implicated the now extensively replicated *TRAF1/C5* locus associated with RA. The initial study included an independent replication, in almost 1,000 seropositive cases and 1,800 controls, resulting in convincing combined statistical association for the SNP tested ($OR = 1.32, p = 4 \times 10^{-14}$). Since this preliminary finding many groups have tested for association with RA at this locus and it has become firmly established as a key susceptibility region for RA [13, 15, 40, 41], although interestingly it is one of the few validated RA regions that has not currently been associated with T1D. The associated SNPs in this region are in linkage disequilibrium with two genes, *TRAF1* (encoding tumor necrosis factor receptor-associated factor 1) and *C5* (encoding complement component 5), both extremely good candidates for an inflammatory disease.

GWAS Follow-up Studies

In reality it is the GWAS follow-up studies that have confirmed many of the RA susceptibility loci.

In this regard a number of different approaches have been utilized to validate the findings from the GWAS. These approaches include:

1. Validation of RA susceptibility loci identified in GWAS in large independent cohorts and confirmation in other populations.
2. Validation of susceptibility loci identified in GWAS of other autoimmune/inflammatory diseases in large independent RA cohorts and confirmation in other populations.
3. Identification of additional susceptibility loci by increasing numbers in the GWAS and meta-analysis of different datasets.
4. Identification of further novel RA susceptibility loci by using novel approaches to interrogate the GWAS data, including pathway-based analysis.

Validation of RA Susceptibility Loci Identified in GWAS in Large Independent Cohorts and Confirmation in other Populations

GWAS, with many thousands of genetic markers and therefore statistical tests, requires large-scale validation studies in order to robustly interrogate the initial findings. The first of these validation studies in RA brought together a UK consortium of more than 5,000 independent cases and around 4,000 controls to investigate the nine tier two SNPs from the WTCCC study [42]. The study confirmed association to a region on 6q23, near the *TNFAIP3* gene, as a true RA susceptibility locus. A US study, published simultaneously, performed a smaller-scale GWAS with 400 cases, 1,200 controls and 110k SNPs with the primary, well-replicated finding from this study confirming unequivocally that this region was important in the genetics of RA [43]. Interestingly, the most associated SNP in the UK population conferred susceptibility to RA, while the strongest effect in the US population was to a totally independent SNP, within the same LD block, but one which conferred protection to RA.

The block of associated SNPs lies very close to the *PTPN11* pseudogene and within 200 kb of *TNFAIP3*, the gene encoding A20, a negative regulator of NF-kB.

This well-powered UK validation cohort, including around 5,000 independent cases and bolstered by upwards of 10,000 controls, has been successful in replicating other SNPs from the original WTCCC GWAS from the tier 2 and tier 3 significance strata [44]. These loci include SNPs on 12q23 near the kinesin heavy chain (*KIF5A*) and phosphatidylinositol-5-phosphate 4-kinase, type II, gamma (*PIP4K2C*) genes, an SNP on 10p15, upstream of the protein kinase gene *PRKCQ*, a gene thought to play a role in T-cell activation and an SNP on 22q13 adjacent to the immune molecule, interleukin receptor *IL2RB*.

Validation of Susceptibility Loci Identified in GWAS of other Autoimmune/Inflammatory Diseases in Large Independent RA Cohorts and Confirmation in other Populations

It was apparent from the very earliest genetic associations that auto-immune diseases shared some genetic overlap. *HLA, PTPN22* and *STAT4* were already established as pan-autoimmune loci even before GWAS suggested other common loci. Testing confirmed or putative loci from T1D, coeliac disease (CD), systemic lupus erythematosus (SLE), and multiple sclerosis (MS) in RA cohorts has led to the identification of further RA susceptibility loci. Indeed using the strategy, a region on chromosome 2q11, mapping to the 5' and promoter region of *AFF3(45)*, a transcription factor implicated in lymphoid development, a further region on 6q25 region implicating *TAGAP* (T-cell activation RhoGTPase activating protein) which is expressed in activated T-cells, as well as *CTLA4* have been established as confirmed loci in RA.

Continuing with the strategy of testing susceptibility loci from one autoimmune disease in all of the others is likely to lead to insights into which pathways may be common and which are unique to different autoimmune conditions.

Identification of Additional Susceptibility Loci by Increasing Numbers in the GWAS and Meta-analysis of Different Datasets

One of the most successful and compelling ways of finding novel disease loci is by increasing the numbers in the initial GWAS by meta-analysis. The NARAC/EIRA GWAS was supplemented with the WTCCC data for the first RA meta-analysis, resulting in four new, robustly replicated loci [46]. The SNPs identified as associated with RA in that study implicated the genes *CD40, CCL21, CDK6* and the *MMEL/TNFRSF14* region. Building on the success of this strategy and using the original NARAC/EIRA study that successfully identified TRAF1/C5 as an RA locus, a further 1,500 and 3,000 controls were tested for association with RA and a meta-analysis with the original data was undertaken [47]. Validation in 2,600 cases and 3,300 independent controls has led to unequivocal evidence that a region on chr2p13 encompassing the *REL* gene, coding for the NF-kB transcription factor c-REL, and the previously reported SLE locus, *BLK*, are two further genes adding to the overall genetic picture in RA.

Identification of Further Novel RA Susceptibility Loci by using Novel Approaches to Interrogate the GWAS Data, including Pathway-based Analysis

Finally, the GWAS data have been interrogated using bioinformatics tools to further increase the number of true RA loci identified. The strategy was predicated on the assumption that some true susceptibility loci lie within the lower tiers of significance in the GWAS, but separating these from false positives presents a challenge. Using a bioinformatics approach (the GRAIL text mining tool) [48] in which known RA loci were used to "seed" the analysis algorithm, less well-statistically implicated loci were graded on their plausibility using likelihood scores generated from text mining. This method generated a list of already moderately associated, biologically plausible candidates, which were then tested directly for association in a large international validation effort [49]. Thus, in over 8,000 independent cases and 11,000 controls from five countries, convincing evidence for replication of association to SNPs adjacent to *CD2/CD58* ($p = 2 \times 10^{-6}$ replication datasets, $p = 2 \times 10^{-9}$ combined analysis), *CD28* ($p = 7 \times 10^{-5}$ replication, $p = 3 \times 10^{-8}$ overall) and *PRDM1* ($p = 2 \times 10^{-5}$ replication, $p = 3 \times 10^{-8}$ overall) was established.

Maximizing Evidence for True Susceptibility Loci

Of course this whole process is an iterative one. Each time a new susceptibility locus is identified, it needs to be confirmed in other large independent cohorts, preferably in multiple populations. Meta-analysis of these multiple datasets can often provide the most convincing evidence of association. Accumulation of evidence, leading to unequivocal statistical evidence ($p < 10^{-10}$, for example), or finding loci with strong evidence of association with no real evidence contradicting the association, can lead to greater confidence in assigning loci to disease. The largest RA meta-analysis to date, incorporating all the published data and a large, pan-European cohort, has provided the best current summary of the state of play for many loci in RA [50]. Table 19.1 summarizes the findings of this paper and provides the most significant evidence available for association for all of the loci described above. In summary, there are 19 loci with compelling evidence to be implicated in RA susceptibility. Of note and in concordance with earlier validation studies, none of the loci currently on this list of confirmed Caucasian RA susceptibility genes are the ones implicated by the earlier, pre-GWAS candidate gene studies (*PADI4*, *FCRL3*, *MHC2TA* and *SLC22A4*).

Determining the True Effect of a Given Susceptibility Locus

Many of the susceptibility loci identified to date appear to confer a relatively low risk of disease. However, initial estimates based on a single associated SNP may well be an underestimate of the real effect of a given locus. There are a number of reasons why this may be the case:

1. The SNP identified may not be the causal variant.
2. There may be multiple effects at a single locus.
3. Interaction between loci may exist.

Identification of Causal Variant

Identification and subsequent validation of an RA-associated SNP is only the first step in identifying the casual variant at that locus. Fine-mapping studies are required to identify all possible causal variants but, because of the linkage disequilibrium in the genome, there are few examples where genetic evidence alone is sufficient to pin-point the causal SNP. One example is the *PTPN22* gene where fine-mapping and haplotype analysis has shown that the exonic R620W polymorphism is likely to be causal [51]. In most cases, however, functional studies will be required to correlate genotype with phenotype and that work is under way.

Multiple Effects at One Locus

It is becoming increasingly apparent that for many loci the association may not be accounted for by a single SNP but by multiple effects at a single locus. Not only may different SNPs be associated with the same disease at a locus but different SNPs at that locus may also be associated with different diseases. This is best illustrated by the association of the 6q23 region harboring the *TNFAIP3* gene with RA, which has a complex genetic architecture. Fine-mapping of the region has found three independent SNP associations with RA, two of which confer susceptibility and one protection. Carriage of the "worst case scenario" alleles, i.e., both susceptibility minus the protective variant results, increases the odds of RA to 1.5, making the locus the third strongest region associated with RA to date [52]. It nicely illustrates the point that many of the risk estimates for the loci identified so far may be underestimates, first, as the causal alleles have not yet been identified and, second, because fine-mapping may reveal further associated variants. This has implications in terms of risk assessment for individuals, assigning a hierarchy of importance to loci, for epistatic interaction analysis and not least for any future functional work in trying to determine the genotype/phenotype relationship.

The 6q23 region has also been associated with T1D [53], psoriasis [54], SLE [55, 56] and celiac [57]. However, different SNPs are associated with the different diseases, illustrating the point that simply testing for association with an SNP previously associated with another disease risks missing potentially important associations with the region (Fig. 19.1). This degree of complexity is not restricted to the 6q23 region; for example, the IL2RA locus has a similar complex association with T1D, MS and JIA [58] where different SNPs show the strongest evidence for association with the different diseases and the same SNP can have opposing risks of susceptibility/protection.

TABLE 19.1   Maximum Evidence of Association for Confirmed RA Susceptibility Loci

SNP	Gene	Case/control	Chi² p	OR (CI 95%)	Ref.
rs5029937	6q23	7,731/7,391	6.84E-09	1.41 (1.26, 1.59)	[50]
rs6920220	6q23	14,567/12,469	1.15E-15	1.19 (1.14, 1.24)	[50]
rs13207033	6q23	13,286/11,909	2.50E-06	0.91 (0.87, 0.94)	[50]
rs1160542	AFF3	10,902/9,247	2.6E-09	1.13 (1.09, 1.18)	[50]
rs27396340	BLK	9692/14508	5.69E-09	1.19(1.12–1.27)	[47]
rs2812378	CCL21	14,147/15,062	2.62E-07	1.10 (1.06, 1.14)	[50]
rs1980422	CD28	11403/24239	2.7E-08	1.12	[49]
rs4810485	CD40	14,031/15,103	2.19E-11	0.88 (0.84, 0.91)	[50]
rs11586238	CD58	11403/24239	1.7E-09	1.13	[49]
rs3087243	CTLA4	12,372/11,360	5.68E-09	0.89 (0.86, 0.93)	[50]
rs6822844	IL2/IL21	12,953/11,512	2.65E-10	0.86 (0.82, 0.90)	[50]
rs2104286	IL2RA	10,112/8,532	1.55E-05	0.90 (0.86, 0.94)	[50]
rs743777	IL2RB	11,038/8,749	5.98E-07	1.28 (1.16, 1.40)	[50]
rs1678542	KIF5A	14,378/15,597	3.49E-09	0.90 (0.87, 0.93)	[50]
rs548234	PRDM1	11403/24239	2.9E-08	1.10	[49]
rs4750316	PRKCQ	14,306/15,575	1.89E-07	0.89 (0.86, 0.93)	[50]
rs2476601	PTPN22	20,079/18,826	5.26E-100	1.61 (1.54, 1.69)	[50]
rs13031237	REL	8652/13526	3.08E-14	1.26(1.17–1.32)	[47]
rs7574865	STAT4	14,394/14,296	2.21E-17	1.19 (1.14, 1.24)	[50]
rs394581	TAGAP	11403/24239	2.3E-06	0.91	[49]
rs2900180	TRAF1/C5	12,358/11,765	2.80E-11	1.14 (1.10, 1.19)	[50]

Interaction between Loci

A third important reason why current estimates of effect sizes for a given locus may be underestimates, is that no account has yet been taken of possible interactions between genes and environmental or other genetic susceptibility factors. In preliminary analysis of the first five loci to be convincingly associated with RA (SE, *PTPN22*, 6q23, *STAT4* and *TRAF1/C5*) carriage of combinations of risk alleles were associated with substantially higher OR than each alone [59]. Furthermore, previous studies have reported epistasis of HLA and smoking in susceptibility to

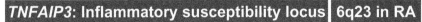

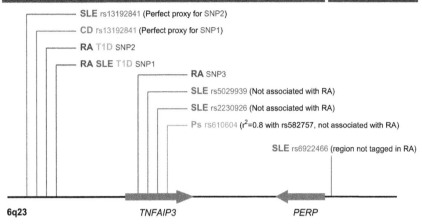

RA: Thomson et al. [42]; Plenge et al. [43]; Orozco et al. [52]
SLE: Graham et al. [55]; Musone et al. [56]
Type 1 Diabetes: Fung et al. [53]
Psoriasis: Nair et al. [54]
Celiac disease: Trynka et al. [57]

FIGURE 19.1

RA [60]. In order to fully understand individual risks, further exploration of these interactions will be required.

Other Genetic Variants and RA

It is unlikely, given all the variants seen in a human genome (which include copy number polymorphism, insertions, deletions, variable number tandem repeats and inversions) that it will be relatively common SNPs that hold all the answers to the genetic component of RA. Attempts to characterize the role of other types of variation contributing to the genetic component in RA are in their relative infancy but are now under way.

Copy Number Variation (CNV) Studies

Tentative, initial, CNV studies have highlighted a number of genomic regions, found for autoimmune diseases, which may be worthy of further investigation in RA. These include FCGR3 [61, 62], CCL3L1 [63], and the defensin gene cluster [64, 65]. Due to the technical difficulties involved in CNV genotyping, the "candidate" approach of the field and the small sample sizes involved in these studies, this area of investigation is still very much analogous to the pre-GWAS SNP era, and results published to date require thorough evaluation.

Rare Variant Studies

One area of interest under current scrutiny is the role that rare SNPs play in disease. Technical advances in genotyping/sequencing coupled with the availability of large, international collections of samples have meant studying SNPs with a minor allele frequency (MAF) of <1% is now a real possibility. A "cluster" analysis approach in RA has implicated TNFAIP3 as harboring more rare variants in RA cases than controls, suggesting an increased mutation load for this gene (E. Zeggini, personal communication). However, again the field is relatively embryonic in its analysis and validation studies.

FUTURE STUDIES

What have the Genetic Studies Shown us to Date?

Genetics allows us to identify potential novel drug targets, define markers for disease prognosis and treatment response, highlight biological pathways shared by diseases, and may help to explain why some people develop specific diseases. A large number of loci implicated in RA mediate their effect through the NF-κB pathway. This has led some to speculate that the survival of CD8 T-cells may play a role, through the effect of c-REL and PRKCQ through the IL2 network of genes [47].

Potentially, strongly omnipotent autoimmune genes, like *HLA*, *PTPN22*, *TNFAIP3*, *CTLA4* and the *IL2* pathway, may increase an individual's risk of developing autoimmune disease, while other genes may determine which course this will take. The RA genetic studies to date suggest that maybe *CD40*, *TRAF/C5* and *CCL21* could be specific to an RA autoimmune response, but extensive resequencing and fine-mapping will be required in all diseases to determine this with any sort of certainty. The findings could eventually lead to novel insights into the specific mechanisms of RA susceptibility.

Clinical Utility

Identification of susceptibility loci will facilitate our understanding of disease mechanisms and may lead to the identification of novel therapeutic targets. However, within the clinic what is important is the ability to predict outcome, including likely response to treatment, as early in the course of disease as possible. To date, very few studies have tested this, but those that have suggest that the genes involved in susceptibility to disease are not necessarily also involved in determining disease course. There are two reports of the *TRAF1/C5* locus being associated with the development of joint erosions while one study found the *CCL21* RA susceptibility allele was also associated with premature mortality from cardiovascular disease in patients with RA. The risks associated with these variants in predicting adverse outcome are not sufficiently large to be useful clinically on their own. Ideally, GWAS in large inception-based cohorts of early inflammatory arthritis with detailed clinical follow-up will be required if we are to meet the goal of identifying prognostic predictors for RA.

Of course as our knowledge of the genetic risk factors for RA increases, we are moving closer to the ultimate goal, the development of predictive algorithms of disease development based on genetic and environmental risk factors, with those at high risk being offered preventive treatments.

OVERALL CONCLUSION

Tremendous progress has been made in our understanding of RA genetics in recent years with at least x loci now confidently confirmed as associated. HLA DRB1 SE remains the largest susceptibility factor but is neither necessary nor sufficient to cause disease. Knowledge of other genetic susceptibility loci has identified key pathways which may be targets for therapeutic intervention. Studies to translate these findings into clinical practice are still in their early stages but remain the ultimate goal.

References

[1] P. Stastny, Association of the B-cell alloantigen DRw4 with rheumatoid arthritis, N. Engl. J. Med. 298 (1978) 869–871.

[2] P.K. Gregersen, J. Silver, R.J. Winchester, The shared epitope hypothesis. An approach to understanding the molecular genetics of susceptibility to rheumatoid arthritis, Arthritis Rheum. 30 (1987) 1205–1213.

[3] N. Bottini, L. Musumeci, A. Alonso, S. Rahmouni, K. Nika, M. Rostamkhani, et al., A functional variant of lymphoid tyrosine phosphatase is associated with type I diabetes, Nat. Genet. 36 (2004) 337–338.

[4] A.B. Begovich, V.E. Carlton, L.A. Honigberg, S.J. Schrodi, A.P. Chokkalingam, H.C. Alexander, et al., A missense single-nucleotide polymorphism in a gene encoding a protein tyrosine phosphatase (PTPN22) is associated with rheumatoid arthritis, Am. J. Hum. Genet. 75 (2004) 330–337.

[5] R.M. Plenge, L. Padyukov, E.F. Remmers, S. Purcell, A.T. Lee, E.W. Karlson, et al., Replication of putative candidate-gene associations with rheumatoid arthritis in >4,000 samples from North America and Sweden: association of susceptibility with PTPN22, CTLA4, and PADI4, Am. J. Hum. Genet. 77 (2005) 1044–1060.

[6] A. Hinks, A. Barton, S. John, I. Bruce, C. Hawkins, C.E. Griffiths, et al., Association between the PTPN22 gene and rheumatoid arthritis and juvenile idiopathic arthritis in a UK population: further support that PTPN22 is an autoimmunity gene, Arthritis Rheum. 52 (2005) 1694–1699.

[7] O.M. van, R.F. Wintle, X. Liu, M. Yazdanpanah, X. Gu, B. Newman, et al., Association of the lymphoid tyrosine phosphatase R620W variant with rheumatoid arthritis, but not Crohn's disease, in Canadian populations, Arthritis Rheum. 52 (2005) 1993–1998.

[8] M.F. Seldin, R. Shigeta, K. Laiho, H. Li, H. Saila, A. Savolainen, et al., Finnish case-control and family studies support PTPN22 R620W polymorphism as a risk factor in rheumatoid arthritis, but suggest only minimal or no effect in juvenile idiopathic arthritis, Genes Immun. 6 (8) (2005) 720–722.

[9] M. Pierer, S. Kaltenhauser, S. Arnold, M. Wahle, C. Baerwald, H. Hantzschel, et al., Association of PTPN22 1858 single-nucleotide polymorphism with rheumatoid arthritis in a German cohort: higher frequency of the risk allele in male compared to female patients, Arthritis Res.Ther. 8 (2006) R75.

[10] G. Orozco, E. Sanchez, M.A. Gonzalez-Gay, M.A. Lopez-Nevot, B. Torres, R. Caliz, et al., Association of a functional single-nucleotide polymorphism of PTPN22, encoding lymphoid protein phosphatase, with rheumatoid arthritis and systemic lupus erythematosus, Arthritis Rheum. 52 (2005) 219–224.

[11] K. Ikari, S. Momohara, E. Inoue, T. Tomatsu, M. Hara, H. Yamanaka, et al., Haplotype analysis revealed no association between the PTPN22 gene and RA in a Japanese population, Rheumatology (Oxford) 45 (2006) 1345–1348.

[12] E.F. Remmers, R.M. Plenge, A.T. Lee, R.R. Graham, G. Hom, T.W. Behrens, et al., STAT4 and the risk of rheumatoid arthritis and systemic lupus erythematosus. N. Engl. J. Med. 357 (2007) 977–986.

[13] A. Barton, W. Thomson, X. Ke, S. Eyre, A. Hinks, J. Bowes, et al., Re-evaluation of putative rheumatoid arthritis susceptibility genes in the post-genome wide association study era and hypothesis of a key pathway underlying susceptibility, Hum. Mol. Genet. 17 (2008) 2274–2279.

[14] G. Orozco, B.Z. Alizadeh, A.M. gado-Vega, M.A. Gonzalez-Gay, A. Balsa, D. Pascual-Salcedo, et al., Association of STAT4 with rheumatoid arthritis: a replication study in three European populations, Arthritis Rheum. 58 (2008) 1974–1980.

[15] M.I. Zervou, P. Sidiropoulos, E. Petraki, V. Vazgiourakis, E. Krasoudaki, A. Raptopoulou, et al., Association of a TRAF1 and a STAT4 gene polymorphism with increased risk for rheumatoid arthritis in a genetically homogeneous population, Hum. Immunol. 69 (2008) 567–571.

[16] H.S. Lee, E.F. Remmers, J.M. Le, D.L. Kastner, S.C. Bae, P.K. Gregersen, Association of STAT4 with rheumatoid arthritis in the Korean population, Mol. Med. 13 (2007) 455–460.

[17] Y.H. Lee, J.H. Woo, S.J. Choi, J.D. Ji, G.G. Song, Association between the rs7574865 polymorphism of STAT4 and rheumatoid arthritis: a meta-analysis, Rheumatol. Int. 30 (5) (2009) 661–666.

[18] A. Suzuki, R. Yamada, X. Chang, S. Tokuhiro, T. Sawada, M. Suzuki, et al., Functional haplotypes of PADI4, encoding citrullinating enzyme peptidylarginine deiminase 4, are associated with rheumatoid arthritis, Nat. Genet. 34 (2003) 395–402.

[19] A. Barton, J. Bowes, S. Eyre, K. Spreckley, A. Hinks, S. John, et al., A functional haplotype of the PADI4 gene associated with rheumatoid arthritis in a Japanese population is not associated in a United Kingdom population, Arthritis Rheum. 50 (2004) 1117–1121.

[20] K. Ikari, M. Kuwahara, T. Nakamura, S. Momohara, M. Hara, H. Yamanaka, et al., Association between PADI4 and rheumatoid arthritis: a replication study, Arthritis Rheum. 52 (2005) 3054–3057.

[21] C.P. Kang, H.S. Lee, H. Ju, H. Cho, C. Kang, S.C. Bae, A functional haplotype of the PADI4 gene associated with increased rheumatoid arthritis susceptibility in Koreans, Arthritis Rheum. 54 (2006) 90–96.

[22] M.L. Burr, H. Naseem, A. Hinks, S. Eyre, L. Gibbons, J. Bowes, et al., PADI4 genotype is not associated with rheumatoid arthritis in a large UK Caucasian population, Ann. Rheum. Dis. 69 (2009) 666–670.

[23] S. Tokuhiro, R. Yamada, X. Chang, A. Suzuki, Y. Kochi, T. Sawada, et al., An intronic SNP in a RUNX1 binding site of SLC22A4, encoding an organic cation transporter, is associated with rheumatoid arthritis, Nat. Genet. 35 (2003) 341–348.

[24] A. Barton, S. Eyre, J. Bowes, P. Ho, S. John, J. Worthington, Investigation of the SLC22A4 gene (associated with rheumatoid arthritis in a Japanese population) in a United Kingdom population of rheumatoid arthritis patients, Arthritis Rheum. 52 (2005) 752–758.

[25] M. Kuwahara, K. Ikari, S. Momohara, T. Nakamura, M. Hara, H. Yamanaka, et al., Failure to confirm association between SLC22A4 polymorphism and rheumatoid arthritis in a Japanese population, Arthritis Rheum. 52 (2005) 2947–2948.

[26] G. Orozco, E. Sanchez, M.A. Gonzalez-Gay, M.A. Lopez-Nevot, B. Torres, D. Pascual-Salcedo, et al., SLC22A4, RUNX1, and SUMO4 polymorphisms are not associated with rheumatoid arthritis: a case-control study in a Spanish population, J. Rheumatol. 33 (2006) 1235–1239.

[27] Y. Kochi, R. Yamada, A. Suzuki, J.B. Harley, S. Shirasawa, T. Sawada, et al., A functional variant in FCRL3, encoding Fc receptor-like 3, is associated with rheumatoid arthritis and several autoimmunities, Nat. Genet. 37 (2005) 478–485.

[28] S. Eyre, J. Bowes, C. Potter, J. Worthington, A. Barton, Association of the FCRL3 gene with rheumatoid arthritis: a further example of population specificity? Arthritis Res. Ther. 8 (2006) R117.

[29] C.B. Choi, C.P. Kang, S.S. Seong, S.C. Bae, C. Kang, The -169C/T polymorphism in FCRL3 is not associated with susceptibility to rheumatoid arthritis or systemic lupus erythematosus in a case-control study of Koreans, Arthritis Rheum. 54 (2006) 3838–3841.

[30] W.G. Newman, Q. Zhang, X. Liu, E. Walker, H. Ternan, J. Owen, et al., Rheumatoid arthritis association with the FCRL3-169C polymorphism is restricted to PTPN22 1858T-homozygous individuals in a Canadian population, Arthritis Rheum. 54 (2006) 3820–3827.

[31] X. Hu, M. Chang, R.K. Saiki, M.A. Cargill, A.B. Begovich, K.G. Ardlie, et al., The functional -169T→C single-nucleotide polymorphism in FCRL3 is not associated with rheumatoid arthritis in white North Americans, Arthritis Rheum. 54 (2006) 1022–1025.

[32] K. Ikari, S. Momohara, T. Nakamura, M. Hara, H. Yamanaka, T. Tomatsu, et al., Supportive evidence for a genetic association of the FCRL3 promoter polymorphism with rheumatoid arthritis, Ann. Rheum. Dis. 65 (2006) 671–673.

[33] M. Swanberg, O. Lidman, L. Padyukov, P. Eriksson, E. Akesson, M. Jagodic, et al., MHC2TA is associated with differential MHC molecule expression and susceptibility to rheumatoid arthritis, multiple sclerosis and myocardial infarction, Nat. Genet. 37 (2005) 486–494.

[34] S. Eyre, J. Bowes, K. Spreckley, C. Potter, S. Ring, D. Strachan, et al., Investigation of the MHC2TA gene, associated with rheumatoid arthritis in a Swedish population, in a UK rheumatoid arthritis cohort, Arthritis Rheum. 54 (2006) 3417–3422.

[35] G. Orozco, G. Robledo, M.V. Linga Reddy, A. Garcia, D. Pascual-Salcedo, A. Balsa, et al., Study of the role of a functional polymorphism of MHC2TA in rheumatoid arthritis in three ethnically different populations, Rheumatology (Oxford) 45 (2006) 1442–1444.

[36] B. Yazdani-Biuki, K. Brickmann, K. Wohlfahrt, T. Mueller, W. Marz, W. Renner, et al., The MHC2TA -168A>G gene polymorphism is not associated with rheumatoid arthritis in Austrian patients, Arthritis Res. Ther. 8 (2006) R97.

[37] P. Harrison, J.J. Pointon, C. Farrar, A. Harin, B.P. Wordsworth, MHC2TA promoter polymorphism (-168*G/A, rs3087456) is not associated with susceptibility to rheumatoid arthritis in British Caucasian rheumatoid arthritis patients, Rheumatology (Oxford) 46 (2007) 409–411.

[38] Anonymous, Genome-wide association study of 14,000 cases of seven common diseases and 3,000 shared controls, Nature 447 (2007) 661–678.

[39] R.M. Plenge, M. Seielstad, L. Padyukov, A.T. Lee, E.F. Remmers, B. Ding, et al., TRAF1-C5 as a risk locus for rheumatoid arthritis – a genomewide study, N. Engl. J. Med. 357 (2007) 1199–1209.

[40] F.A. Kurreeman, L. Padyukov, R.B. Marques, S.J. Schrodi, M. Seddighzadeh, G. Stoeken-Rijsbergen, et al., A candidate gene approach identifies the TRAF1/C5 region as a risk factor for rheumatoid arthritis, PLoS. Med. 4 (2007) e278.

[41] F.A. Kurreeman, D. Rocha, J. Houwing-Duistermaat, S. Vrijmoet, V.H. Teixeira, P. Migliorini, et al., Replication of the tumor necrosis factor receptor-associated factor 1/complement component 5 region as a susceptibility locus for rheumatoid arthritis in a European family-based study, Arthritis Rheum. 58 (2008) 2670–2674.

[42] W. Thomson, A. Barton, X. Ke, S. Eyre, A. Hinks, J. Bowes, et al., Rheumatoid arthritis association at 6q23, Nat. Genet. 39 (2007) 1431–1433.

[43] R.M. Plenge, C. Cotsapas, L. Davies, A.L. Price, P.I. de Bakker, J. Maller, et al., Two independent alleles at 6q23 associated with risk of rheumatoid arthritis, Nat. Genet. 39 (2007) 1477–1482.

[44] A. Barton, W. Thomson, X. Ke, S. Eyre, A. Hinks, J. Bowes, et al., Rheumatoid arthritis susceptibility loci at chromosomes 10p15, 12q13 and 22q13, Nat. Genet. 40 (2008) 1156–1159.

[45] A. Barton, S. Eyre, X. Ke, A. Hinks, J. Bowes, E. Flynn, et al., Identification of AF4/FMR2 family, member 3 (AFF3) as a novel rheumatoid arthritis susceptibility locus and confirmation of two further pan-autoimmune susceptibility genes, Hum. Mol. Genet. 18 (2009) 2518–2522.

[46] S. Raychaudhuri, E.F. Remmers, A.T. Lee, R. Hackett, C. Guiducci, N.P. Burtt, et al., Common variants at CD40 and other loci confer risk of rheumatoid arthritis, Nat. Genet. 40 (2008) 1216–1223.

[47] P.K. Gregersen, C.I. Amos, A.T. Lee, Y. Lu, E.F. Remmers, D.L. Kastner, et al., REL, encoding a member of the NF-kappaB family of transcription factors, is a newly defined risk locus for rheumatoid arthritis, Nat. Genet. 41 (2009) 820–823.

[48] S. Raychaudhuri, R.M. Plenge, E.J. Rossin, A.C. Ng, S.M. Purcell, P. Sklar, et al., Identifying relationships among genomic disease regions: predicting genes at pathogenic SNP associations and rare deletions, PLoS Genet. 5 (2009) e1000534.

[49] S. Raychaudhuri, B.P. Thomson, E.F. Remmers, S. Eyre, A. Hinks, C. Guiducci, et al., Genetic variants at CD28, PRDM1, and CD2/CD58 are associated with rheumatoid arthritis risk, Nat. Genet. 41 (2009) 1313–1318.

[50] D. Plant, F. Cornelis, S. Rantapää-Dahlqvist, G. Goulielmos, M.L. Hetland, L. Klareskog, et al., Investigation of potential non-HLA RA susceptibility loci in a European cohort increases the evidence for 10 markers, Arthritis Rheum. 60 (2009) S278.

[51] D.J. Smyth, J.D. Cooper, J.M. Howson, N.M. Walker, V. Plagnol, H. Stevens, et al., PTPN22 Trp620 explains the association of chromosome 1p13 with type 1 diabetes and shows a statistical interaction with HLA class II genotypes, Diabetes 57 (2008) 1730–1737.

[52] G. Orozco, A. Hinks, S. Eyre, X. Ke, L.J. Gibbons, J. Bowes, et al., Combined effects of three independent SNPs greatly increase the risk estimate for RA at 6q23, Hum. Mol. Genet. 18 (2009) 2693–2699.

[53] E.Y. Fung, D.J. Smyth, J.M. Howson, J.D. Cooper, N.M. Walker, H. Stevens, et al., Analysis of 17 autoimmune disease-associated variants in type 1 diabetes identifies 6q23/TNFAIP3 as a susceptibility locus, Genes Immun. 10 (2009) 188−191.

[54] R.P. Nair, K.C. Duffin, C. Helms, J. Ding, P.E. Stuart, D. Goldgar, et al., Genome-wide scan reveals association of psoriasis with IL-23 and NF-kappaB pathways, Nat. Genet. 41 (2009) 199−204.

[55] R.R. Graham, C. Cotsapas, L. Davies, R. Hackett, C.J. Lessard, J.M. Leon, et al., Genetic variants near TNFAIP3 on 6q23 are associated with systemic lupus erythematosus, Nat. Genet. 40 (2008) 1059−1061.

[56] S.L. Musone, K.E. Taylor, T.T. Lu, J. Nititham, R.C. Ferreira, W. Ortmann, et al., Multiple polymorphisms in the TNFAIP3 region are independently associated with systemic lupus erythematosus, Nat. Genet. 40 (2008) 1062−1064.

[57] G. Trynka, A. Zhernakova, J. Romanos, L. Franke, K.A. Hunt, G. Turner, et al., Coeliac disease-associated risk variants in TNFAIP3 and REL implicate altered NF-kappaB signalling, Gut 58 (2009) 1078−1083.

[58] L.M. Maier, C.E. Lowe, J. Cooper, K. Downes, D.E. Anderson, C. Severson, et al., IL2RA genetic heterogeneity in multiple sclerosis and type 1 diabetes susceptibility and soluble interleukin-2 receptor production, PLoS Genet. 5 (2009) e1000322.

[59] A. McClure, M. Lunt, S. Eyre, X. Ke, W. Thomson, A. Hinks, et al., Investigating the viability of genetic screening/testing for RA susceptibility using combinations of five confirmed risk loci, Rheumatology 48 (2009) 1369−1374.

[60] H. Kallberg, L. Padyukov, R.M. Plenge, J. Ronnelid, P.K. Gregersen, A.H. van der Helm-van Mil, et al., Gene-gene and gene-environment interactions involving HLA-DRB1, PTPN22, and smoking in two subsets of rheumatoid arthritis, Am. J. Hum. Genet. 80 (2007) 867−875.

[61] T.J. Aitman, R. Dong, T.J. Vyse, P.J. Norsworthy, M.D. Johnson, J. Smith, et al., Copy number polymorphism in Fcgr3 predisposes to glomerulonephritis in rats and humans, Nature 439 (2006) 851−855.

[62] M. Fanciulli, P.J. Norsworthy, E. Petretto, R. Dong, L. Harper, L. Kamesh, et al., FCGR3B copy number variation is associated with susceptibility to systemic, but not organ-specific, autoimmunity, Nat. Genet. 39 (2007) 721−723.

[63] C. McKinney, M.E. Merriman, P.T. Chapman, P.J. Gow, A.A. Harrison, J. Highton, et al., Evidence for an influence of chemokine ligand 3-like 1 (CCL3L1) gene copy number on susceptibility to rheumatoid arthritis, Ann. Rheum. Dis. 67 (2008) 409−413.

[64] K. Fellermann, D.E. Stange, E. Schaeffeler, H. Schmalzl, J. Wehkamp, C.L. Bevins, et al., A chromosome 8 gene-cluster polymorphism with low human beta-defensin 2 gene copy number predisposes to Crohn disease of the colon, Am. J. Hum. Genet. 79 (2006) 439−448.

[65] E.J. Hollox, U. Huffmeier, P.L.J.M. Zeeuwen, R. Palla, J. Lascorz, D. Rodijk-Olthuis, et al., Psoriasis is associated with increased [beta]-defensin genomic copy number, Nat. Genet. 40 (2008) 23−25.

Index

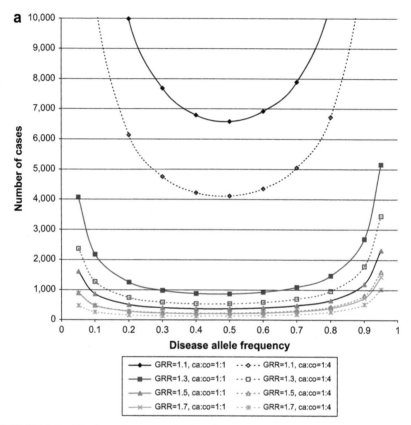

FIGURE 3.2 Number of cases required to detect a variant conferring genotype relative risks between 1.1 and 1.7 (under a multiplicative model) to a disease with a population prevalence of 5%, with 80% power, for (a) a candidate gene study involving 20 independent tagSNPs (SNP-based type I error = 0.0025); (b) a GWA study (SNP-based type I error = 5.7×10^{-7}) (14). Adapted from [58]. (Please refer to Chapter 3, page 41).

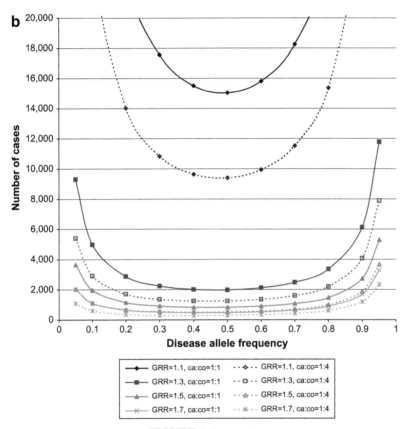

FIGURE 3.2—*Cont'd.*

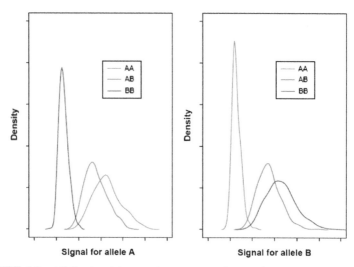

FIGURE 5.2   Allelic signal from each genotype. (Please refer to Chapter 5, page 81).

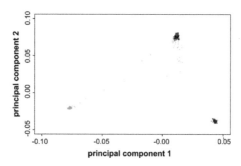

FIGURE 7.3   Identification of individuals with discordant ancestry using principal components analysis. Individuals of known ancestry from the HapMap (CEU (European) – **Red**, YRI (African) – **Green** and CHB&JPT (Asian) – **Purple**) have been used to seed a PCA analysis to detect individuals in of non-European ancestry in a large GWA study of European cases and controls. Crosses indicate GWA-samples removed from further study due to evidence of non-European ancestry and black circles indicate samples remaining under analysis. (Please refer to Chapter 7, page 103).

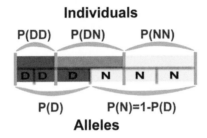

FIGURE 8.1 Example of a Manhattan plot of genome-wide association p-values. Taken from [17]. (Please refer to Chapter 8, page 119).

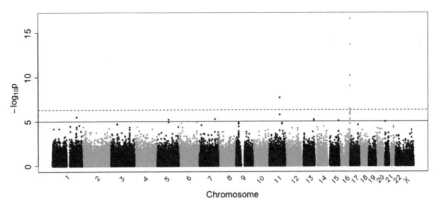

FIGURE 9.1 Genotypic and allelic frequency distribution in a population; $q = P(D) = P(DD) + \frac{1}{2} \cdot P(DN)$. (Please refer to Chapter 9, page 128).

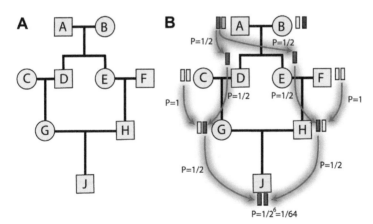

FIGURE 9.2 Inbred family structure (A) and probability of individual "G" being autozygous for the "Red" ancestral allele. (Please refer to Chapter 9, page 129).

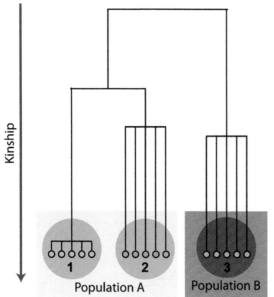

FIGURE 9.3 Three samples from two populations. 1: Family-based sample; 2, 3: random sample of "independent" people. (Please refer to Chapter 9, page 146).

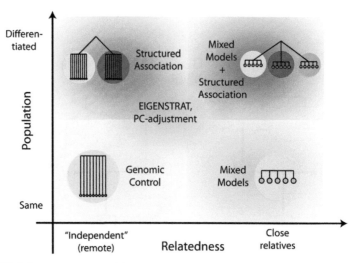

FIGURE 9.4 Applicability of different methods for association analysis. (Please refer to Chapter 9, page 154).

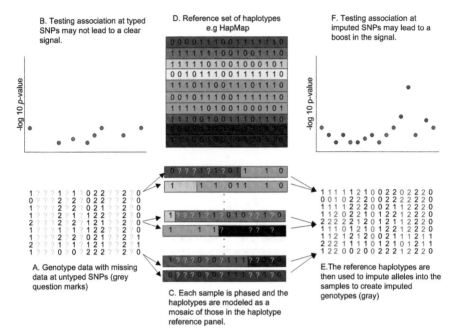

B. Testing association at typed SNPs may not lead to a clear signal.

D. Reference set of haplotypes e.g HapMap

F. Testing association at imputed SNPs may lead to a boost in the signal.

A. Genotype data with missing data at untyped SNPs (grey question marks)

C. Each sample is phased and the haplotypes are modeled as a mosaic of those in the haplotype reference panel.

E. The reference haplotypes are then used to impute alleles into the samples to create imputed genotypes (gray)

FIGURE 10.1   The figure illustrates the idea of genotype imputation in a sample of unrelated individuals. The raw data consists of a set of genotyped SNPs with a large number of SNPs without any genotype data (A). Testing for association at just these SNPs may not lead to a significant association (B). Imputation attempts to predict these missing genotypes. Algorithms differ in their details but all essentially involve phasing each individual in the study at the typed SNPs. The figure highlights three phased individuals (C). These haplotypes are compared to the dense haplotypes in the reference panel (D). Strand alignment between datasets is an important practical task that must be done before the comparison takes place. In the figure the phased study haplotypes have been colored according to which reference haplotypes they match. This highlights the idea implicit in most phasing and imputation models that the haplotypes of a given individual are modeled as a mosaic of haplotypes of other individuals. Missing genotypes in the study sample are then imputed using those matching haplotypes in the reference set (E). In real examples, the genotypes are imputed with uncertainty and a probability distribution over all three possible genotypes is produced. It is necessary to take account of this uncertainty in any downstream analysis of the imputed data. Testing these imputed SNPs can lead to more significant associations (F) and a more detailed view of associated regions. (Please refer to Chapter 10, page 159).

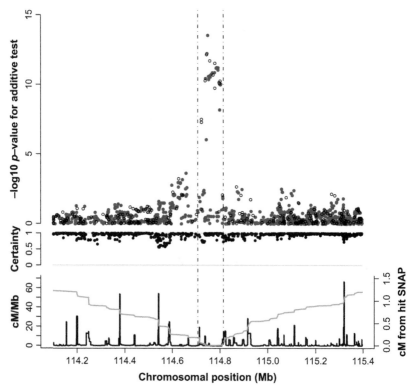

FIGURE 10.2 The results of imputation in and around the TCF7L2 gene in the WTCCC GWAS for type 2 diabetes. The upper part of the plot shows the log10 P-value for the additive model versus a model of no association. The p-values were calculated using called genotypes (black circles) and imputed genotypes (gray circles), called at a threshold of 0.9. The middle panel shows a measure of certainty for each SNP, which is defined as the average maximum posterior genotype call probability. The lower panel shows the fine-scale recombination rate across the region (dark gray) and the cumulative recombination rate measured away from the most highly associated genotyped SNP (horizontal gray). The vertical dashed lines on the plot delineate the main region of association. The largest log10 P value at a genotyped SNP (rs4506565) is 12.25, whereas the largest log10 P-value at an imputed SNP (rs7903146) is 13.57 (taken from Marchini et al. [10]). (Please refer to Chapter 10, page 161).

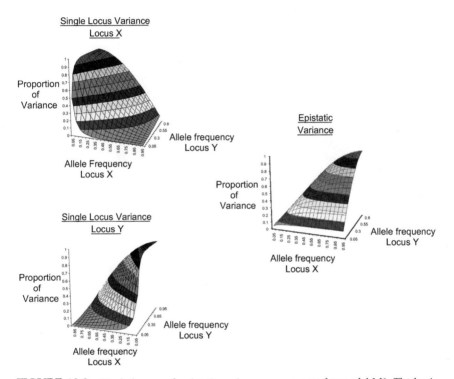

FIGURE 12.2 Single-locus and epistatic variance components for model M1. The horizontal axes on the graphs represent allele frequencies at the first and second loci, whereas the vertical axes give the proportion of the total genetic variance due to each component. (Please refer to Chapter 12, page 204).

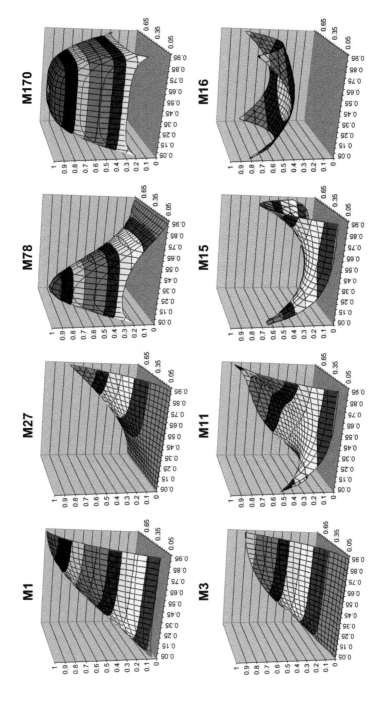

FIGURE 12.3 The proportion of the total genetic variance within the epistatic variance component for eight models from Fig. 12.1. The vertical axis denotes the proportion of variance, whereas the horizontal axes denote allele frequencies at each locus. For each model, there are portions of the parameter space where much of the variance is contained within the epistatic component. (Please refer to Chapter 12, page 205).

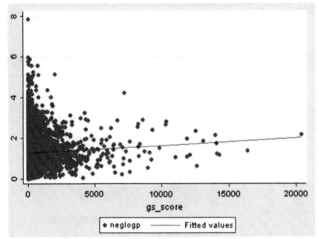

FIGURE 15.1 A plot of the negative-log (*p*-value) of the lowest *p*-value for association with type 2 diabetes per gene region plotted against GeneSniffer candidacy score for type 2 diabetes gene candidacy. The red line shows there was a subtle but significant correlation (coefficient = 0.000326, $p < 10^{-3}$). (Please refer to Chapter 15, page 252).

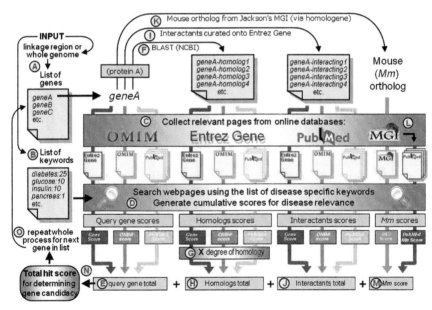

FIGURE 15.2 Flow diagram of GeneSniffer. (Please refer to Chapter 15, page 255).

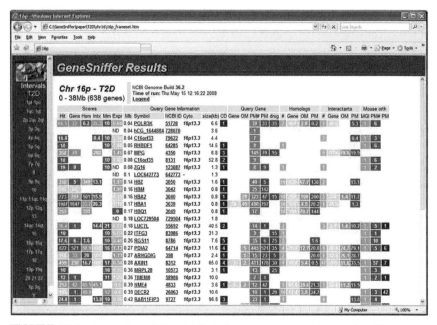

FIGURE 15.3 Screenshot of GeneSniffer's frontpage summarising results. (Please refer to Chapter 15, page 257).

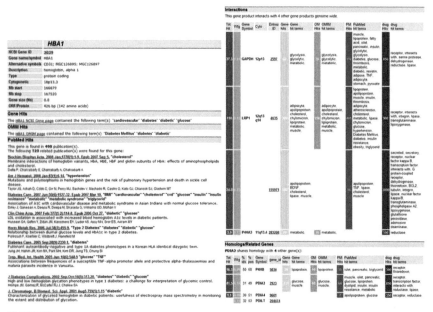

FIGURE 15.4 Screenshots of GeneSniffer's gene results page. (Please refer to Chapter 15, page 257).

Printed and bound by CPI Group (UK) Ltd, Croydon, CR0 4YY

03/10/2024

01040415-0008